Business Resilience System (BRS): Driven Through
Boolean, Fuzzy Logics and Cloud Computation

(Vigilant Eye of the Sphinx)

Bahman Zohuri • Masoud Moghaddam

Business Resilience System (BRS): Driven Through Boolean, Fuzzy Logics and Cloud Computation

Real and Near Real Time Analysis
and Decision Making System

 Springer

Bahman Zohuri
Galaxy Advanced Engineering, Inc.
Albuquerque, NM, USA

Masoud Moghaddam
Galaxy Advanced Engineering, Inc.
Albuquerque, NM, USA

ISBN 978-3-319-85148-8   ISBN 978-3-319-53417-6   (eBook)
DOI 10.1007/978-3-319-53417-6

Printed on acid-free paper

This Springer imprint is published by Springer Nature
The registered company is Springer International Publishing AG
The registered company address is: Gewerbestrasse 11, 6330 Cham, Switzerland

This book is dedicated to my daughter Natalie Zohuri (M.B.A.).

—Bahman Zohuri

I would like to dedicate this book to my parents who did not hesitate anything and sacrificed so many things in their lives so that I can achieve the best in my life. Also to my two lovely brothers Soheil and Hassan who always have encouraged me and supported me by all means.

—Masoud Moghaddam

Preface

From the ancient history, we know that the Babylonians had risen to power in the late seventh century and were heirs of the urban traditions, which had long existed in southern Mesopotamia. They eventually ruled an empire as dominant in the Near East as that held by the Assyrians before them. This period is called the Neo-Babylonian Empire because Babylon had also risen to power earlier and became an independent city-state, most famously during the reign of King Hammurabi (1792–1750 BC). With the recovery of Babylonian independence under Nabopolassar, a new era of architectural activity ensued, and his son Nebuchadnezzar II made Babylon into one of the wonders of the ancient world.

King Nebuchadnezzar II (605 BC–562 BC) ordered the construction of the Ishtar Gate in about 575 BC and was part of his plan to beautify his empire's capital. It was under his rule that Babylon became one of the most splendid cities of the ancient world. He ordered the complete reconstruction of the imperial grounds, including rebuilding the Etemenanki Ziggurat (the Temple of Marduk), and is also credited for the construction of the Hanging Gardens of Babylon—said to have been built for his homesick wife Amyitis.

The Magnificent Ishtar Gate of Babylon and the protecting guards in front of these gates in the form of a lion's body with a human head were a symbolization of their defense system against intruders. The Ishtar Gate was the eighth gate of the city of Babylon (Iraq, in present day) and was the main entrance into the great city. It was a sight to behold; the gate was covered in lapis lazuli-glazed bricks, which would have rendered the façade with a jewel-like shine. Alternating rows of bas-relief lions, dragons, and aurochs representing powerful deities formed the processional way. The message of course was that Babylon was protected and defended by the gods, and one would be wise not to challenge it. The magnificent gate, which was dedicated to the Babylonian goddess Ishtar, was once included among the Seven Wonders of the Ancient World until it was replaced by the Lighthouse of Alexandria in the third century BC.

Later on, the Egyptians in their old civilization presented the same symbol, by building their Sphinx statues, and one of the famous ones that is remaining today is the Great Sphinx.

For our modern world data and information that are flying around at almost speed of light, existence of Business Resilience System to protect our enterprise against any man-made or natural threat is a must thing.

To understand resilience with more precision, we need to understand the concept of the system state first. A system state is the general configuration of the system. For example, if we think of a glass jar as being a system, then smashing the jar into little pieces would be a change to the system state, or, if we think of a farm as being a system, then neglecting the farm for so long that it grows into a forest would be a change to the system state.

What qualifies as a state change depends on how we define the system. There are often many ways of defining a system. Therefore, there will also be many ways of defining its states and changes to them. We should have the mental flexibility to imagine systems and states in different ways, so that we can define them in ways that are helpful for our purposes.

Given this understanding of the system state, we can now define resilience with more precision. Resilience is the ability of a system to maintain certain functions, processes, or populations after experiencing a disturbance. The more a system is able to maintain its functions and components after a disturbance, the greater its resilience to that disturbance will be.

For these authors to establish such resilience as it is suggested in this textbook, we rely on events and using event management and best practices, known as Redbook written by IBM corporation, in particular in Chaps. 1 and 2. With their permission, we have replicated Chaps. 1 and 2 of the book into Chaps. 3 and 4 of this present textbook; however, we encourage the readers to read through that book as well.

Pursuit of sustainable development requires a systems approach to the design of industrial product and service systems. Although many business enterprises have adopted sustainability goals, the actual development of sustainable systems remains challenging because of the broad range of economic, environmental, and social factors that need to be considered across the system's life cycle. Traditional systems engineering practices try to anticipate and resist disruptions but may be vulnerable to unforeseen factors. An alternative is to design systems with inherent "resilience" by taking advantage of fundamental properties such as diversity, efficiency, adaptability, and cohesion.

In summary, "resilience" first conjures up in the mind pictures of bouncing back from adversity, and it is about improved prevention measures; restoring power on the network after a major outage is an example. This captures some of the essentials, with an emphasis on flexibility, coping with unexpected and unplanned situations, and responding rapidly to events, with excellent communication and mobilization of resources to intervene at the critical points. However, we would argue that we should extend the definition a little more broadly, in order to encompass also the ability to avert the disaster or major upset, using these same characteristics. Resilience then describes also the characteristic of managing the organization's activities to anticipate and circumvent threats to its existence and primary goals.

Albuquerque, NM, USA   Bahman Zohuri
Albuquerque, NM, USA   Masoud Moghaddam

Acknowledgments

I am indebted to the many people who aided me, encouraged me, and supported me beyond my expectations. Some are not around to see the results of their encouragement in the production of this book, yet I hope they know of my deepest appreciations. I especially want to thank my friend Bill Kemp, to whom I am deeply indebted, for continuously giving his support without hesitation. He has always kept me going in the right direction.

Above all, I offer a very special thanks to my late mother and father and to my children, in particular, my son Sasha. They have provided constant interest and encouragement, without which this book would not have been written. Their patience with my many absences from home and long hours in front of the computer to prepare the manuscript is specially appreciated.

Contents

About the Authors

Bahman Zohuri is currently at the Galaxy Advanced Engineering, Inc., a consulting company that he started himself in 1991 when he left both semiconductor and defense industries after many years working as a chief scientist. After graduating from the University of Illinois in the field of physics and applied mathematics, he joined Westinghouse Electric Corporation where he performed thermal hydraulic analysis and natural circulation for inherent shutdown heat removal system (ISHRS) in the core of a liquid metal fast breeder reactor (LMFBR) as a secondary fully inherent shutdown system for secondary loop heat exchange. All these designs were used for nuclear safety and reliability engineering for self-actuated shutdown system. He designed the mercury heat pipe and electromagnetic pumps for large pool concepts of LMFBR for heat rejection purpose for this reactor around 1978 where he received a patent for it. He then was transferred to the defense division of Westinghouse later, where he was responsible for the dynamic analysis and method of launch and handling of MX missile out of canister. He later on was a consultant at Sandia National Laboratory after leaving the US Navy. Dr. Zohuri earned his bachelor's and master's degrees in physics from the University of Illinois and his second master's degree in mechanical engineering as well as his doctorate in nuclear engineering from the University of New Mexico. He has been awarded three patents and has published 26 textbooks and numerous other journal publications.

Recently he has been involved with cloud computation, data warehousing, and data mining using fuzzy and Boolean logic.

Masoud Moghaddam got his master's degree in business administration and has been a programmer and developer for over 25 years now. He became an ISP (Internet service provider) on the early days of Internet popularity and has a great knowledge and experience in IP-driven and Web applications, Intranet services, cyber security, database programming, and graphical user interface (GUI) design.

His years of experience and knowledge of networking and cyber security involved him on many projects in the past 20 years. In addition, he has managed many total solution projects in networking and software development, while he was working as the director of IT for Galaxy Advanced Engineering, and he was involved in many scientific and technical projects to date.

Masoud also has deep views about the future of networking, artificial intelligence (AI), and organic computing where he is currently working on the fundamentals and methodology of those new concepts.

Chapter 1
Resilience and Resilience System

Resilience thinking is inevitably systems thinking, at least as much as sustainable development is. In fact, "when considering systems of humans and nature (social-ecological systems) it is important to consider the system as a whole." The term "resilience" originated in the 1970s in the field of ecology from the research of C.S. Holling, who defined resilience as "a measure of the persistence of systems and of their ability to absorb change and disturbance and still maintain the same relationships between populations or state variables." In short, resilience is best defined as "the ability of a system to absorb disturbances and still retain its basic function and structure."

1.1 Introduction

In the earliest of Egyptian times, men saw and knew about the power of beasts and seem to have envied them. There is a sense that humans, at the dawn of civilization, were subject to, and seemingly inferior to, the world's more feral inhabitants. However, as man's intellect grew, together with his ability to control, or at least defend himself from wild beasts, so too did his confidence. Many scholars believe that mixed images such as the *Sphinx* symbolize humankind's domination over wild beasts and over chaos itself. Such images as the Great Sphinx may very well represent animal power tamed by human intelligence and thus transformed into divine calm. Traditionally, mixed or composite images were always seen as divine. One way or another, what could be more dangerous and powerful or more self-assured than the king of the jungle with the mind of a human king?

In today's world of cyber war, where power of Internet allows the variety of cyber attack have power to prevent them ahead of the time even by seconds is an advantage to the enterprises. Other threats and countering them with properly require different kind of Sphinx that defend the enterprises and organization against

© Springer International Publishing AG 2017
B. Zohuri, M. Moghaddam, *Business Resilience System (BRS): Driven Through Boolean, Fuzzy Logics and Cloud Computation*,
DOI 10.1007/978-3-319-53417-6_1

these threats whether it is man-made or a natural disaster. We must protect ourselves in fast pace ever changing computer world by taking the advantages of vast world of structured or unstructured data or other means such as information coming out from geopolitical or human intelligence perspective, to be able to make right choice of action or interaction to our advantages.

Early Egyptian where symbolizing these Sphinx with symbolization of human head on body of a lion, as it is evident from remaining of Great Sphinx Today as it is illustrated here in Fig. 1.1. The human head and eyes in it was presentation of **Vigilant Eye of the Sphinx** to protect and guard a king and his/her kingdom against any adverse events that might have a threat to their kingdom. Such evidence can be found in ancient Babylon and Persian kingdom in the form of standing lion with human head known as Ishtar and its eight gates to inner city.

The Ishtar Gate was the eighth gate of the city of Babylon (in present-day Iraq) and was the main entrance into the great city. It was a remarkable sight; the gate was covered in Lapis lazuli-glazed bricks, which would have rendered the façade with a jewel-like shine. Alternating rows of bas-relief lions, dragons, and aurochs representing powerful deities formed the processional way. The message of course was that Babylon was protected and defended by the gods, and one would be wise not to challenge it. The magnificent gate, which was dedicated to the Babylonian goddess Ishtar, was once included among the Seven Wonders of the Ancient World until it was replaced by the Lighthouse of Alexandria in the third century BC. Today, a reconstruction of the Ishtar Gate, using original bricks, is located at the Pergamon Museum in Berlin.

Considering the fact that Babylon civilization goes beyond any other ancient kingdom, they were the pioneer in the symbolization of Ishtar.

For us in modern world computation and handling of information in the form of cloud system, existence of an autonomous and intelligence system in place is a necessity, and we like to call it a Business Resilience System (BRS), which is the foundation of this text write-up.

Fig. 1.1 Remaining of Great Sphinx in Egypt

1.2   Resilience and Stability

Stability and sustainability in any point of life cycle of a system in place or a process within a system are very critical and important for performance and driving efficiency of that system or process from day-to-day operations. Therefore, having a resilience and stability idea within system or process to carry capacity is closely related to the idea of sustainability. Here we are going to explore another closely related idea of resilience. Resilience is a property of all systems and is related to how a system responds to a disturbance or stressor. In rough terms, the more resilient a system is, the larger a disturbance it can handle [1].

To understand resilience with more precision, we need to first understand the concept of system state. A system state is the general configuration that system is currently in it. For example, if we think of a glass jar as being a system, then smashing the jar into little pieces would be a change to the system state. On the other hand, as an example, if we think of a farm as being a system, then neglecting the farm for so long that it grows into a forest would be a change to the system state.

What qualifies as a state change depends on how we define the system. There are often many ways of defining a system, so there will also be many ways of defining its states and changes to them. We should have the mental flexibility to imagine systems and states being defined in different ways, so that we can define them in a more helpful way for our purposes.

Given this understanding of system state, we can now define resilience with more precision.

- *Resilience is the ability of a system to maintain certain functions, processes, or reactions after experiencing a disturbance.*

Let us continue with the jar metaphor and imagine that the jar is a system for holding sand. The system components would then be the glass jar, the lid, and the sand and air inside the jar. If our glass jar system is thrown at a wall with enough force, it will smash into little pieces and will no longer be able to perform its principal function of holding sand, or anything, for that matter. However, what if the force of impact was only strong enough to crack the jar without breaking it apart? In this case, one of the system components—the jar—is changed, but the system can continue to hold sand, and thus the system state remains essentially the same. The jar system's resilience, then, is the size of the impact it can withstand without smashing to pieces. Remember that disturbances *always change systems* in some way (otherwise, we would not call them disturbances). The more a system is able to maintain its functions and components after a disturbance, the greater its resilience to that disturbance.

With simple systems like the glass jar filled with sand, resilience can be (and often is) represented using the metaphor of a ball in a basin. If the ball is pushed a little bit, it will return to the bottom of the basin, i.e., to its initial state. If the ball is pushed hard enough, it will leave the basin and eventually settle somewhere else, i.e., in an additional state. The height of the basin thus corresponds with resilience: the higher the basin, the harder of a push the ball can withstand and still return to its initial state. Of course, this metaphor becomes less helpful with more complex systems, where we are having many constituents, processes, and functions in effect at the same time. In reality, most systems are only relatively resilient to most disturbances. Most complex systems are able to maintain some, but not all, constituents, processes, and functions after any given disturbance (as long as it is not catastrophic). In other words, resilience in real-world systems is usually *relative* to the type of disturbance and specific constituents, processes, and functions. Figure 1.2 is illustration of resilience and state for metaphor of a ball in a basin.

Resilience is often viewed as a good thing. If an ecosystem is resilient, or if human society is resilient, then they will be quite capable of withstanding the disturbances that they face. For any system to sustain any particular state, the system cannot experience any disturbances that exceed its resilience for that state. Thus, resilience, like carrying capacity, is closely related to sustainability. This is why we see efforts to enhance resilience from groups like the Resilience Alliance [2]. They would like our human–environment systems to be sustained.

However, whether or not resilience actually is good is an ethical question, and the answer is not automatically yes. We will discuss ethics further in Module 3, but, for now, consider this. A terrorist network might be resilient if it can withstand many attacks or other efforts to destroy or disrupt it. Likewise, a dangerous pathogenic virus might be resilient if it can withstand many antiviral medicines or other measures that we take to curtail the virus. In these two cases, resilience is not a good thing. At least, we can imagine that, in these cases, some people might reasonably consider resilience to be bad. Therefore, while resilience is certainly an important concept and may often be considered as a good thing, we should not blindly assume that it always is.

As far as the stability of resilience system is concerned, one important concept related to resilience is its stability. Stability is the opposite side of the disturbances a system faces. If there are few disturbances or small disturbances, then the system is relatively stable. If there are many disturbances or large disturbances, then the system is relatively unstable.

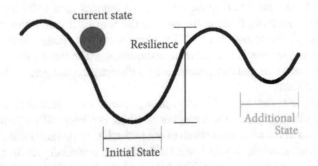

Fig. 1.2 Resilience and state (Courtesy of Yooinn Hong)

Stability is a very important concept in agriculture as an example. We would very much like it if our farms would yield (produce) about the same amount of food each year, because in general we eat about the same amount of food each year. If there is an unusually large food yield in one year, this can cause complications but is typically not a huge problem.

However, if there is an unusually small food yield in one year, then this can be a larger problem, and it may cause a huge shortage in products and increase in prices or even chaos. In the agriculture module, we will examine yield stability in more detail. There, we'll discuss the Irish Potato Famine, which occurred in the mid-1800s. This was a case of extreme instability in food yield, which had disastrous consequences.

One might think that a resilient system would be one with more stability, but this is not always the case. Sometimes, some instability or disturbance can help to increase resilience. This occurs when the disturbances increase the system's ability to respond to further disturbances.

For example, think of our bodies as systems. If we do not exercise regularly, then we gradually lose our ability to withstand against disturbances which have negative effect on our bodies like joint and muscle weakness which leads to further instability and finally collapsing. Then such instability sends our body to a different weaker state. On the contrary, as we get more exercise, then there is an increase in disturbance and instability in our body in a positive way which increases our ability to withstand further exercise without collapsing and therefore more stability of our body at the end. In this example, the exercise is a positive disturbance, and instability, and, as we exercise more, our bodies get less stability but more resilience. This often happens with other systems, too [2].

Resilience and intelligent systems along with their engineering are becoming a new paradigm for complex system performance and maintenance, decision-making for survivability of organization and enterprise, business process management (BPM), and business continuity management (BCM).

The concept of resilience was introduced by Holling [3] in the field of ecology and has been well documented in ecological and social literature and in some management cases. The initial definition of resilience is that it determines the persistence of relationships within systems and is a measure of the ability of these systems to absorb changes of state in variables, driving variable and parameters to the desired state and yet to be able to function accordingly and still persist.

Also, other definitions can be described and included like:

- The potential of particular configuration of a system to maintain its structure and functionality in the face of any disturbance
- To have the ability in the system to reorganize following disturbance-driven change and measured by size of "stability domain," and "the capacity" of a system to absorb disturbance and reorganize while undergoing change so as to still retain essentially the same function, structure, identity and feedback.

An intelligent and autonomous resilience system built on data infrastructure should be able to handle its functional fidelity and to be able to take a proper countermeasure facing unpredictable or predictable events, no matter if it is man-made or a natural disaster. For resilient system to have such fiduciary responsibility and to be persistent, there is a demand of real-time accessibility to trusted data coming from various directions based on cloud computation and real-time analysis based on fuzzy logic. However, more research has been done on the concept of resilience and its applicability to ecological, social, and business systems in comparison to engineered systems.

Resilience engineering or resilience business system in case of main subject of this book represents a major step forward by proposing a completely new vocabulary instead of adding one more concept to an existing lexicon. Although various definitions of resilience exist that are dependent on the subject area, resilience in infrastructure system, such as energy system or financial system (banking), has a different concept.

Such systems driven by trusted data in real time are able to take proper measures, to adapt and recover from external shocks, including natural, artificial, and technological disasters and failures.

On the other side, poor infrastructure and design ultimately affect the smooth and efficient operation of systems and may have a serious interruption in existence of enterprise BPM and BCM accordingly. However, such affect in day-to-day operations of the system demands a shift of processes, tactics, strategies, coordination, and measurement utilizing tools such as cloud computation to collect the right information from right and trusted data.

This information may come into dash-port monitoring screen in the infrastructure systems by triggering the decision-making points that are set per Service Level Agreement (SLA) in place.

Infrastructure systems in most cases are interconnected via enterprise service bus (ESB) or cloud system (CS). Therefore, analysis of the system should consider interdependency properties, because of both dependencies and interdependencies connectivity and interoperability. Due to these circumstances, there are various types of effects, and few can be named here as follows:

1. **Cascading effect:** When disruption in one infrastructure causes disruption in a second
2. **Escalating effect:** When disruption in one infrastructure exacerbates an independent disruption of a second infrastructures
3. **Common cause effect:** When a disruption of two or more infrastructures occurs at the same time

The last is more prevalent during natural disasters. The interactions create a very delicate spider web effect between infrastructures and feedbacks and complex topologies at different levels.

Therefore, it is nearly impossible to analyze the behavior of any infrastructure in isolation of its environment and surroundings [3].

The basic elements of all the infrastructures are vulnerable to physical and natural disruptions and technogenic disasters. The many interrelationships among the infrastructures call for analysis in which various system components are interrelated and for management strategies that allow easy adjustment as more information and data become available. Many different characteristics are shared among the interdependent infrastructure systems, including but not limited to the following:

1. They are large-scale dynamics and nonlinear, spatially distributed "system of systems" with various components.
2. They are administered by different organization/agencies with different objectives.
3. They have multiple decision-makers/stakeholders (i.e., man in the loop) and sometime conflicting and competing objectives.

Considering the general characteristics of these infrastructures, including their inherent complexity, each has a unique field of research. The traits of "system of systems" will include systems that follow different deterioration patterns; hence they fall under different monitoring and maintenance policies and governing rules, defined by different SLA base on stakeholders.

A critical concept and model of infrastructure dependencies have to address both the level and the interconnectedness. In most cases, some of the analysis and the methods are coincided at various levels. One notable characteristic of the hierarchy is that at the top level, socioeconomic, gaming, scenario techniques are used, and at the lower level more experimental and technical simulation are used.

For instance, if we consider a business continuity plan (BCP), which is a plan to help ensuring that business processes can continue during a time of emergency or disaster. Such emergencies or disasters might include a natural event such as earthquake, fire, or man-made catastrophes such as terrorist acts or any other similar case, where business is not able to function under normal conditions. Thus, having a reliable and intelligent Business Resilience System (BRS) in place could be the most proactive protection of shareholder value against the adverse impact of business disruption at any scale and is the next step beyond BCP.

These very effective risk management approaches can be shown as follows and supported by the illustration in Fig. 1.3 as business resilience life cycle.

Steps involved in effective risk management are:

- Enables an organization to become resilient to small and large disruptions
- Defines an early warning approach to identifying and dealing with disruptions versus having to recover from potentially disastrous events
- Drives systematic identification and management of critical business processes, assets, and risks
- Provides systematic monitoring and alerting of risk to critical business processes and assets
- Allows for proactive, automated, and informed responses to mitigate the impact from disruptions
- Emphasizes mission resource dependencies equally—not just technology

Fig. 1.3 Business
resilience life cycle

The fundamental challenge now is to develop resilience indices and workflow that can be used within the "system of systems" framework and use cases, given the complexity and some properties of interdependencies of different infrastructures. Such indices built in an autonomous resilience system should be capable of analyzing the resiliency of the overall system.

Furthermore, inclusion of uncertainties that require the opinion of experts can substantially increase the complexity of a particular decision, location, and form of disruption in many situations. Although we commonly listen to and deal with experts, in most cases, these subject matter experts are knowledgeable in the field but may have limited information by not having enough trusted data. Therefore, imperfect information can still be used to come up with the optimal decisions for disaster recovery (DR) as an example.

In summary, we can show an optimal Business Resilience System (BRS) overview, with components offered by Fig. 1.4.

It is notable to state that there are several topical areas related to control resilience of a system. These complement the fundamental concept of dependable or reliable computing by characterizing resilience in regard to the particular control system concerns, including design considerations that provide a level of understanding and assurance in the safe and secure operation of enterprise or organization. These areas are presented below with discussion to characterize the basis for consideration as an area of resilience.

1. *Human systems*

   The human ability to quickly understand novel situations by employing heuristics and analogy can provide additional control to system's resilience. On the other hand, there are situations in which we may have a general inability to reproducibly predict human behavior.

   This may be true in situations of fatigue or high stress or decision-making under high levels of uncertainty. Bayesian methods provide one method by

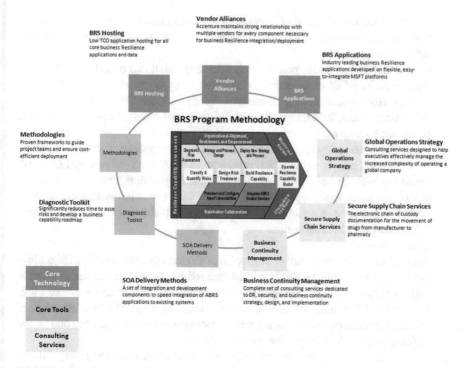

Fig. 1.4 BRS offering overview components

which to take into account evidence regarding human response, but this is one among many approaches. The literature in human reliability analysis provides an orientation regarding ergonomics, workload, complexities, training, experience, etc., which may be used to characterize and quantify human actions and decisions.

Digital technology, used to benefit control system interaction, can come from the operators' perspective and provide additional complexity. For example, more information can be presented to the human operator to form a response. However, the response could be completely automated and act as an autonomous, human manipulated, or a combination of both. The dependencies and rules for these complex interactions, or mixed initiative, are not necessarily well defined or clear.

Resiliency results from understanding of this complexity by adding human factor and designing an error-tolerant control system which complements perception, fusion, and decision-making.

2. *Complex cloud network*

As control systems become more decentralized, the ability to characterize interactions, performance, and security becomes more critical to ensuring resilience. While more decentralization can provide additional reliability due to implicit redundancy and diversity, it may also provide more avenues or vectors to cyber attack. Therefore, the design of complex networks needs to consider all factors that influence resilience and optimize the flow of information for multiple considerations [4].

Global stability is often, perceived as something that can be achieved by local minimization of all process unit operations, many of which are contained in a facility. However, there is no assurance that global stability can be achieved in this manner, and, in addition, this philosophy maintains a reactionary control paradigm by its nature.

However, considering the latencies in digital control systems, there is a tendency as well as a desire to provide faster responses when the feedback and response occur close to the point of interaction with the application. Therefore, it is suggested that a true global optimization coupled with a local interaction can achieve both the assurance of a global minima and an acceptable response when designing control system architecture.

3. *Cyber awareness*

   Because of the human element of a malicious actor, traditional method of achieving reliability cannot be used or implemented to characterize cyber awareness and resilience. Dynamic mechanisms of probabilistic risk analysis that can link human reliability with the system state are still maturing. The intellectual level and background of the adversary make stochastic methods unusable due to the variability of both the objective and the motives.

   In addition, the strength of the adversary is increased because the existing control system architecture is not random, and response characteristics are reproducible. Therefore, a resilient design can find strength in similar fashion by becoming a typical normal control system architectural design and appearing random in response and characteristics to the adversary.

The above discussion provides a holistic and conceptual framework and brief overview of the architectural considerations of control system as part of subcomponent and component of a smart business resilience system, and more details will be provided in the following chapters and sections of this book.

1.3 Reactive to Proactive Safety Through Resilience

In a world of finite resources, irreducible uncertainty, and multiple conflicting goals, safety is created through proactive resilient processes rather than through reactive barriers and defenses. Building a solid infrastructure based on different facets of resilience as the ability of systems to anticipate and adapt to the potential for surprise and failure, or unpredictable events, whether man-made or natural disaster, is a fundamental requirement for today's organizations and enterprises. These fundamental requirements guarantee the survivability of organizations and enterprises from business process management (BPM) and business continuity management (BCM) perspective.

Until recently, the dominant safety paradigm was based on searching for ways in which limited or erratic human performance could degrade a well-designed and "safe system." Techniques from many areas such as reliability engineering and management theory were used to develop "demonstrably safe" systems. The assumption seemed to be; once that safety is established, it can be maintained by requiring that human performance stays within prescribed boundaries or norms. Since "safe" systems needed to include mechanisms that are guarded against people as unreliable components, understanding of how human performance which could stray outside these boundaries could become important.

According to this paradigm, "error" was something that could be categorized and counted. This led to numerous proposals for taxonomies, estimation procedures, and ways to provide the much needed data either structured or unstructured, for error tabulation and extrapolation. Studies of human limits became important to guide the creation of remedial or prosthetic systems that would make up for the deficiencies of people.

Since humans (as unreliable and limited as system components) were assumed to degrade what would otherwise be flawless system performance, this paradigm often prescribed automation as means to safeguard the system from the people in it. In other words, in the "error counting" paradigm, work on safety is comprised of protecting the system from unreliable, erratic, and limited human components (or more clearly protecting the people at the blunt end—in their roles as managers, regulators, and consumers of systems—from unreliable "other" people at the sharp end, who operate and maintain those systems) [5].

Efforts to improve the safety of systems have often (some might say always) been dominated by hindsight, both in research and in practice which perhaps is more surprising in the former than in the latter.

The practical concern for safety is usually driven by events that have happened, either in one's own company or in the industry as such. There is a natural motivation to prevent such events from happening again. In concrete cases they may incur severe losses of equipment and/or of a life in general cases because they may lead to new demands for safety from regulatory bodies, such as national and international administrations, organizations, business enterprises, and agencies. New demands are invariably seen as translating into increased costs for companies and are for that reason alone undesirable.

However, this is not an inevitable consequence, especially if the enterprise takes a longer time perspective. Indeed, for some businesses, it makes sense to invest proactively in safety, although such cases are uncommon. The reason for this is that sacrificing decisions usually are considered over a short-time horizon in terms of months rather than years or in terms of years rather than decades.

Finding the best optimum and intelligent BRS requires fundamental research and engineering design for software and hardware to interact with each other in rather real-time circumstances, where cloud computation based on data is involved.

In the case of research, i.e., activities that take place at academic institutions rather than in industries and are driven by intellectual rather than economic motives, the effects of hindsight ought to be less marked. Research by its very nature should be looking to problems that go beyond the immediate practical needs and hence address issues that are of a more principal nature.

When researchers in the early 1980s began to reexamine human error and collect data on how complex systems had failed, it soon became apparent that people actually provided a positive contribution to safety through their ability to adapt to changes, gaps in system design, and unplanned situations.

Many studies of how complex systems succeeded and sometimes failed found that the formal descriptions of work embodied in policies, regulations, procedures, and automation were as incomplete as models of expertise and success. Analysis of the gap between formal work prescriptions and actual work practices revealed how people in their various roles throughout systems always struggled to anticipate paths toward failure, to create and sustain failure-sensitive strategies, and to maintain margins in the face of pressures to increase efficiency.

Overall, analysis of such circumstances taught us that failures represented breakdowns in adaptations directed at coping with complexity, while success was usually obtained as people learned and adapted to create safety in a world fraught with hazards, trade-offs, and multiple goals [6].

In summary, these studies revealed [5]:

- How workers and organizations continually revise their approach to work in an effort to remain sensitive to the possibility for failure
- How distant observers of work, and the workers themselves, are only partially aware of the current potential for failure
- How "improvements" and changes create new paths to failure and new demands on workers, despite or because of new capabilities
- How the strategies for coping with these potential paths can be either strong and resilient or weak and mistaken
- How missing the side effects of change is the most common form of failure for organizations and individuals
- How a culture of safety depends on remaining dynamically engaged in new assessments and avoiding stale, narrow, or static representations of the changing paths (revising or reframing the understanding of paths toward failure over time)
- How overconfident people have already anticipated the types and mechanisms of failure, and the strategies they have devised are effective and will remain so
- How continual effort after success in a world of changing pressures and hazards is fundamental to creating safety

In the final analysis, safety is not a commodity that can be tabulated. It is rather a chronic value "under our feet," which infuses all aspects of practice. Safety is, in the words of Karl Weick [7], a dynamic nonevent. Progress on safety therefore ultimately depends on providing workers and managers with information about changing vulnerabilities and the ability to develop new means for meeting these.

The initial steps in developing a practice of business resilience system have focused on methods and tools:

- To analyze, measure, and monitor the resilience of organizations in their operating environment
- To improve an organization's resilience vis-à-vis the environment
- To model and predict the short- and long-term effects of change and line management decisions on resilience and therefore on risk

This book charts an effort by these authors to lay out a practical and intelligent Business Resilience System driven through Boolean, fuzzy logics, and cloud computation.

1.4 The Business Resilience System Backdrop

Backdrop for BRS can be summarized as follows:

Organizations have significant capability in risk management and BCP, but the focus is on a narrow subset of high-level risk, and they often fail to actively address operational risks that may result in degradation rather than "destruction" of operational capabilities.

A "BRS" seeks to extend organizations' risk management and BCP efforts to address threats to the processes that support their primary operational capabilities—the primary source/driver of business (shareholder) value:

- Research has shown that even the result of degraded operations has significant impact on shareholder value.
- Organizations generally monitor operational capabilities using KPIs which are lagging indicators.
- BRS identifies and monitors leading indicators of potential impacts to resources that support critical business processes and drive shareholder value.
- BRS is a holistic, enterprise approach to proactively protect shareholder value against the adverse impact of business disruption at any scale, thereby allowing organizations to become more resilient to service degradation without having to invoke BCP steps because of unplanned, destructive events.
- Becoming more resilient allows an organization to rapidly recover from a disruption and resume normal operations with limited revenue, cost, and time impacts.

The importance of business resilience cannot be overstated, and it can be depicted as Fig. 1.5 in case of let say stock market analysis, when a supply chain malfunction is announced, stock prices plunge on average, and shareholder wealth decreases by certain amount per company.

Current business continuity plans are fragmented and do not include a holistic approach to identifying and effectively managing risk to the enterprise. This statement can be seen in Fig. 1.6 that illustrates the key characteristic of historical path and current and future elements of business resilience to a futuristic and intelligent form.

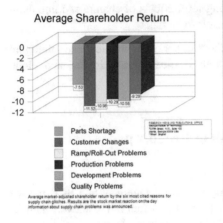

"When a supply chain malfunction is announced, stock prices plunge an average of 8.62%, and shareholder wealth decreases by $120M or more per company."*

Average Shareholder Return

- Parts Shortage
- Customer Changes
- Ramp/Roll-Out Problems
- Production Problems
- Development Problems
- Quality Problems

Average market-adjusted shareholder return by the six most cited reasons for supply chain glitches. Results are the stock market reaction on the day information about supply chain problems was announced.

A recent study of increased investment in supply chain security found additional benefits to that area of business activity for the 11 manufacturing companies participating in the study:

- 38% reduction in lost cargo
- 37% reduction in product tampering
- 14% reduction in excess inventory
- 47% reduction in delivery times
- An increase in customer satisfaction leading to 26% reduction in customer attrition
- 20% increase in new customers

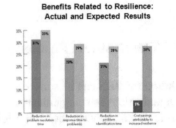

Benefits Related to Resilience: Actual and Expected Results

Source: The Manufacturing Institute and Stanford University, July 2006

Fig. 1.5 Illustration of BRS importance

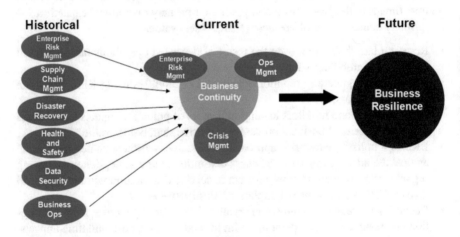

Fig. 1.6 Business resilience system historical path

We can also present this historical path in the following form as Fig. 1.7, as a summary to be:

In today's world of cyber threat and cyber attack, we are in desperate need of a smart business resilience system to be in place. Thus, we can suggest a holistic BRS that has capability of reducing the magnitude and duration of a business disruption as Fig. 1.8, presented in following illustration.

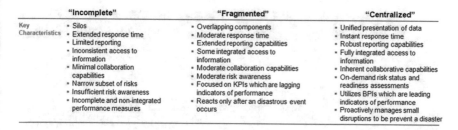

	"Incomplete"	"Fragmented"	"Centralized"
Key Characteristics	• Silos • Extended response time • Limited reporting • Inconsistent access to information • Minimal collaboration capabilities • Narrow subset of risks • Insufficient risk awareness • Incomplete and non-integrated performance measures	• Overlapping components • Moderate response time • Extended reporting capabilities • Some integrated access to information • Moderate collaboration capabilities • Moderate risk awareness • Focused on KPIs which are lagging indicators of performance • Reacts only after an disastrous event occurs	• Unified presentation of data • Instant response time • Robust reporting capabilities • Fully integrated access to information • Inherent collaborative capabilities • On-demand risk status and readiness assessments • Utilizes BPIs which are leading indicators of performance • Proactively manages small disruptions to be prevent a disaster

Fig. 1.7 Key characteristic of BRS

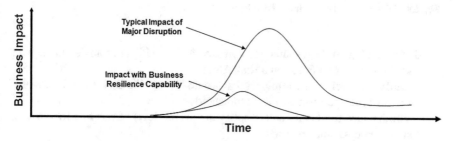

Fig. 1.8 Illustration of business impact versus time

Such BRS infrastructure will be responsive to today's threat of cyber attack, a need for homeland security, and business continuity of enterprise from BCM, BPM, and BCP point of view.

The granular step by step of plot depiction of Fig. 1.8 could breakdown to Fig. 1.9, where we can see tabulation of Typical Impact and Resilient Impact.

As we stated before, the suggested Business Resilience System (BRS) is the proactive protection of shareholder value against the adverse impact of business disruption at any scale and is the next step beyond BCP, in support of BCM and BPM. This was holistically presented in Fig. 1.3, and then it was expanded to Fig. 1.4 sort of showing the offering overview of BRS components.

The secret sauces to the suggested BRS approach by these authors involve the identification and utilization of the what we call "Risk Atom." The Risk Atom is comprised of various interrelated and continually moving and interacting components that are arranged in orbits surrounding a process data point (a PDP—the nucleus) and helps a business organization maintain resilient processes through an effective resource response to direct and indirect threats manifested by some preceding event.

This approach helps an organization to:

• Better identify, quantify, and respond to risks at the business process level
• Define the Risk Atoms that are appropriate to a specific business process (a business process may contain one or multiple Risk Atoms)
• Identify and quantify those events and threats that could "force" the movement of a Risk Atom across identified and calculated performance measure thresholds over time

Latent Risk	Threat Exposes Vulnerability	Disruption Identified	Business Responds	Business Recovers	Business Adapts
Typical Impact Risks are not clearly assessed, quantified and categorized	Limited ability to monitor threats and vulnerabilities on a real time basis	Reactive notification and awareness of disruptions	Slow and uncoordinated response	Inefficient recovery model leads to excessive costs	Limited processes for assessing performance and improvement
Resilient Impact Risks are continuously evaluated and managed centrally	24/7 monitoring and alerting capability provides early warning	Early detection arms the right people with the information they need to act	Pre-determined collaboration and action plans are put in motion	Business recovers rapidly as planned	Performance metrics are reviewed and used to make improvements

Fig. 1.9 Tabulation of Typical Impact and Resilient Impact

- Identify and quantify resource responses to avoid, mitigate, transfer, or recover from the impact of a threat on a Risk Atom
- Quantify the level of threat impact that would "force" the Risk Atom to traverse through various performance level thresholds
- Determine how to identify and quantify the overall level of risk to a Risk Atom, business process, and enterprise
- Set the stage for the establishment of an "early warning" approach that would enable an organization to respond to threats and their impact before a catastrophic situation materialized

The Risk Atom concept can be applied to all business processes and is applicable to just about any "system" that must be resilient and to have a solid BCM and BPM in place. The Risk Atom is the foundation upon which the BRS solution is built and determines the business actions to be taken upon threat manifestation. This solution is illustrated in Fig. 1.10 as follows where data processing through cloud computation as nucleus of this Risk Atom is either fuzzy logic or Boolean logic or combination of both.

The BRS Risk Atom is composed of those interrelated business intelligence "particles" that help a business define the Processing Data Points (PDPs) to be monitored, the impact upon the business and critical business processes as those PDPs pass through various threshold levels, and the business' response to a situation where the PDP is affected.

The Risk Atom is comprised of four (4) orbits of company-specific intelligence which is needed to effectively manage a critical business process (CBP) as follows:

- **First orbit:** Processing Data Point (PDP—the nucleus), with real-time or manual reporting capabilities
- **Second orbit:** Business processes and accompanying resources
- **Third orbit:** Threats to the PDP and impacts to the business and process
- **Fourth orbit:** Business risk elements and responses to the threats acting upon the PDP

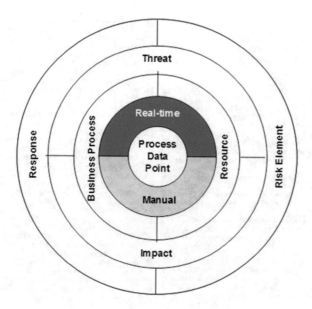

Fig. 1.10 The BRS Risk Atom

For any critical business process, there may be one or multiple Risk Atoms, but any Risk Atom must reflect a critical business process measure that, when "tipped," it will begin degrading process capabilities, and, if left unchecked, it will result in a disaster/destruction situation requiring the invocation of a Business Continuity Process (BCP).

A PDP can "move" through various levels of thresholds (as a result of threat manifestation) which will determine the type of business activities to be performed to remedy any foreseeable process degradation before it becomes process destruction.

To fully understand the Risk Atom and its function within BRS, component parts must be identified and defined. These component parts are depicted in Fig. 1.11 and they are explained to some details.

1.5   Risk Atom Key Concept

Key to using the Risk Atom concept is establishing performance thresholds, which set parameters defining when targeted response activities are enacted. The Risk Atom key concept is movement through performance thresholds and is illustrated in Fig. 1.12.

Schematic of Fig. 1.12 for Risk Atom and key concept is summarized as follows:

- A Risk Atom begins its journey at Point A—stasis or equilibrium.

The PDP (Process Data Point)
- A Key Performance Indicator (KPI) or Business Process Influencer (BPI) that could have a direct and negative impact to a company's "bottom line" if that KPI or BPI significantly missed a performance target.
- Most typically a BPI (a leading indicator) that must be identified after an in-depth analysis of the business process, versus a current KPI that is a lagging indicator.
- Can be quantified via different measures (e.g. days, percent, $, etc.)
- May or may not be unique to an industry, client-type or business process.

The First Orbit
- A PDP can be measured and monitored on a **Manual** (via a human interface) or **Real-time** (via a system output) basis.

The Second Orbit
- **Business Process** represents those discrete business processes that an organization wants to make more resilient.
- **Resource** defines into which one of the resource categories a PDP can be placed (i.e. **people, process, technology, network, data or facilities**.)

The Third Orbit
- **Threats** can be segregated into **environmental, supply, demand, process or controls**, and are typically, but not always, preceded by an event (e.g. event = earthquake, threat = tsunami).
- **Impact** represents the financial ($), tangible or intangible impact to the business should a threat materialize because of a preceding event thereby causing the RiskAtom to "move" from its stasis or equilibrium point through a performance threshold.

The Fourth Orbit
- **Response** defines the activities that a business must perform in order to respond to a manifested threat that causes the RiskAtom to "move" across performance thresholds over time.
- A response may be enacted to avoid, mitigate, transfer or recover from a particular situation and would encompass activities in one or all of the following areas: people, process, technology, network, data and facilities.
- **Risk Element** identifies those elements of the business that an organization wishes to "guard" in order to protect concepts such as customer goodwill, labor productivity, market capitalization or brand (for example) – things that could be irrevocably destroyed or have a severe impact on the organization's external standing in the business community if the organization was not prepared and/or resilient.
- A risk element may or may not be unique to a particular organization, and a single RiskAtom may encompass multiple risk elements.

Fig. 1.11 Overview of Risk Atom orbits

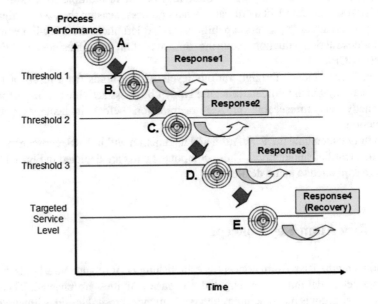

Fig. 1.12 Risk Atom movement through Performance Thresholds. (This figure does not show business impact by threshold.) *Note*: (1) Thresholds are set in accordance with process needs. (2) Threshold measures are user selectable

- A threat impacts the Risk Atom and its performance measure drops until it passes Threshold 1 and finds itself at Point B. At that point the first response is activated in the hopes of potentially avoiding any further performance degradation to the business process.
- The Risk Atom continues to fall and passes through Threshold 2. At Point C the second response is activated in an attempt to mitigate any impact from the threat.
- The scenario continues to Point D where the third and final "resilience" response is activated.
- If the Risk Atom performance continues to fall toward the targeted service level, the final response is to recover from the threat situation which means that the previous three responses did not rectify the fall of the Risk Atom and there could be a direct and negative impact on the company's "bottom line."
- The concept of a BRS Risk Atom is to provide and act upon threats before they critically impact the business and cause potentially irreparable harm (i.e., become more resilient).

To reduce potential business impact and response duration, BRS helps drive proactive and timely response activities according to performance threshold levels. This is schematically illustrated in Fig. 1.13 here.

As part of our discussion in *Risk Atom*, we can summarize them here as:

- With BRS, the magnitude and duration of a disruption can be reduced as the business is receiving early warnings that a potential disastrous scenario could be imminent and that targeted responses are being invoked to resume normal operations earlier in the degradation situation.
- The setting of thresholds allows the business to adequately respond to a degradation situation before it becomes a disaster or destruction scenario.
- A Risk Atom will "move" because a manifested threat causes a degradation in the level of service that the Risk Atom represents.
- As a Risk Atom "moves," overall status may or may not change. For example, if the level of degradation is very small at Threshold 1, overall status may remain

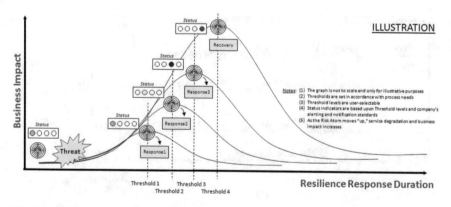

Fig. 1.13 Schematic of resilience response duration

"green" and require simple monitoring of the situation. However, as the Risk Atom passes through the remaining thresholds, failure to adequately respond could drive the organization to recovery via the appropriate BCP, thereby incurring significant impacts to the business and shareholder value.

From a functional process and system tool standpoint, BRS follows a linear track with the focus of reducing enterprise risk and increasing business resilience. This statement is schematically depicted in Fig. 1.14.

Using fuzzy logics mathematical modeling to streamline the BRS risk management and business resilience approach, a high-availability toolset could be developed that suggests and consists of the following components:

- PDP identification, threshold, and response management database
- External and internal monitoring of threats and events
- BRS response management unit
- BRS action response system
- BRS dashboard and reporting system
- Proactive prevention and diagnostic tool

The above steps schematically are depicted in Fig. 1.15, which is presentation of BRS application architecture overview.

As authors of this book, we are suggesting a holistic approach to simple yet intelligent business resilience system, architecture configuration here. Of course by no

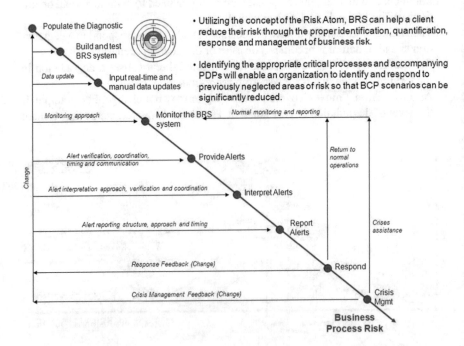

Fig. 1.14 BRS functional process and system tool standpoint

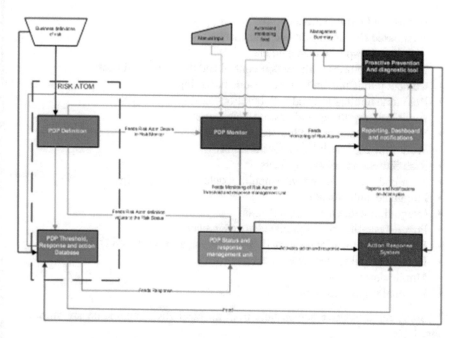

Fig. 1.15 BRS application architecture overview

means this is a final and bulletproof approach. In today's computational technology along with autonomous functionality built on artificial intelligence and cloud computation availability and yet continuously thriving, a better and more robust system could be configured and designed. However, all futuristic BRS design will have one thing in common, and that is the information that comes to them via PDP in Risk Atom nuclei. For the PDP to be processed in real time, wave of information and data via cloud computation should come in a very unidirectional or omnidirectional mode, if not real-time at least in near real-time basis. Considering the growth size of data in today's business, we certainly need to cut back on processing time of these data to be able to reduce them to some trusted values that PDP can process them and be able to send the notification to stakeholder and key decision-maker in organizations and enterprises. By having a very well-thought trigging threshold point within dashboard of BRS system, per Service Level Agreement in place for the given application, alarming threats can then be identified. This way a measure can be taken with appropriate time frame to counter the threat and react accordingly.

1.6   Business Resilience System Features

Business Resilience System (BRS) is a multifeatured application that is designed to use the latest available Web-based technology with features such as:

- Highly configurable architecture
- Web-based risk definition and prioritization
- Web-based response management
- Document libraries with version control and document roll back
- Content publishing and online content authoring
- Real-time notification and collaboration
- Audit trail, reporting, and custom dashboards
- Web-based business intelligence
- Work flow engine and task coordination
- Rich Site Summary (RSS) feeds
- High-end search engine with inbuilt people search capability
- Pluggable authentication
- Integration with Office applications
- Policies auditing and compliance management
- Enhanced notifications—e-mail, Short Message Service (SMS), and voice
- Document and folder level access control
- Mobile device support
- Personal digital assistant (PDA) devices

Some representative screenshots from an ideal BRS application are shown in Fig. 1.16:

To help analyze, define, and implement a BRS solution, the BRS functional stack in Fig. 1.17 will prove instructive as well as provide the project building blocks.

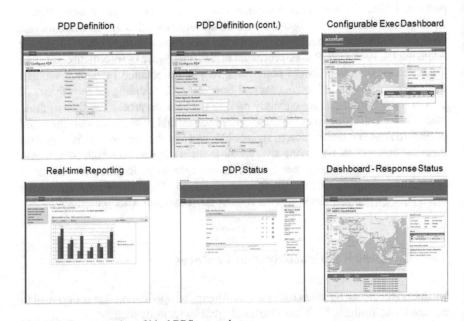

Fig. 1.16 Representation of ideal BRS screenshots

Fig. 1.17 The BRS functional stack

The functional stack that is illustrated in Fig. 1.17 is summarized below, with description of each element of stack.

- The footings for the BRS "stack" are an organization's current critical business, risk management, and BCP processes.
- Enhancing these capabilities requires a diagnosis of the current process environment to identify, document, and quantify the key performance indicators (KPIs) and/or business process influencers (BPIs) that have a direct impact on shareholder value.
- Following diagnosis, planning must be performed to identify and develop appropriate threshold and situational responses as well as determine the degree to which any identified gaps from the diagnosis would be filled.
- Results from the diagnosis and planning steps would be configured within the BRS tool to allow monitoring and presentation of threats, process/facility readiness, response activities, and external information data feeds.
- The next layer refers to the activities required within each resource set to mitigate the impact from a specific threat scenario. In turn, these are assembled into appropriate responses based upon attained thresholds.
- As understanding of the organization's risk, response, and mitigation capabilities matures, the BRS risk management and notification environment will be continually improved.

Business Resilience System (BRS) offering is briefly illustrated in Fig. 1.18.

Identifying the correct critical business processes (CBPs) to monitor and the Process Data Points (PDPs) that support them is the foundation of proactive business resilience, and that is shown here in Fig. 1.19 as well.

Functional model of BRS offering at top level where fuzzy logic plays a major role is given in the following format as it can be seen in Fig. 1.20.

BRS portal technology—phase I to start with in order to put the design of such system into action—is schematically offered here in Fig. 1.21.

Business Resilience System (BRS) portal—high-level system diagram—is depicted in Fig. 1.22 as follows.

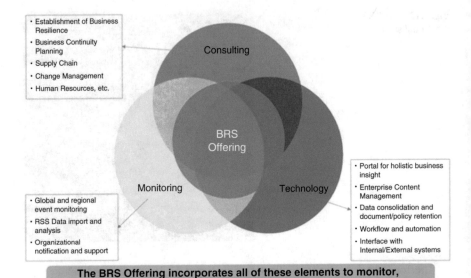

The BRS Offering incorporates all of these elements to monitor, maintain and merge corporate risk at all levels of the organization.

Fig. 1.18 BRS offering

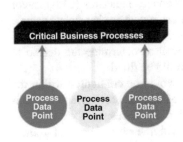

- The foundation of BRS is the organizations ability to identify the Critical Business Processes (CBP's) that support the organization or a particular component of the organization

- The Process Data Points (PDP's) that support these will be identified and thresholds and actions will be determined

- The PDP thresholds will be the basis for determining the appropriate reactive or proactive response that will need to be taken to mitigate or avoid the risk

- These thresholds can be changed or modified over time to optimize the response to meet the needs of the organization

- The PDP's and Critical Business Processes will be determined during the diagnosis and analysis processes

Fig. 1.19 Illustration of critical business processes

The technology overview as part of BRS offering feature is briefed here as follows:

- **Crisis preparedness**
  Crisis preparedness means knowing what to do and how to do it with a team that is practiced and ready. Safeway Business Resilience System (BRS) has particular expertise in developing policy, plan, and procedures in support of corporate

The complete BRS offering then rolls in all of the CBP's and CBA's that need to be monitored and managed through a single portal.

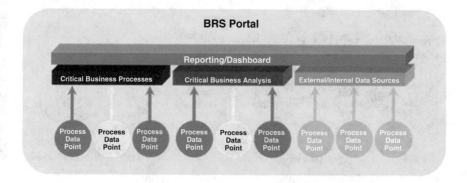

- The portal becomes the business dashboard for executives to proactively manage their organization
- External and Internal Data sources (ERP, MRP, etc.) combined with the accumulation of CBP's and CBA's create a holistic view of business resilience

Fig. 1.20 Business Resilience System portal configurations

The BRS Portal will include the following components for Phase 1.

Phase 1 Components:
- Monitoring and emergency call support
- Delegated user administration and defined user access
- Manual configuration and set-up
- Profiling and push capability
- Defined relationships between data points
- Logic supported for limited number of PDPs and associated threats
- Meta-data model by document and document type
- Rules for user-based review of activities
- Search capabilities based on use cases
- Enterprise Content Management
- Data Mining and Data Analysis (What If and Fuzzy Logic Functionality) for PDPs
- Real Time Communication (i.e. IChat, Web Mail)

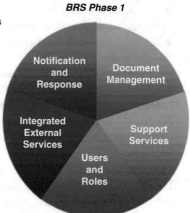

Fig. 1.21 BRS phase I portal

crisis management, supply chain risk management, expatriate risk management, and site emergency management. We also collaborate with leading risk consulting companies such as Marsh Risk Consulting and Kroll to deliver a comprehensive program tailored to each business-specific needs.

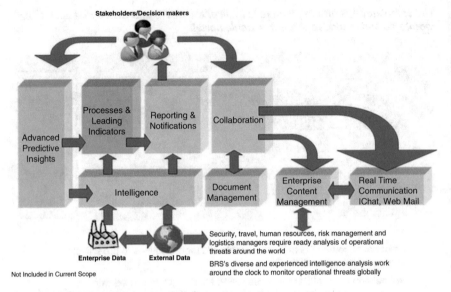

Fig. 1.22 BRS portal high-level system diagram

- **Integration services**
  BRS can help you integrate data from your crisis/emergency management systems, employee data systems, and supply chain data systems as well as other related data systems (i.e., Teradata, master data management).
- **Implementation services**
  BRS has successfully completed hundreds of complex, client activation projects and refined the implementation process.
- **Special projects**
  BRS uses its extensive source network to produce detailed risk assessments and specialized intelligence on countries, cities, and, often, specific neighborhoods. Driven by each client's needs, these detailed intelligence reports range from on-the-ground crime, kidnapping, and other security threats to geopolitical conditions, natural disaster, competitive business environments, and transportation logistics.
- **Training programs**
  BRS offers a variety of specialized training programs that detail the travel security and safety issues that international travelers face, as well as the measures that corporate security; travel and risk professionals should take to ensure the safety of executives and corporate employees as they operate in the global marketplace.
- **Technology approach**
  BRS approach is based on real and total Web-based solution utilizing .NET and J2EE as open source which is totally independent of OS clients for its implementation crosses the Internet and intranet.

Top Ten Questions to Think About

- Does our organization face rising stakeholder and shareholder expectations for its **ability to sustain** operations despite disruptive events?
- Does our company employ a **comprehensive event response capability**, using responses to routine disruptions to build skills and experience that will apply to disaster recovery?
- Have large or small disruptive events (e.g., blizzards, transit strikes, delivery delays, West Coast Port embargo, a power or network outage, computer virus, etc.) **materially impaired** your organization's delivery capability?
- Are "routine" disruptive events (e.g., power outages, network downtime, seasonal flu epidemics, hurricane evacuations, supply chain delays, etc.) considered part of **"normal" operations** not requiring a specific event response plan and capability?
- Have we analyzed our organization's **annual quality and financial costs** due to "routine" disruptive events?
- Can **external events** (e.g., regulatory changes, public panics) cause spikes or troughs in citizen demand for organization services? Does your organization have response plans for these disruptive demand changes? Would **planning and preparation** improve the organization's effectiveness in responding to such demand changes?
- Does our organization systematically monitor its environment for **predictable, high-impact disruptive events** and have **automated response capabilities** in place to rapidly communicate status and begin response implementation?
- Have we established **clear organizational accountability** for our response capability?
- Does our organization's disaster recovery plans focus **only on technology recovery or on full business** capability (people, process, facilities, technology) recovery?
- Have our organization's operations (processing centers, call centers, etc.) **quickly bounced back** after Hurricane Katrina (or other hurricanes/disasters)? Are our clients even more dependent on your services during a disaster than during routine operations?"

1.7   Business Resilience System Project Plan

A sample of high-level project implementation timeline in phase I is illustrated in Fig. 1.23 with project overview in Fig. 1.24.

The assumption in case of phase I implementation timeline was that:

Assumptions

- Pricing is subject to additional discussions with business side and due diligence.

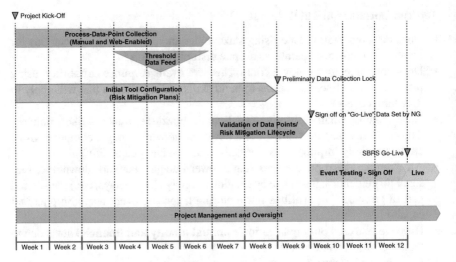

Fig. 1.23 BRS implementation timeline in phase I

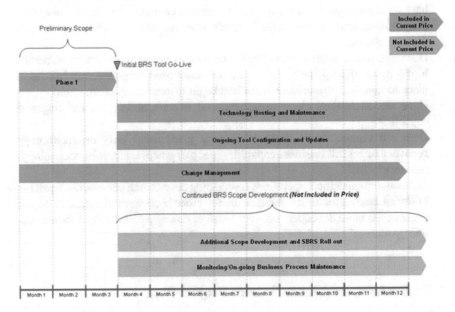

Fig. 1.24 BRS project overview for phase I

- BRS platform is configurable for up to 50 users.
- Does not include the cost of interface development for enterprise ERP or legacy systems into BRS.
- Pricing includes travel and expenses for some external team.
- Pricing does not include fixed (FX) and cost-of-living adjustment (COLA) for year 2017

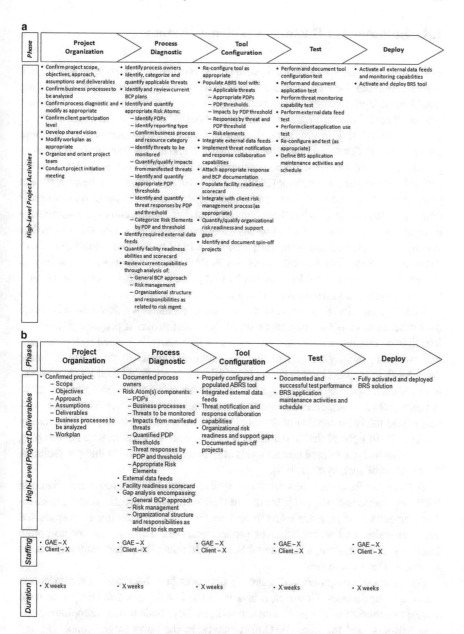

Fig. 1.25 (**a**) Illustration of a high-level project plan. (**b**) Illustration of a high-level project plan

A holistic overview of work-plane as part of project management processes is suggested by these authors here and is shown in Fig. 1.25a, b, as well.

See Appendix A for more details of project plan.

1.8   Summary of Business Resilience System

In summary, Business Resilience System (BRS) is offering some type of system that is built around the best artificial intelligence (AI) concept and approach. Old saying that "Knowledge is Power" works perfectly, if our knowledge has real-time (at least near real) feed from vast data of information floating around us, and that comes to us from every direction providing that we can trust them no matter what structure they possessed. For this AI machine to be in place and act autonomously as much as possible, float of data needs to be continuous to its core of BRS Risk Atom, as we defined it Processing Data Point (PDP) in Fig. 1.10. For this core to be actively and proactively effective and to function properly against incoming cyber threats or any defined threat, building an intelligent model from data mining and expert knowledge is a must and inevitable. A look at some fundamental principles related to this matter is subject of next and following chapters. For a PDP to have an impact and effectively manage the four key orbits defined in Risk Atom, requires its capability to absorb the data processing driven either by fuzzy or Boolean logic as part of its function and data mining, which is also subject later on chapters of this book.

Bear in your mind that the property that has given humans a dominant advantage over other species is not strength or speed, but intelligence. If progress in artificial intelligence continues unabated, AI systems will eventually exceed human in general reasoning ability. A system that is "superintelligent" in the sense of being "smarter than the best human brains in practically every field" could have an enormous impact upon humanity [8]. Just as human intelligence has allowed us to develop tools and strategies for controlling our environment, a superintelligent system would likely be capable of developing its own tools and strategies for exerting control [9]. In light of this potential, it is essential to use caution when developing AI systems that can exceed human levels of general intelligence or that can facilitate the creation of such systems [10].

A Business Resilience System also requires a true Incident Response Planning (IRP), to enable and prepare it for the inevitable threat or threats against organizations or enterprise that are using it and to give enough warning to prepare the decision-maker and stakeholder for proper measures. Thus, as part of corporate Internet security system, a "Computer Security Incident Response Planning" seems a fundamental requirement.

As we know, computers and computer networks have been part of the corporate landscape for decades. However, it is only in the last 5 years that companies have started to connect these systems and networks to the outside world—suppliers, business partners, and the Internet. Unfortunately, in the hurry to get connected and jump on the e-business bandwagon, computer security is frequently given short shrift, placing corporate assets at risk.

As an example of assets at risk, we can look at the media filled with accounts of recent Internet security problems, including the denial of service attacks against Yahoo!, eBay, Amazon, CNN, and other several instances of data theft involving credit cards or personal information, and the "I Love You" virus/worm. Although the press devoted many column inches and on-air minutes to these stories, they focused primarily on the exciting topic of "the chase" to catch the perpetrators and generally ignored the more important topics of how frequently computer security incidents occur, how many companies' data is at significant risk, and the potentially devastating impact of computer security incidents on their victims.

A super smart Business Resilience System will provide a good and impenetrable line of defense to surround an organization or enterprise by providing advanced enough warning and put them ahead the ball.

In summary also, to help you determine if BRS may be right for your critical business processes or organization, ten questions are presented for your consideration as reader or folks that are interested in placing a more detail-oriented BRS in place. These holistic questions were imposed in above, and we repeat it here again.

These holistic questions are:

- Does your organization face rising stakeholder and shareholder expectations for its ability to sustain operations despite disruptive events?
- Does your company employ a comprehensive event response capability, using responses to routine disruptions to build skills and experience that will apply to disaster recovery?
- Have large or small disruptive events (e.g., blizzards, transit strikes, delivery delays, West Coast Port embargo, a power or network outage, computer virus, etc.) materially impaired your organization's delivery capability?
- Are "routine" disruptive events (e.g., power outages, network downtime, seasonal flu epidemics, hurricane evacuations, supply chain delays, etc.) considered part of "normal" operations not requiring a specific event response plan and capability?
- Have you analyzed your organization's annual quality and financial costs due to "routine" disruptive events?
- Can external events (e.g., regulatory changes, public panics) cause spikes or troughs in citizen demand for organization services? Does your organization have response plans for these disruptive demand changes? Would planning and preparation improve the organization's effectiveness in responding to such demand changes?
- Does your organization systematically monitor its environment for predictable, high-impact disruptive events and have automated response capabilities in place to rapidly communicate status and begin response implementation?
- Have you established clear organizational accountability for your response capability?
- Does your organization's disaster recovery plans focus only on technology recovery or on full business capability (people, process, facilities, technology) recovery?
- Have your organization's operations (processing centers, call centers, etc.) quickly bounced back after Hurricane Katrina (or other hurricanes/disasters)? Are your clients even more dependent on your services during a disaster than during routine operations?"

References

1. Published on GEOG 030: Geographic Perspectives on Sustainability and Human-Environment Systems. (2011). https://www.e-education.psu.edu/geog030/node/327
2. http://www.resalliance.org/
3. Holling, C. S. (1973). Resilience and stability of ecological system. *Annual Review of Ecology and Systematic, 4*, 1–23.
4. Wang, F.-Y., & Liu, D. (2008). *Networked control systems: Theory and applications*. London: Springer.
5. Woods, D. D., & Hollnage, E. (2006). *Resilience engineering: Concepts and precepts*. Boca Raton: CRC Press.
6. Rasmussen, J. (1997). Risk management in a dynamic society: A modelling problem. *Safety Science, 27*(2–3), 183–213.
7. https://en.wikipedia.org/wiki/Karl_E._Weick
8. Bostrom, N. (2014). *Superintelligence: Paths, dangers, strategies*. New York: Oxford University Press.
9. Muehlhauser, L., & Salamon, A. (2012). Intelligence explosion: Evidence and import. In A. Eden, J. H. Soraker, J. H. Moor, & E. Steinhart (Eds.), *Singularity hypotheses: A scientific and philosophical assessment. The frontiers collection*. Berlin: Springer.
10. Soares, N., & Fallenstein, B. (2014). Agent Foundations for Aligning Machine Intelligence with Human Interests: A Technical Research Agenda. Machine Intelligence Research Institute.

Chapter 2
Building Intelligent Models from Data Mining

In order for Business Resilience System to function and induced advanced warning for proper action, the BRS Risk Atom and, in particular, the fourth orbit have to stay in stable status, by assessing the risk elements and responses. Thus, looking at risk assessment and understanding it are an essential fact, and it is inevitable; therefore, we need to build an intelligent model to invoke information from variety of data available to us at more than terabyte from around the globe we are living on. These data need to be processed by Process Data Point in the core of Risk Atom, either real time or manually. The feed point for PDP is structured on fuzzy or Boolean logic as suggested by us authors in this book. This chapter under lays foundation for Risk Atom by discussing the risk assessment and goes through process of building intelligent models, along with data mining and expert knowledge and a look at some fundamental principles that can interact with the Risk Atom.

2.1 Introduction

Information has become an organization's most precious asset. Organizations have become increasingly dependent on information, since more information is being stored and processed on network-based systems. A significant challenge in providing an effective and efficient protective mechanism to a network is the ability to detect novel attacks or any intrusion works and implement countermeasures. Intrusion detection is a critical component in securing information systems and a subset of Business Resilience System (BRS). Intrusion detection is implemented by an Intrusion Detection System (IDS), which as we said could be a subset of BRS. Today, we can find many commercial Intrusion Detection Systems available in the market, but they are restricted in their monitoring functionality and they need frequent updates and patches. The widespread use of e-commerce, for example, has

© Springer International Publishing AG 2017
B. Zohuri, M. Moghaddam, *Business Resilience System (BRS): Driven Through Boolean, Fuzzy Logics and Cloud Computation*,
DOI 10.1007/978-3-319-53417-6_2

increased the necessity of protecting the system to a very high extend. Confidentiality, integrity, and availability of information are major concerns in the development and exploitation of network-based computer systems. Intrusion Detection System can detect, prevent, and react to the attacks. Intrusion Detection has become an integral part of the information security process. However, it is not technically feasible to build a system with no vulnerabilities; as such, intrusion detection continues to be an important area of research.

Obviously, it is immanent that we have an Internet security system and as a result a computer security incident response plan that is prepared for the inevitable as part of fundamental of BRS infrastructure. For such infrastructure to be super smart and fully inherent and autonomous, one needs a good knowledge base (KB) and, thus, indexing of incoming data moment by moment, driven by an intelligent models from data mining and expert knowledge as part of master data management (MDM) repository. This repository is the first line of defense and layer that Processing Data Point (PDP) in nucleus of Risk Atom strives from it.

As part of orbital layer of Risk Atom, we need to have an excellent understanding of each orbit, dealing with risk element and responses, threat and its impact, business processes and resources, and last but not least real-time and manual engagement where core Process Data Point is built on. This chapter will take us through these sequences.

2.2 Risk Assessment and Risk Element

Risk assessment is a step in the risk management process. Risk assessment is measuring two quantities of the risk R, the magnitude of the potential loss L, and the probability p that the loss will occur. This statement is schematically illustrated in Fig. 2.1 as:

Fig. 2.1 Typical risk assessment illustration

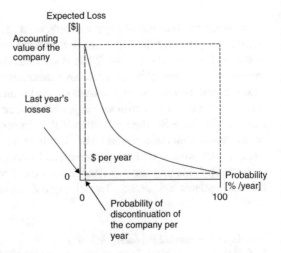

Risk assessment may be the most important step in the risk management process and may also be the most difficult and prone to error. Once risks have been identified and assessed, the steps to properly deal with them are much more programmatical by its nature.

Part of the difficulty of risk management is that the measurement of both of the quantities in which risk assessment is concerned can be very difficult itself. Uncertainty in the measurement is often large in both cases. In addition, risk management would be simpler if a single metric could embody all of the information in the measurement. However, since two quantities are being measured, this is not possible. A risk with a large potential loss and a low probability of occurring must be treated differently than one with a low potential loss but a high likelihood of occurring. In theory both are of nearly equal priority in dealing with first, but in practice it can be very difficult to manage when faced with the scarcity of resources, especially time, in which to conduct the risk management process.

Mathematically, the management process can be expressed as:

$$R_i = L_i p(L_i)$$
(Eq. 2.1)

In addition, form of total situation the Eq. 2.1 can expand to a general form as follow:

$$R_{total} = \sum_i L_i p(L_i)$$
(Eq. 2.2)

Both Eqs. 2.1 and 2.2 induced based on a financial risk assessment such as insurance enterprise.

Financial decisions, such as insurance entity, often express loss terms in dollars. When risk assessment is used for public health or environmental decisions, there are differences of opinions as to whether the loss can be quantified in a common metric such as dollar values or some numerical measure of quality of life. Often for public health and environmental decisions, the loss term is simply a verbal description or possible Service Level Agreement (SLA) in place of the outcome, such as increased cancer incidence or incidence of birth defects. In that case, the "risk" is expressed as:

$$R_i = p(L_i)$$
(Eq. 2.3)

If the risk estimate takes into account information on the number of individuals exposed, it is termed a "population risk" and is in units of expected increased cases per a time period. If the risk estimate does not take into account the number of individuals exposed, it is termed an "individual risk" and is in units of incidence rate per a time period. Population risks are of more use for cost/benefit analysis; individual risks are of more use for evaluating whether risks to individuals are "acceptable."

In the context of public health, risk assessment is the process of quantifying the probability of a harmful effect to individuals or populations from certain human activities. In most countries, the use of specific chemicals or the operations of specific facilities (e.g., power plants, manufacturing plants) is not allowed unless it can be shown that they do not increase the risk of death or illness above a specific threshold. For example, the American Food and Drug Administration (FDA) regulates food safety through risk assessment [1]. The FDA required in 1973 that cancer-causing compounds must not be present in meat at concentrations that would cause a cancer risk greater than one in a million lifetimes.

However, if we expand our argument on the example of insurance entity, in the estimation of the risks, three or more steps are involved, requiring the inputs of different disciplines.

2.2.1   How the Risk Is Determined

The first step, *hazard identification*, aims to determine the qualitative nature of the potential adverse consequences of the contaminant (chemical, radiation, noise, etc.) and the strength of the evidence it can have with that effect. This is done for chemical hazards, by drawing from the results of the sciences of toxicology and epidemiology. For other kinds of hazard, engineering or other disciplines are involved.

The second step for chemical risk assessment is determining the relationship between dose and the probability or the incidence of effect (dose–response assessment). The complexity of this step in many contexts derives mainly from the need to extrapolate results from experimental animals (e.g., mouse, rat) to humans and/or from high to lower doses. In addition, the differences between individuals due to genetics or other factors mean that the hazard may be higher for particular groups, called susceptible populations. An alternative to dose–response estimation is to determine an effect unlikely to yield observable effects. In developing such a dose, to account for the largely unknown effects of animal to human extrapolations, increased variability in humans, or missing data, a prudent approach is often adopted by including safety factors in the estimate of the "safe" dose, typically a factor of 10 for each unknown step.

The third step, exposure quantification, aims to determine the amount of a contaminant (dose) that individuals and populations will receive. This is done by examining the results of the discipline of exposure assessment. As different location, lifestyles, and other factors likely influence the amount of contaminant that is received, a range or distribution of possible values is generated in this step. Particular care is taken to determine the exposure of the susceptible population(s).

The results of the three steps above are then combined to produce an estimate of risk. Because of the different susceptibilities and exposures, this risk will vary within a population. The decisions based on the application of risk assessment are sometimes based on a standard of protecting those most at risk. This problem raises the question of how small a segment of a population must be protected. What if a risk is very low for everyone but 0.1% of the population?

A difference exists whether this 0.1% is represented by:

- All infants younger than X days
- Recreational users of a particular product

If the risk is higher for a particular subpopulation because of abnormal exposure rather than susceptibility, there is a potential to consider strategies to further reduce the exposure of that subgroup. If an identifiable subpopulation is more susceptible, due to inherent genetic or other factors, there is a policy choice whether to set policies for protecting the general population that are protective of such groups. It is currently done for children, when data exists or is under the Clean Air Act for populations such as asthmatics, or whether if the group is too small or the costs too high. Sometimes, a suitable position is to at least limit the risk of the more susceptible to some risk level above which it seems too inequitable to leave them out of the risk protection.

2.2.2 Acceptable Risk Increase

The idea of not increasing lifetime risk by more than one in a million has become a commonplace in public health discourse and policy. How consensus settled on this particular figure is unclear. In some respects, this figure has the characteristics of a mythical number. In another sense, the figure provides a numerical basis for what to consider a negligible increase in risk. In part, the one in a million benchmark arose early in public health risk assessment history when risk assessment was a tempering analysis to existing statutory language such as the Delaney Clause prohibition on any use of introduced carcinogens or where environmental statues were using a "best technology" decision rule. Some current environmental decision-making allows some discretion to deem individual risks potentially "acceptable" if below one in ten thousand increased lifetime risk. Low-risk criteria such as these do provide some protection in the case that individuals may be exposed to multiple chemicals (whether pollutants or food additives, or other chemicals). Nevertheless, both of these benchmarks are clearly small relative to the typical one in four lifetime risks of death by cancer (due to all causes combined) in developed countries.

Individuals may be tempted to advocate the adoption of a zero-risk policy. After all the one in a million policy would still cause the death of hundreds or thousands of people in a large enough population. In practice, however, a true zero risk is possible only with the suppression of the risk-causing activity. More stringent requirements, or even the one in a million policy, may not be technologically feasible at a given time or so expensive as to render the risk-causing activity unsustainable.

In the interest of public health, the risks vs. benefits of the possible alternatives must be carefully considered. For example, it might well be that the emissions from hospital incinerators result in a certain number of deaths per year. However, this risk must be balanced against the available alternatives. In some unusual cases, there are significant public health risks, as well as economic costs, associated with all options. For example, there are risks associated with no incineration (with the potential risk for spread of infectious diseases) or even no hospitals. But, often further investigation identifies further options, such as separating noninfectious from infectious wastes, or air pollution controls on a medical incinerator, that provide a broad range of options of acceptable risk—though with varying practical implications and varying economic costs. Intelligent thought about a reasonably full set of options is essential. Thus, it is not that unusual for there to be an interactive process between analyses, consideration of options, and then further analysis.

2.2.3 Acceptable Risk in Auditing

In auditing, risk assessment is a very crucial stage before accepting an audit engagement. According to ISA 315 Understanding the Entity and its Environment and Assessing the Risks of Material Misstatement, "the auditor should perform risk assessment procedures to obtain an understanding of the entity and its environment, including its internal control" [2].

The main purpose of risk assessment procedures is to help the auditor understand the audit client. Aspects like client's business nature, management structure, and internal control system are good examples. The procedures will provide audit evidence relating to the auditor's risk assessment of a material misstatement in the client's financial statements. Then, auditor obtains initial evidence regarding the classes of transactions at the client and the operating effectiveness of the client's internal controls.

In auditing, audit risk includes:

1. Inherent risk
2. Control risk
3. Detection risk

Each of the above definition can be describe and expand up, but it is beyond the scope of this book; thus, we encourage the readers to do their own investigation and research. However, these types of information become part of database of a knowledge base (KB) as part of Risk Atom that can be processed by Processing Data Point (PDP) as the first line of defense. KB allows for PDP to invoke pervious known knowledge from central database or master data management (MDM), if you will to enhance the triggering point at the dashboard of Business Resilience System (BRS) to set the advanced warning to prevent for event and disaster to take place.

Some researchers and subject matter experts argue and have expressed concerns that risk assessment tends to be overly quantitative and reductive. For example, they argue that risk assessments ignore qualitative differences among risks. Some charge that assessments may drop out important nonquantifiable or inaccessible information, such as variations among the classes of people exposed to hazards. O'Brien further claims that quantitative approaches divert attention from precautionary or preventative measures [3].

The above process and infrastructure easily could be expanded across all types and classes of organization and enterprises, and based on their classification and type of business they are in, their own risk assessment can be defined and proper Service Level Agreement (SLA) set in place.

In conclusion, the risk analysis process is an important aspect of business recovery planning. The probability of a disaster occurring in an organization is highly uncertain. Organizations should also develop written, comprehensive business recovery plans that address all the critical operations and functions of the business.

The plan should include documented and tested procedures, which, if followed, will ensure the ongoing availability of critical resources and continuity of operations. A business recovery plan, however, is similar to liability insurance. It provides a certain level of comfort in knowing that if a major catastrophe occurs, it will not result in financial disaster for the organization.

Insurance, by itself, does not provide the means to ensure continuity of the organization's operations and may not compensate for the incalculable loss of business during the interruption or the business that never returns.

2.2.4   Risk Analysis Process

A critical aspect of the risk analysis process is to identify the preparedness and preventive measures in place at any point in time. Once the potential areas of high exposure to the organization are identified, additional preventative measures can be considered for implementation.

Regardless of the prevention techniques employed, possible threats that could arise inside or outside the organization need to be assessed. Although the exact nature of potential disasters or their resulting consequences are difficult to determine, it is beneficial to perform a comprehensive risk assessment of all threats that can realistically occur to the organization. Regardless of the type of threat, the goals of business recovery planning are to ensure the safety of customers, employees, and other personnel during and following a disaster.

The relative probability of a disaster occurring should be determined. Items to consider in determining the probability of a specific disaster should include, but not be limited to:

1. Geographic location
2. Topography of the area
3. Proximity to major sources of power
4. Bodies of water and airports

5. Degree of accessibility to facilities within the organization
6. History of local utility companies in providing uninterrupted services
7. History of the area's susceptibility to natural threats
8. Proximity to major highways, which transport hazardous waste and combustible products

The above are few natural disasters that one can think of it and we are certain there are more to think about it. However, the human threat is a different ball game that our intelligent communities under the umbrella of Department of Homeland Security need to be involved and their data need to be built in a secure master data management (MDM) system with strong firewall, which prevents any hacker to penetrate the security of this firewall from cyber attack perspective.

Potential exposures may be classified as natural, technical, or human threats. Examples include:

Natural threats: internal flooding, external flooding, internal fire, external fire, seismic activity, high winds, snow and ice storms, volcanic eruption, tornado, hurricane, epidemic, tidal wave, and typhoon

Technical threats: power failure/fluctuation, heating, ventilation or air conditioning failure, malfunction or failure of Central Processing Unit (CPU), failure of system software, failure of application software, telecommunications failure, gas leaks, communication failure, and nuclear fallout

Human threats: robbery, bomb threats, embezzlement, extortion, burglary, vandalism, terrorism, civil disorder, chemical spill, sabotage, explosion, war, biological contamination, radiation contamination, hazardous waste, vehicle crash, airport proximity, work stoppage (internal/external), and computer crime

All locations and facilities should be included in the risk analysis. Rather than attempting to determine exact probabilities of each disaster, a general relational rating system of high, medium, and low can be used initially to identify the probability of the threat occurring.

The risk analysis also should determine the impact of each type of potential threat on various functions or departments within the organization.

2.3 Incident Response

In the event that the security of a system has been compromised, an incident response is necessary. It is the responsibility of the security team to respond to the problem quickly and effectively. *Incident response* is an expedited reaction to an issue or occurrence. Pertaining to information security, an example would be a security team's actions against a hacker who has penetrated a firewall and is currently sniffing internal network traffic. The incident is the breach of security. The response depends upon how the security team reacts, what they do to minimize damages, and when they restore resources all while attempting to guarantee data integrity.

Think of your organization and how almost every aspect of it relies upon technology and computer systems. If there is a compromise, imagine the potentially devastating results. Besides the obvious system downtime and theft of data, there could be data corruption, identity theft (from online personnel records), embarrassing publicity, or even financially devastating results as customers and business partners learn of and react negatively to news of a compromise.

Research on past security breaches (both internal and external) shows that companies can sometimes be run out of business as result of a breach. A breach can result in resources rendered unavailable and stolen or corrupted data. However, one cannot overlook issues that are difficult to calculate financially, such as bad publicity. An organization must calculate the cost of a breach and how it will detrimentally affect an organization, both in the short and long term.

2.3.1   *What Is Incident Response*

As businesses and other institutions increase their online presence and dependency on information assets, the number of computer incidents also rises. Consequently, these organizations are finally increasing their security postures. This is accomplished in three stages. Firstly, organizations must develop and implement security plans and controls in a proactive effort. Secondly, they must work to ensure their plans, and controls are effective by continually reviewing and adapting them to ensure that appropriate security is always in place. Finally, when controls are bypassed either intentionally or unintentionally, organizations must be prepared to act quickly and effectively to minimize the impact of situations. The goal is to prevent an operational security problem from becoming a business problem that impacts revenue. This book provides guidelines to help organizations plan their responses to incidents and minimize any negative impact to their businesses.

All too often, when organizations develop information security programs, they treat security issues as a simple "check-box" on the list of required corporate functions. After giving security its due attention (often times very little), senior executives happily and honestly check the box indicating they have somehow dealt with information technology (IT) security and then move on to the next issue. Many of these organizations assume that once the security program is established (the box is checked, remember), they are assured of complete security and that—like hanging a painting on the wall—once in place it requires little further attention. Nothing could be farther from the truth [7].

For one thing, there is no such thing as total security. Good security controls keep folks honest and make it so challenging for an adversary to get around such controls that they will give up and move on to an easier target. Security supports business operations and ensures uptime and efficiency of mission-critical systems needed by the business in its daily operations to generate revenue and profit. From that perspective, security is as critical to business operations as the reliability and stability of the company's networks, servers, and phone lines. However, what happens when something unexpected happens or someone manages to get around the established security controls in a manner that threatens business operations, and subsequently revenue and profitability?

When responding to an incident, there are a number of considerations:

- What happened?
- What has been damaged?
- What business processes are being impacted and how do we minimize those impacts?
- Who did it and what did they do?
- How did they do it?
- Can and should any forensic information be preserved?
- What are the legal issues?
- Is the damage continuing or has the activity been contained?

Waiting until the incident occurs is too late to begin planning how to address the situation. Incident Response Planning requires both administrative and technical roles. The relationship between these roles takes time to flourish if they are to be truly effective during an incident. Both parties need to be familiar with the other's role, responsibilities, and capabilities. Administrators and managers need to understand the value and function of an incident response team in order to support the technical people regardless of whether an incident is actually occurring. In addition, the technical people need to have a current understanding both of the specific environment of the organization and of the current state of the attacks that are likely to be experienced [7].

In summary, *incident response* is a vital part of any successful IT program and is frequently overlooked until a major security emergency has already occurred, resulting in untold amounts of unnecessary time and money spent, not to mention the stress associated with responding to a crisis. In the most basic terms, an incident is a situation in which an entity's information is at risk, whether the situation is real or simply perceived. Common examples of incidents include the following, by no means complete list of incident types:

- A company's web site is defaced by an intruder. The company seeks to find the perpetrator and recoup financial damages for tarnishing the company's reputation.
- An employee at a company is believed to be selling trade secrets to a competitor.
- A rival corporation is believed to be dialing into a company's computing systems and downloading financial performance data.
- A computer virus is spreading among employees by way of infected Microsoft Word document files shared over e-mail.

These situations are serious incidents that could easily result in significant impact to a company if not handled appropriately. Clearly, it is that level of impact, which is most important to a business. To a typical corporation, the most severe type of incident is one that adversely affects a business process. Any company that does not understand the potential impact of an information system security incident need only ask its senior business managers what the impact would be if their business functions were delayed, halted, or otherwise diminished. To exacerbate the situation, business managers and many senior executives are generally not technologists or experts in the underlying IT infrastructure supporting their business process.

2.3.2   Real-Life Incidents

Contrary to public perceptions, not all incidents have dramatic dollar losses or make the front-page news in sensational stories of computer terrorists wreaking havoc around the cyber world. Rather, most incidents rarely get a passing glance from even the most investigative reporter and are often rather mundane and uninteresting for people outside of the affected area or company. As this book is not sensational and does not make unrealistic claims of gloom and doom, let us look at some typical situations that incident responders deal with on an almost-daily basis [7].

2.3.3   Incident Response Planning Program

With so many choices, how does an organization begin to set up an incident response program? As with most journeys (or corporate actions), making the decision to support the idea is the crucial first step. It is absolutely vital to get senior executive-level support for developing your incident response program. One thing is for certain that anyone proposing this to senior management must be able to make a compelling business case. Although much of the information in this book is meant to make that process easier, we cannot possibly give readers all the materials required to take their case to management. A lot of the business case comes down to a cost–benefit analysis presented so the executive can quickly see the benefit of having an incident response capability established. After all, capability and management interest already exist in being able to respond to fire, theft, burglary, or medical emergencies.

This chapter covers the high-level, administrative, management, political, and operational issues of setting up an incident response program along with predictive functionality and its model as part of Incident Response Planning program. Naturally, the specifics vary by organization, but there should also be a great deal of common ground.

2.3.3.1   Establishing the Incident Response Program

Once an organization has made the decision to proceed with an incident response program, the fun really begins for those charged with leading the undertaking. Be prepared to spend time documenting technical and managerial procedures, defining staff roles and responsibilities, staffing requirements, fighting for adequate funding, identifying and purchasing tools, training your staff, and taking care of a whole slew of other administrative details specific to your situation and organization.

On the bright side, the more attention you pay to detail now, the more likely you succeed in the future. The more you train, test, document, validate, optimize, and document again, the smoother things are likely to run during a real incident.

2.3.4   Tools of the Trade

Anyone who reads an information security magazine or spends time on the Internet quickly realizes that there are a myriad of security tools available. These tools range from small, simple freeware security utilities to comprehensive professional security suites of tools designed to solve all your security problems. Because there are a vast number of tools, it can be difficult for the incident response professional to select the right tools for the job. However, contrary to vendor claims, no panacea or silver bullet exists that can solve all your problems. More importantly, very few, if any, security tools are designed from the ground up to the incident response tools. There are tools for encryption, authentication, intrusion detection, and so on, but there are arguably no tools designed specifically for incident response. Many of the available tools are useful for incident response activities, but were not designed for that purpose. Thus, it is up to the incident response practitioner to study the available tools and their applicability to incident response operations and then to adapt to the tools they deem appropriate for the task at hand.

Though certainly not all—of the currently available tools and discusses each tool's strengths and weaknesses specifically with, regards to its utility and applicability as an incident response tool. The discussions about the tools are based, in almost every case, on our experiences using them in actual incidents. The operational aspects of the tool, such as the vendor's ability to support it, are stressed in the discussions here. We do not seek to endorse or condemn any tools only to show the novice incident responder the capabilities, pros, and cons of various tools you will encounter. Where one tool comes up short, you can bet others will fill in the gaps nicely.

2.3.5   What Is Out There

As we have already indicated, the spectrum of available tools is broad. Even under ideal situations, not all of them are applicable to incident response operations. Although some can be considered system administration tools as well, we are going to focus on the tools that can be used as part of incident response operations. The list of tools we have assembled has been put together through actual incident response operations. As such, it has been growing, changing, shrinking, and updating for some time now.

2.3.6 *Risk Assessment and Incident Response*

It is clear why a company should invest the resources to establish an incident response program: consider the results and impact on a corporation that suffers a disaster without having prepared for it! In other words, what level of risk is a company willing to accept on its information resources and businesses?

This is addressed through the concept of risk management or when senior management conducts a cost–benefit analysis to weigh the pros and cons of implementing various security countermeasures such as an incident response program. Risk management defines levels of risk by examining the types and probabilities of threats and vulnerabilities associated with a given organization and balance those findings against the costs associated with protecting against such potential problems. These assessments help senior management decide the level of risk they and the company are willing to accept as a result of implementing (or not implementing) specific countermeasures to potential security problems. For example, not having an incident response process may mean extended periods of downtime and confusion that could affect business operations or revenue, just as not having a properly configured firewall increases the probability of a network being compromised.

- Threat intent to do harm (e.g., crackers, viruses, criminals)
- Vulnerability: weaknesses (e.g., no firewalls, outdated antivirus tools)

While many resources provide in-depth details of risk management, here are some points to ponder in assessing the levels of risk for enterprise information resources. Truthful answers to questions like these will help determine how robust an incident response capability may be required at a given company and thus the level of resources needed to make it happen:

- What business processes are dependent upon the proper functioning of IT systems?
- To what level has the company entrusted its IT staff to access these critical IT systems?

These are types of questions that are handled by an intelligent and autonomous Business Resilience System in place in any organization of enterprise.

2.4 Predictive Modeling

As part of BRS component and feature of smartness, we need some predictive modeling. We need to examine the nature of this model and review its benefits for enterprise management and business processing orbit for PDP. Our focus in this section is directed toward the use of prediction in the overall enterprise management scheme

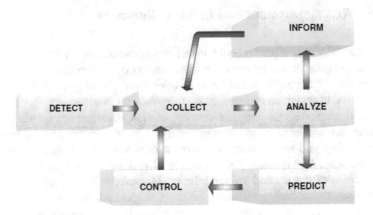

Fig. 2.2 The enterprise performance cycle

and how various kinds or modes of predictive model can benefit the ordinarily inelastic management techniques used in conventional analysis and reporting processes. Thus, looking at the general nature of the enterprise management cycle is a necessity for BRS and its PDP nucleus.

2.4.1 Prediction in Enterprise Management

Prediction is an essential component of Business Resilience System and of any comprehensive enterprise architecture and survivability of enterprise as its day-to-day Business Continuity Process, since it couples analysis to control of business continuity management (BCM) and business process management (BPM). This coupling forms part of basic management and analytical phases of BRS driven that a mature enterprise should incorporate to deeply understand and administer the performance of their system. Figure 2.2 here is a depiction of a high level of this process as subset of their Business Resilience System.

As it can be seen in Fig. 2.2 schematic, prediction sits on six different nodes as part of the analysis and control cycle in the management architecture. Uber prediction model depicted here depends on analysis and is controlled by these nodes as part of prediction process. Each node is explained here as follows [4]:

1. **Direct**
   Detection is the process of recognizing and quantifying the physical and logical, i.e., software elements of an enterprise, and extends both vertically and horizontally across the IT platforms as well as the conventional lines of business requirements.

2. **Collect**

   Once the physical assets comprising the enterprise have been identified, we need to collect information about their operations—typically, this is in the form of performance measurements—transactions per second, response times, machine utilization statistics, on-hand inventory levels, point of sales completions, and so forth.

3. **Analyze**

   The core of nearly all performance management tools is the set of analyses operations that are applied to the data. In most systems, this consists of standard statistical measures, correlation analysis, rates of change management, and the detection, through statistical means of the normal operating characteristics of machines, databases, web sites, and client applications. These normal operating characteristics are then used to detect anomalous behaviors.

4. **Inform**

   Almost all the produced reports for upper management and operational D- and C-level executives derive from the analysis component. Analysis is not only the process of statistical measurements; it is also the process of filtering, aggregation, higher-dimensional organization, and generating new values from the mathematical treatment of many other values. Reports reflect these analyses either as formal documents or through graphical visualizations.

5. **Predict**

   Prediction comes in a variety of flavors, and some are easier to understand and implement than others. The basic principle of prediction, however, remains the same: forecasts future behavior by understanding, as deeply as possible, past behavior. Prediction differs from analysis in its goals. Analysis attempts to determine the interplay of various components and how and to what degree; they comprise a working system. Prediction takes one or more of the performance indicators and predicts, with associated degree of confidence, their values in the near or far future.

6. **Control**

   Control is the ultimate goal of enterprise management—the ability to fuse analysis, prediction, and autonomic computing into a self-monitoring and self-healing environment. In order to control a process, fundamental control theory tells us that we need to understand the system dynamics and we need to predict future behaviors so that anomalous performance drift can be detected and corrected. Control of all but simple systems is currently beyond modern enterprise management capabilities [5], but limited types of control, mostly concerned with recovery and load balancing, are supported through forms of fault tolerance and redundant fail-safe measures.

The flowcharts of processes, illustrated in Fig. 2.2, are naturally, high-level perspectives on the functional and operational services in a real system, and the boundaries between many processes are not so clearly defined. Within a complete system, analysis, reporting, and prediction are often combined to handle root cause analysis, capacity planning, Service Level Agreements (SLA), multidimensional visualization, and so forth.

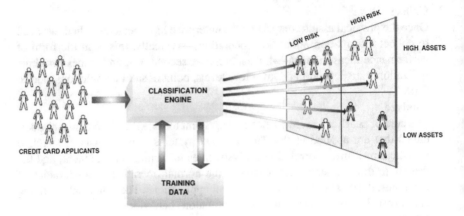

Fig. 2.3 Credit card classification model [4]

The nature of prediction mathematically has two distinct types of activities. These two types are identified and each one description is as follows:

1. **Prediction by Classification**

   In prediction by classification situation, a value is assigned one of *N* possible categories. The categories are set up by the model designer and values are placed in a category based on their attributes. This assignment process comes from a set of training data in the form of Boolean logic (i.e., true or false) that examines historical data and their classification by model builder. As a rather simplified example of prediction by classification, we can look at a company who is issuing credit cards. The issuing credit company might like to predict the probability or risk that an applicant will default on their payments. Furthermore, they would like this risk broken down by applicant's total liquid assets. If the model divides applicants into two risk categories such as low and high and two assets categories such as high and low, then the classification, as illustrated in Fig. 2.3, supports 2 × 2 outcome space.

   This model assigns an incoming applicant to one of the categories and predicts whether or not the individual will be a high or low risk within the applicant's asset class. One way to build this model is through a machine learning technique such as neural networks or decision tree algorithms like the Business Resilience System (BRS). These approaches discover the behavior of an individual in a class from the behavior of the class member. By collecting a historical sample of good, moderate, acceptable, and bad credit card holders, the objective function, with their known attributes, will function. The variable of this are elements, such as age, occupation, income, residence and job longevity, home ownership, and their total liquid assets minus home valuation; a model can assign a new individual to a class based on how well he or she matches the properties of a class.

   In this way, the model is making a prediction in risk of default or late payments for a class of individuals predicated on economic, demographic, employment,

Table 2.1 Dependent data (i.e., Boolean-type data)

Cases	Independent		Dependent
	x_1	x_2	y
1	10.5	0.03	30
2	8.7	0.07	54
3	12.0	0.01	16
4	9.4	0.02	18
.	.	.	.
.	.	.	.
n	8.3	0.06	48

and other factors. Of course, a more straightforward model could simply categorize candidates into two classes:

(a) *Accept*
(b) *Reject*

These two above steps are part of the workflow and engine rule implemented into the model. Other forms of classification models can work on continuous representations, so they can classify candidate by his or her degree of risk or degree of assets.

2. **Prediction by Forecasting**

Prediction by forecasting is primarily used to forecast the values of other indicators in the enterprise performance measure based on the values of other indicators, which we hope *independent* variables affect the outcome of these key indicators. As an example, Table 2.1 shows a set of independent and dependent variables.

The purpose of a predictive model is to develop a method of predicting "y" using the values of "x_1" and "x_2." The accuracy of this prediction that is called the standard error of estimate is a measure of the model's reliability in estimating the dependent variable for new cases. There are many types of predictive models, but nearly all of them work on the same principle:

Discover and Quantify past Behavior and Habit in order to Forecast Future Behavior

The above message is the basic and principle of foundation of *Business Resilience System* that has been called **BRS** so far. Past *behavior* and *habit* are the databases which are the information that we need to be able to predict, to classify, and to forecast the future behavior. Such databases if in the form of fuzzy or Boolean format allow the PDP to inform stakeholders in organization or enterprise the possible occurrences of an inevitable event. The unwanted threat, either man-made or natural, political or ecumenical, or otherwise, that has adverse effects can be prevented if they are able to make the right decision for right countermeasure, so the BCM, BPM, or BCP is not going to be interrupted. Thus, continuous flow of data in a trusted format of Fuzzy and Boolean combined, from every possible direction driven by cloud computation needs to be accessible by the PDP of risk atom within BRS, if not real-time, but at least near real-time is an ideal concept.

The need for information available to PDP in the form of data is a must principle; thus having an intelligent model from data mining and thus expert knowledge is required for BRS to be functional and carry on the task that is assigned to it for prediction and prevention of hostile threat. The threat can be defined per application, where BRS needs to be operational and requirements assigned to it from SLA point of view.

Note that the technique used to create a forecasting model is called *regression analysis*. Among the most common forms is the linear regression, which attempts to fit a straight line through historical data points in such a way that it minimizes the error between the line and the actual data points. A linear regression model assumes that the underlying equation for the line looks something like the following form:

$$y = a + bx_1 \pm \varepsilon \qquad \text{(Eq. 2.4)}$$

The equation is for a model with one outcome-dependent variable and one independent variable.

The linear regression analysis predicts a value for "y" by finding the value for "a," called the intercept, and "b" which is the coefficient of the slop and measures the rate of change in the data. In practice the equation also includes a measure of random noise ε, which is a measure of the uncertainty or error in the estimate. Note that the name simple linear regression is somewhat misleading. It is the model, not the method of fitting, that is simple.

By using various kinds of polynomials, a regression analysis can also find the equation for nonlinear data such as data that has periodic or seasonal behaviors. Figure 2.4 illustrates the concepts associated with regression forecasting. In simple term, then, a prediction is an estimate of a future value from a collection of historical values. Nevertheless, not any future value will do. The value must be a credible

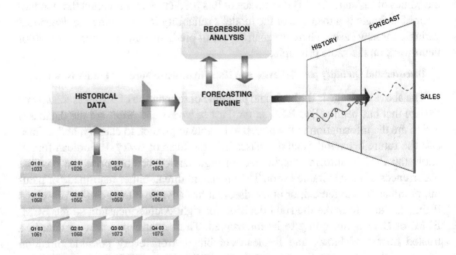

Fig. 2.4 Time series prediction model

value based on sound statistical or mathematical foundations. However, the prediction must be liable to assessment by the user in terms of its utility and its accuracy. In this case, every point in a prediction can be viewed as an occurrence in three-dimensional space as (x_i, c_i, p_i) [4], where x_i is the value; c_i is the certainty that this value is valid, based on the underlying evidence; and p_i is the periodicity of the model for which the prediction was made. Understanding the nature of a prediction is crucial in making sound judgments from the model. This is important whether the outcome is interpreted by a human or is used by a predictive component embedded in a large system.

2.4.2 The Benefits and Predictive Modeling Basics

Predictive modeling is a name given to a collection of mathematical techniques having in common the goal of finding a mathematical relationship between a target, response, or "dependent" variable and various predictor or "independent" variables with the goal in mind of measuring future values of those predictors and inserting them into the mathematical relationship. This allows to predict future values of the target variable. Because these relationships are never perfect in practice, it is desirable to give some measure of uncertainty for the predictions, typically a prediction interval that has some assigned level of confidence like 95%.

Building an intelligent predictive model as part of BRS infrastructure requires some understanding of functionalities that support this infrastructure and features and concepts of predictive models that need to be layout driven in examining the ways in which organizations or enterprises approach the model building their process. A wide variety of techniques emerge, and often the model centers around the kinds that we need and how the model addresses such issues as the nature of the data we receive on minute-by-minute fed to Processing Data Point (PDP) of Business Resilience System (BRS). This model should utilize the reasoning processes that manipulate the data to correct trigger point warning displays in BRS dashboard and finally the type of outcomes the model can produce for stakeholder to take the proper measurements.

Another task in the process is model building. Typically, there are many available potential predictor variables, which one might think of in three groups: those unlikely to affect the response, those almost certain to affect the response and thus destined for inclusion in the predicting equation, and those in the middle, which may or may not have an effect on the response. For this last group of variables, techniques to test whether to include those variables have been developed, and research on this "model building" step continues today. This paper addresses some basic predictive modeling concepts and is meant for people new to the area. Predictive modeling is arguably the most exciting aspect in the emerging and already highly sought-after field of data analytics. It is the way in which big data, a current buzzword in business applications, are used to guide decisions for smart business operations.

The benefits of predictive modeling have both tangible and intangible positive impact to the organization and enterprise. The intangible benefits include yet not limited to a sense of empowerment and a sense of cohesion in the organization and enterprise which will bring an ability to look into future at level that is more granular. Justifying predictive models, however, rests on a collection of tangible benefits and requires collection of trusted databases for stakeholder to set the right Service Level Agreement (SLA) in place. This SLA then allows the predictive model to feed the right information to BRS in real-time mode. Some other aspects of tangible benefits are returns that are associated with both the performance of enterprise management system and the various line of business process management (BPM) that rely on the management system.

Reviewing at high level some of the benefits of the model and the basics of creating and using a predictive model could be pointed out here, and they are as follows [4]:

1. **Deeper understanding of system behavior**
   A deeper understanding and awareness can be gained through an analysis of key indicator predictions. This knowledge allows a better understanding of how the system performs over time (i.e., behavior) and under varying kinds of stress. With this gained knowledge and understanding, we are able to fine-tune the system behavior. System may be defined as political threat, economical threat, and natural threat or man-made, including terrorist threats. Tuning a system improves performance and throughput with a resulting reduction in long-term acquisition and support costs. To understand the system behaviors deeper will allow both analysts and management to isolate business, machine, staffing, and financial anomalies that, cumulatively, impair customer service; increase machine, support, and operational staffing costs; affect vendor relationships; and add an increased level of uncertainty to the day management of the machine resources supporting vital of business in the enterprise.

2. **Improved Service Level Agreement compliance**
   Understanding the behavior of serve systems and their application mix at various levels of load and predicting the throughout performance under conditions of increased pressure provides key indicators in forecasting the ability to meet various kinds of Service Level Agreement (SLA). Because an SLA is often accompanied by cost penalties, predictive models can forecast windows of opportunity in shifting loads, predict situations when the current job mix will exceed SLA maximums, and provide justification for the acquisition or upgrade of equipment such as servers, network routers, or load balancers.

3. **Higher fault tolerance and system reliability**
   Predictive models, combined with statistical models of normal behavior, can predict critical threshold violations. If these thresholds are based on the availability of limited resources or tied to the maximum carrying capacity of networks or processors, these predictions can be used to interdict faults before they occur, shift loads in anticipation of bottlenecks, and change the job mix before a critical, limiting resource constraint is exceeded.

4. **Better load balancing**

   Related to improved service level compliance, predictive models keyed to forecasting the periodicity of service loads can provide an excelled means of identifying way of distributing work over server clusters or over distributed systems. Better load balancing not only improves overall system performance but also increases total capacity and provides better computing resource utilization. By reducing the task or job queue on server, the average wait time for jobs is reduced and the throughput is improved. This is the result of Little's theorem [6].

   This theorem relates averaged parameters of a queue in the steady state. It is especially useful, because it makes no assumptions about the arrival pattern or service time distribution. In another word, it can be stated as:

   "The average number of customers is equal to the arrival rate multiplied by the average resident time. This theorem can be applied equally well to the queuing system as a whole, which means queue plus server or just the queue by itself" [4].

5. **Faster error diagnosis, recovery, and error aversion**

   In the same way that modeling can isolate overloads and bottlenecks caused by thresholds tied to limiting resources, the forecasts can quickly identify the causative agents in machine, database, and application failures facilitating a rapid restart. The same forecasts can be used in a diagnostic tool; they can also be used as an early warning system to avoid potential errors. This use of prediction allows a high availability of machine and other computing resources, improving performance and overall throughput. Rapid error diagnosis and repair also reduces staff time, limits cost for overtime, and allows operation staff to remain focused on important business objectives.

6. **Deeper understanding of business objectives and relationships**

   Perhaps one of the most important benefits of predictive modeling is its ability to give line of business managers and executive a deeper insight into the behavior of customers and vendors as well as a more thorough understanding of how business policies and markets affect such important bottom-line issues. These issues are including events, such as sales, inventory levels, shipping, or in general supply chain, as part of BCM and BPM and finally return. Forecasting can, as an example, identify the buying habits and behaviors of important customers based on seasonal or other kinds of periodicity cycles. Amazone.com is a good example. Knowing the cumulative or aggregate buying habits or behaviors of major customers can provide the intelligence necessary to plan campaign, synchronize vendor shipments and inventory stocks with buying behaviors, reduce or increase product lines, assess sales efficiencies, assign or redistribute sales territories, and judge the effectiveness of marketing strategies. This naturally allows a better supply chain in place for organization in support of their different stores based on their Key Performance Indicator (KPI) and allows a better Price Promotion Optimization (PPO) in place both in front stores or online and e-commerce point of view.

7. **Ability to address and answer business strategy decisions**

   Predictive models provide an extended glimpse into the future and offer important insight into the behavior of customers and vendors. As a result, they are able to answer a number of difficult questions that analytical methods cannot completely address. Cast as questions about future revenues and future attrition rates, or future actions, predictive models are well equipped to answer questions such as:

- Is this customer or prospect, or Web browser likely to buy my product?
- Is this customer likely to respond to my credit offer or PPO?
- Is this a profitable area for an expansion of existing or new services?
- What is the potential lifetime value of this customer (as part of Customer Relation Management (CRM))?
- What other services or products can I sell to this existing customer (i.e., PPO)?
- How many widget will be purchased in the next months, which falls in supply chain process, inventory control, and at higher-level as part of business continuity management (BCM), business process management (BPM) process, and eventually Business Continuity Process (BCP) process?

8. **Better and more reliable strategic planning**
   Strategic planning is corrected with the long-term stability and growth of the organization, the acquisition of new lines of business, and the exploitation of existing business opportunities. Predictive modeling at the application and resource consumption level provides executive and their planning staff with the insights they need to make sound judgments about the allocation of capital, staff, and other forms of renewable and nonrenewable resources.
   The benefits derived from predictive modeling derive not only from the ability to look into the future but also from the underlying nature of time series and behavior-based models. A strong data management and analytical infrastructure must be present in any predictive modeling in cases where the forecasting abilities of a model are seriously compromised. Much of this is discussed in the following sections later on.

In summary, the benefits derived from predictive modeling derive not only from the ability to look into the future, but also from the underlying nature of time series and habit and behavior-based models. A strong data management, data mining through Boolean and fuzzy logics, and analytical infrastructure must be present in any predictive modeling in order to insure that the predictions are based on solid evidence. Without these foundations, the forecasting abilities of a model are seriously compromised. The data modeling, data mining, and data management will be discussed in the next chapter.

2.4.3 Basic Infrastructure of Modeling

This section will provide the foundations or very high-level background that we need to have in order to have some idea about the concepts and features, which possibly drives the basics of predictive models. In examining the ways in which

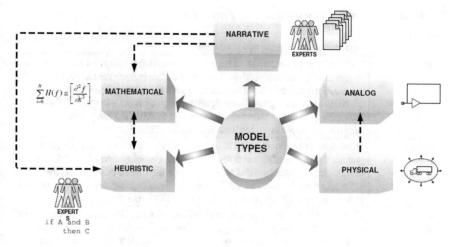

Fig. 2.5 Top view of types of models

organizations or enterprises approach the model building process, for their Business Continuity Process (BCP) and a wide variety of techniques that is emerging from it.

These techniques often center on the kind of models we need and how the model addresses such issues as the nature of the data, the reasoning processes that manipulate the data, and the type of outcome the model can produce. Now review, at a high level, the basics of creating and using a model, by first interpreting the model from an engineering point of view.

Systems and knowledge engineers use the word "model" in a variety of contexts, but nearly all of them refer to some digital implementation of a well-defined process. However, the world of models and model building encompasses a wide variety of representations. Although we are primarily concerned with basic design of system models, the evolution of such models often passes through or encompasses other modeling organizations or enterprises.

Figure 2.5 is a top view of the basic model types and some of their possible interconnection components that are considered.

Naturally, neither the model taxonomies and data governess nor the model boundaries are absolutes for any organization or enterprises. From each entity point of view, these rules, taxonomies, governess, and boundaries will need to be modified, so it fits to that particular entity or organization. As dashed lines in Fig. 2.5 illustrates, one model may be prelude to another. We often develop a narrative model before expanding our ideas into a mathematical or system heuristic model.

Further the classification of a model into one class or the other is not always possible—the boundaries are very permeable.

Table 2.2 is summarizing how these models differ and how they are used.

The last two taxonomic classes—mathematical and heuristic—form the focus of nearly all business prediction models. These are roughly classed as symbolic models and generally incorporate both algebraic as well as intellectual relationships.

Table 2.2 Model categories, taxonomies, and government [4]

Nature of model	Organization/enterprise and used cases
Native	These models often consist of experts collaborating on a formal representation of some highly focused system. Examples of this include the Delphi method [1] and many aspects of market research. In such models a collection of questionnaires are repeatedly analyzed among the experts to typically predict such critical issues as vulnerabilities, market trends, long-range sales and margins, and new line of business penetration strategies
Physical	These models are constructed to test or evaluate some essentially physical system. Examples of physical models include precision aircraft models for use in wind tunnels, architectural renderings of building and bridges, and the molecular construction kits used in organic chemistry and genetics to represent such things as benzene rings and the helical DNA molecule
Analog (simile)	These models combine the properties of one system to describe the behavior of another system using different physical forms. Thus they are *similes*—one systems function like another systems. Analog models work because there is a similarity or parallelism between the underlying forces that drive both models. Until the recent accessibility of the personal computer, electrical and mechanical analog systems were routinely used to model complex networks such as process plants and highway traffic flows
Mathematical	Mathematical models and statistical models in this category came into their own right with the common availability of digital computers through timesharing in early days and now routinely on desktop and laptop personal computers. A mathematical model consists of equations, often with interdependencies. The ubiquitous spreadsheet is a prime example of a mathematical model. Control engineering systems and embedded controllers are also examples of mathematical models. Many of the models used in knowledge discovery and data mining are also mathematical models. These include neural networks and other forms of classification schemes, such as decision trees [1], classification and regression trees (CART), and chi-squared automatic induction (CHAID) algorithms
Heuristic	The recent rise in machine intelligence and expert system technologies has introduced another type of model into the mix—the heuristic model. These are often called **if-then-else** production systems and form the core knowledge repository of today's decision-support and expert systems. These models depend solely on the high speed, high computational capabilities of the computer. Heuristic models embody "rules of thumb" and other business processes. We often refer to them as policy-based models, and heuristic representations comprise the vast majority of modern Business Process Models (BPMs)

In the modern sense of word model, we generally combine the two taxonomies into a single or hybrid class: the *knowledge-base model*.

2.4.4 Models and Event State Categorization

There is, however, a more fundamental classification of models based the way model states are generated and how model variables are handled. Table 2.3 shows the partitioning—into discrete or continuous and into deterministic or stochastic.

Table 2.3 Model classification by internal structure [4]

		Event state	
		Discrete	Continuous
Model state	Deterministic	Spreadsheets	Dynamic flow
		Econometric	Process control
		Budget	Differential equations
		Inventory control	
	Stochastic	Quality measurement	
		Queuing models	
		Market share	
		Price positioning	

From technical point of view, the classification of models as discrete or continuous refers to the model's composite variable organization, but in actual fact and practice, the term is used to describe the model's treatment of time. Many models have clear and practice the term is used to describe the model's treatment of time. Many models have clear and precise demarcations, which break up the system into regular intervals, such as many queuing applications, production or project schedules, traffic analysis models, portfolio safety and suitability models, etc.

Other models have a horizon that varies smoothly across time. These continued models are rare in the business world, although they do occur in nonlinear random walk models that attempt to follow the chaotic trends of the stock market. In any event, of course, few continuous models are actually implemented continuously but use a form of periodic or random data sampling.

The model state categorization—and a fundamentally important perspective on how the model is constructed—reflects the ways in which the underlying model relationships are or can be described. Outcomes in a deterministic model can be predicted completely if the independent variables like input values and the initial state of the model are known. This means that a given input always produces a given output. The outcome for a stochastic system is not similarly defined. Stochastic models have an intrinsic degree of randomness coupling the variables with the ongoing state of the model.

2.4.5 Model Type and Outcome Categorization

From a knowledge-base and systems engineer's perspective, the most critical model classification isolates the kind of outcome. There are two broad types:

1. **Predictive**

   A predictive model generates a new outcome in the form of the value of the dependent variables based on the previous state of independent variables. Regression analysis, as an example, is a predictive time series model. Given a least squares linear interpretation of a data points $x_1, x_2, x_3, \ldots, x_n$, the regression model predicts the value of point x_{n+1} and a vector of subsequent points with varying degrees of accuracy.

2. **Classification**

On the other hand, classification models analyze the properties of a data point and assign it to a class or category. Cluster analysis is a classification model. Neural networks are also predominantly classifiers—they activate an outcome neuron based on the activation of the input neurons. Such taxonomies, however, are hardly rigid. A rule-based predictive model can, under some circumstances, be viewed as a classifier. Moreover, many classification models, especially those based on decision tree algorithms, can generate rules and governess that produce a model prediction.

Like taxonomic partitioning, these types are hardly "pure" and models often involve both. Table 2.4 shows model families sorted into cluster according to the underlying model taxonomy and the outcome type.

Now, we need to examine the principal techniques used to build predictive models in a business enterprise. Each of the approaches is summarized in terms of its features and approach to classification and prediction. Robust predictive models form the core of the analytical tool kit for most modern decision-making. As such, it is critical that the decision-maker understand the nature of the underlying model and the trade-offs that are made between functionality, simplicity, and accuracy.

Nearly all useful models are predictive. In fact, the rationale behind constructing a model must in some either specific or general way be the forecast of some future properties of the system based on an understanding of how the model is working in the present. In general, if we account for the mechanics underlying the model itself, there are three basic types of predictive models, each derived from their fundamental relationship between how data is understood and used in the prediction.

Table 2.4 Model classification by outcome [4]

		Outcome	
		Predictive	Classifier
Model type	Knowledge based	Expert systems	Dynamic flow
		Fuzzy systems	Process control
		Evolutionary	Differential equations
	Mathematical	Regression analysis	Cluster analysis
		Statistical leaning	Classification and regression tree (CART)
		Correlation	Self-organizing maps
		Radial basis	
		Functions	
		Adaptive	

These types are:

1. **Qualitative Methods and Techniques**
A qualitative or methodological model is a form of the narrative approach to model formulation. Experts, market segments such as focus groups, and knowledge about special situations are formally organized in order to understand the problem and make a prediction. Quantitative approach may or may not consider the past in their prediction. Although a qualitative model may rely on statistical and mathematical information, it is, by its nature, a static representation of the narrative at some point in time.

2. **Time Series Analysis and Regression**
Time series or regression analysis is one of the oldest and most frequently used of the mathematical models. A time series model is almost completely dependent on historical data. The regression analysis finds a line through the data that represents the underlying trend. From this trends future data points can be predicted with varying degrees of certainty. Figure 2.6 illustrates a simple regression model of sales over time.

Time series analysis is capable of modeling highly complex data. Advanced forms of regression that combine nonlinear analysis with statistical learning theory can uncover and model the periodicity of the data. That is, the model can find and separate out patterns involved in daily, weekly, monthly, or other seasonal variations and cyclical patterns that repeat on a more or less regular basis, as well as the rates of change of each trend.

3. **Causal or Behavior Models**
Behavior models that also should include habits from these authors' point of view in particular when human being is involved are concerned with discovering, quantifying, and exploiting the relationships between system components and are almost always powerful enough to account for anomalous changes in the model environment.

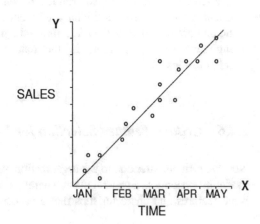

Fig. 2.6 Time series regression line

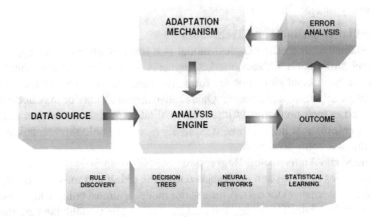

Fig. 2.7 Behavior models with a feedback mechanism

Since they are generally based on an understanding of changes over time and over other dimensions, the behavior model relies a great deal on historical data. Behavior models are extremely powerful ways of uncovering and modeling highly complex relationships in data. Unlike regression analysis, which lacks any connection with the "laws" of the underlying system, a causal or behavior model can work in very sophisticated ways—as an example illustrated in Fig. 2.7, behavior models can involve adaptive feedback, the ability of the model to learn from its mistakes and change to new outside pressures.

Behavior models differ from their regression model types in a more robust and insightful way. Their underlying analysis engine often incorporates a variety of techniques drawn from the fields of artificial intelligence (AI) and machine learning built in AI system from its inventor, the mankind, and its knowledge-based information. These techniques include *rule discovery*—the process of discovering if-then-else rules—which can explicitly define the conditional behavior of a system, *decision trees*.

Tree-like classifiers that also produce if-then rules; *neutral networks*—high non-linear classifiers from artificial intelligence that work in ways similar to the human brain; and *statistical learning*—the process of uncovering and quantifying time varying patterns in the data with their possible lead and lag relationships.

2.4.6 Critical Feature Selection for Predictive Models

Now we turn our attention to the organization, properties, and features of predictive modeling systems as a general class. Therefore, in this section, we will examine the basic features and characteristics that a particular predictive model must have in order not only to be trustworthy but also to remain robust and trustworthy. We also examine some of the external pressures, impacts, and misconceptions that contribute to problems in the design and use of a predictive model. The list of those features crucial to the success of any forecasting model is listed in the following contents in this section.

Generally speaking there are no such things as a "correct" model representation for a particular system, although there is obviously a wide range of incorrect models. In selecting a model approach, the designer must weigh the objectives of the modeling project against the available modeling selection process approach properly; the main concern of the decision-maker becomes one of model validity. This problem of validity is complicated and is a central issue when executives and stakeholder use models to set policy or launch organizations on new course of action, such as entering a new line of business. Over all, model validity depends on a number of factors and they are:

- Insuring that the modeling approach is appropriate for both the study as well as the underlying system.
- Insuring that the model incorporates all the necessary components to adequately represent the portion of the real-world system under study.
- Insuring that the components are organized and connect properly.
- Insuring that the model behaves at the level of granularity and over the required time frame in a manner close to the behavior of the real world. This involves not only a measure of the model's error rate but also an understanding of its limitations and how quickly, as example, does the error of estimate increase as the predictions move into the future.

As a result, validity is often a question of how well the final model satisfies its design criteria. Since model always expects an abstraction of the underlying real-world system, it can never be expected or should never precisely match the real world. We can, however, improve the validity of our predictive models by understanding the properties of various model types and what features should be central to any predictive model.

2.4.7   Core Features of Robust Predictive Models

Predictive models represent a particular class of computer models. They have special architectural, functional, and operational requirements that extend beyond spreadsheet, simulation, and optimization models. We can conclude this discussion of critical feature selection with a list of the core features necessary to support a reliable and trustworthy model.

1. **A Concern with Evidence**

   A predictive model must be concerned with two forms of evidence. The first is the amount of historical data used to make its predictions. The higher the number of dimensions in the model as, for an example, the number of separate periodicity cycles that the model must consider, the more data the model will require. A lack of sufficient data reduces the evidence associated with the underlying regression. This is the difference between forecasting the future with three data points or three hundred data points. The second form of evidence is the amount of certainty associated with each forecasted data point. The further into the future a prediction looks, the less certainty is associated with each predicted point. A model must provide a measure of the certainty for each predicted point.

2. **The Ability to Tolerate Noise**

   Noise is a measure of information entropy—which is a measure of the amount of randomness in the data. This randomness is introduced into data through a variety of causes such as:

   - Sensor failures.
   - Transcription errors.
   - Missing data.
   - Illegal data types and entrusted data.
   - Numeric overflow or underflow.
   - The appearance of overlaying patterns (i.e., a time point t_n along the horizontal axis). This time point has several data points for product sales at this time interval—some belong to regular customers. However, some belongs to a client that places large orders for the product at regular intervals, and t_n just happens to be at one of these interval points.

     Robust predictive models must tolerate noise, but must also consider the ratio of noise to normal data in assessing the utility of the data. Increasing noise levels affect the underlying evidence associated with a prediction.

3. **A Concern for Self-Measurement**

   Predictive behavioral models must be concerned with the level of compatibility between their rules and their predictions. This is especially true for deployed operational models that have been successfully validated. Because rule-based models do not reply solely on historical data to generate a prediction, their forecast can become increasingly error prone if the data begins to shift in value and the model does not use an adaptive scheme to relearn its rules. A behavior model can treat the average degree of evidence in each prediction as a data point for that execution of the model. By performing a time series regression on its own collection of evidence, a model can measure its own reliability and detect an unacceptable increase in the standard error, that is, a high positive or negative rate of change for the trend line through the evidence value indicates a lack of trustworthiness in the model.

4. **Ability to Generate Nonlinear Regressions and Predictions**

   Finally, a good predictive model, whether qualitative, time series, or behavioral, must be able to evaluate and use highly nonlinear data. Nonlinearity is not the same as non-deterministic nor is the same as stochastic (random). Nonlinearity means that the outcome of a process is not linearly correlated with the inputs, so a small change in one or more of the model properties can sometimes produce an unexpectedly large change, or no change, or a rapid deceleration in the outcome. All but the simplest business and public-sector problems involve nonlinear elements. Models that consider market growth or penetration rates; competition pricing, including Price Promotion Optimization (PPO); capital budgets; and risk assessments, especially risk models that use non-equilibrium game theory, are all faced with nonlinear predictions.

2.4.8 Model Type Selection

Selecting the proper type of predictive model depends on the goals of the study. The model type may also be influenced by practical considerations—the availability of historical data, the accessibility of experts, and, in some instances, the political pressures. There are, however, some basic factors that mitigate for and against the selection of a particular model type.

1. **The Qualitative or Methodological Model**
   Either model based on the knowledge of experts directly with the use of experts in evaluating questionnaires or surveys can provide a powerful tool when data is scarce, such as new product market strategies, mergers and acquisitions. This includes project risk assessment or when the competitive and marketing pressures of the outside world are undergoing rapid changes. These changes are such as the emergence of the personal computer and the rapid emergence of the Internet. These models fuse human judgment and various forms of rankings to transform qualitative knowledge into quantitative predictions.

   Qualitative models however are often subject to high degrees of error, usually stemming from the corporate culture itself. There are several common factors that lead to this error rate, and they are:

   - First, if the study team is dominated by a perceived "expert" in the field, the expert's influence prevents the other study team members from challenging the expert's view of the world.
   - Second, when the experts are under political pressure to view a particular process in a positive or negative light or to confirm a judgment made by senior executives, the model is likely to become little more than an exercise in creative writing.
   - Third, when a study team is composed of true experts, who are often senior executives, they frequently come to view the study as an unwelcome diversion from their management responsibilities. This attitude has serious implications on the depth and validity of the final study outcome.

   Qualitative models incorporate a systematic and formal methodology for collecting information, evolving hypotheses, and generating forecasts. They are excellent modeling tools for multiple collaborating experts and are very good, when used in an open and undirected way, at forming a consensus from conflicting experts in a variety of related disciplines. Qualitative models are used effectively in market research, new line of business exploration, new product development, risk assessment, forecasts of short- and long-term sales, and market barrier penetration.

2. **The Time Series Analysis and Regression Model**

   All statistical models are predicated on the assumption that the future is like the past. When this assumption is more or less true, a time series forecast provides a powerful method for uncovering not only general trends but also the rates of change in this trend and how fast the trend is moving either up or down, over some interval of time. It is this up or down rate of change, which forms the basis for making a forecast from a time series.

   Using a time series analysis requires a sufficient level of historical data—usually several years worth of data since it is important to have enough data to account for multiple data points in each repeating time period over the model's time horizon. A time series model must be validated by comparing its predictions against a set of known values. Thus, the collection of historical data is divided into a training set and a validation set. The training set is usually the larger of the two and, as the name implies, is used to train the model. The model is then run against the validation set, and the difference between its predictions and the actual values is measured. This difference, called the standard error, is a measure of the model's accuracy.

   Time series models are simple to understand and, due to the availability of regression software in Visual Basic, C++, and Java, are certainly the most common class of predictive business model. The fact that they are widely used does not mean, however, that they are used correctly. In practice, they are difficult to use properly.

   - First, a difficulty arises from a lack of data. If we want to predict monthly sales, having data for 12 months (1 year) is insufficient. You would need several years of data to be sure that you have enough points for each January, each February, each March, etc.
   - Second, it is generally difficult to make accurate predictions from raw data. The underlying trends and rates of change are often confused with noise (transcription errors, missing data points, etc.) and seasonal and other regular periodic variations, as well as occasional distortions due to marketing campaigns or severe, short-term economic downturns.
   - Third, time series analyses tend to be heavily influenced by a few data points that are significantly different from the normal data in any time period. These are called outliers. Sometimes outliers are recording errors, sometimes they are simple anomalies, but other times they might represent a long-term periodicity that is unrecognized by the level of granularity in the current data.

Time series and regression models are very good models for estimating and predicting the future from extensive historical data. A time series model can decompose data into its underlying periodicity curves, isolate irregular patterns, and develop both a periodic-based trend and the rate of change along each of the trend lines. Time series prediction is used in the forecasting of both short- and long-range product sales, anticipated margins, production and inventory levels, cash flows, cash balances, capital budgets, and threshold violation warnings.

3. **The Causal or Behavior Model**

While behavior models can be mathematical (such as economic models), the current generation of behavior models employs statistical, time series, and advanced heuristic techniques to capture the deep nature of very complex business systems. It is the core set of heuristic or knowledge-base rules that allows a behavior model to represent highly complex and often (nonlinear systems). Behavior models come in several flavors.

- The most common is the knowledge base or expert system—a collection of if-then-else rules describing the system's behavior and derived from one or more experts. Expert system capabilities not only provide the behavior model with easy-to-understand rules, but they directly incorporate facilities for goal seeking, scheduling, and configuration analysis.
- Another behavior model, usually more appropriate to classification than prediction, is the neural network. This model learns behavior patterns from historical data in a way similar to time series regression. It is validated in the same way as well. A neural network trained to recognize, as an example, applicants that are likely to default on a mortgage or become a bad credit card risk can "predict" a mortgage failure by classifying a candidate in the probable foreclosure category. Neural networks have the added disadvantage of requiring numeric input, so that categorical string data must be converted to some kind of enumeration sequence.
- Another form of behavior modeling involves the construction of decision trees—decision trees like neural networks are generated from historical data and validated in the same way. A decision tree creates implicit if-then rules that can classify a model state with a minimum amount of information. Like neural networks, this classification can also serve as a prediction.

More comprehensive and advanced behavior models combine expert rules with rules discovered from the data. Thus they incorporate a concept known as data mining or knowledge discovery with the model itself. The data mining feature allows the model to find and quantify the behavior rules of a system directly from the data that flows into a system and the outcomes generated by the system. The same data mining capability means that behavior models, especially those that use learning and adaptive feedback, can modify their behavior based on changes so that their view of the future is not patterned exclusively on the past. Figure 2.8 shows a schematic representation of a typical rule-based behavior model with rule discovery and an adaptive feedback loop.

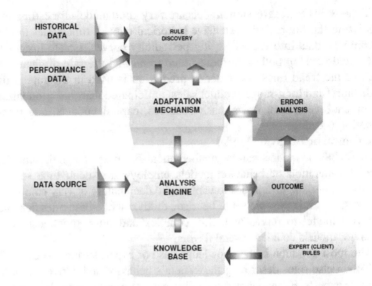

Fig. 2.8 A behavior model with rule discovery

    With their ability to combine both expert rules and rules derived from data, behavior-based models are ideal when you lack of sufficient historical data; have access to committed, available experts; or are dealing with rapidly changing and evolving environments. A behavior model is a good combination of the qualitative or methodology approach and the analytical capabilities of a time series and mathematical model. For all their robustness and power, however, predictive behavior models are often subject to modeling shortcoming that has little to do with their expressive power.

    There are a number of critical issues confronting analysts who select behavior models.

- First, they require a rather deep understanding of the interplay between rules and supporting technologies such as statistical learning and time series regression.
- Second, the model designer must have a way of either acquiring the rules from a set of experts or sufficiently understand the nature of rule discovery so that rules derived from data can be seamlessly integrated into the model's logic.
- Third, understanding the mechanics of an adaptive feedback mechanism is a nontrivial exercise. In the case of a rule-based system, adaptive feedback means generating new rules from the data and then merging these in some way with existing rules. Some mechanism must be in place to decide which rules to retain from one adaptation cycle to the next.
- Fourth, behavior models that incorporate time series analysis into their reasoning engines are subject to the same constraints on the data as time series regression models discussed in the previous section.

Behavior or causal model is excellent tools for modeling, studying, and predicting the functional relationships between sales and other economic, competitive, and external factors, modeling interindustry, cross-sector, and interdepartmental flows of goods, services, inventories, and products within a changing economy and markets. Behavior models are also sensitive to the rules and constraints placed on the underlying representation, so they are ideal choices for sensitivity analysis and goal seeking. Because of their complexity, behavior models are also well suited to predictive economic, multiple constraint regression, diffusion index, and leading indicator models.

2.4.9 Basic Principles of Predictive Models

Model selection, discussed in the previous section, is concerned with mapping the proper analytical approach to the purpose and scope of the study itself. But underlying any predictive mode is a set of principles that should guide us in the formulation of the model regardless of the analysis approach. These principles take two forms, those that interfere with the design and implementation of a model and those that define the overall nature and characteristics of a workable model. Like all principles that involve human decision-making, they are somewhat fuzzy in their constraints.

1. **Definition and Support Principles**

   - **The Principle of Similarity:** A good model design follows Occam's razor—a model should include just those factors and features necessary to bring the model into step with reality. Important properties should be included and unimportant ones omitted. A simple model not only encourages comprehension and understanding but also makes model less prone to introduced error during maintenance and modification.
   - **The Principle of Extensibility:** A sound predictive model must be extensible—this principle goes hand in hand with simplicity. A model must support changes to its capabilities, changes in data, and data in its overall functionality. This extensibility should also extend to the level of expertise of the model user.
   - **The Principle of Robustness:** A good predictive model makes it difficult for a user to produce a bad answer. In a predictive model, assuming that the model itself is working correctly, the bad idea centers around a prediction that lacks sufficient evidence. Hence, a robust predictive model should assign a degree of evidentiary support to each forecasted point or automatically adjust and limit its forecast horizon based on the amount of evidence.
   - **The Principle of Fault Tolerance:** A well-designed and well-built predictive model should be tolerant of erroneous data as well as faulty or improper property settings. This reduces the overall brittleness and fragility of the model.

Even some commercial forecasting models do not have reasonableness checks for incoming data, throw fatal exceptions on minor violations of data ranges, fail to provide reasonable defaults for model properties, or do not check that properties set by the user make sense and are not in conflict.

- **The Principle of Ease of Control**: A well-designed predictive model has well-thought-out ergonomics supporting effective and easy-to-use properties. The most powerful and robust model is of little use if its operating properties, data sources, tracing facilities, and outcome requirements cannot be easily understood and easily set. This is also related to the Law of Least Work—a model should only ask the user for those elements it cannot determine, calculate, deduce, or infer through the current model state and its properties.
- **The Principle of Completeness:** For all critical modeling issues, the model must be complete. Completeness is always a trade-off between comprehensiveness, functionality, granularity, and comprehension. A sound predictive model must couple the outcome to all the necessary causal relationships. If the connections are not made, then the model's predictions cannot be trusted.

2. **Interference Principles**

The support principles are concerned with the design and nature of the model itself, while the interference principles are concerned with way in which management and model users interact with the model as an artifact of planning as a component of the organization's strategic planning process. Both of these must be considered as factors in selecting and implementing an effective predictive model.

- **The Lack of Knowledge Principle:** An incorrect perspective on the purpose and value of predictive modeling
- **The Lack of Concern Principle:** A rudimentary or naïve appreciation of and concern for the role of modeling in the planning process, which forms the infrastructure of the predictive model itself. An isolation of the model development from the overall strategic planning process
- **The Lack of Definition Principle:** An insufficient definition and understanding of the objectives in the forecasting model
- **The Lack of Engineering Principle:** An oversimplification of the causal relationships in the model
- **The Lack of Responsibility Principle:** A lack of appreciation for the fundamental constraints placed on model accuracy due to lack of data, lack of expertise, lack of design and development time, lack of client involvement, and similar factors

References

1. Merrill, R. A. (1997). Food safety regulation: Reforming the Delaney Clause. *Annual Review of Public Health, 18*, 313–340. This source includes a useful historical survey of prior food safety regulation.

2. AICPA Statement on Auditing Standards No. 109.
3. Commoner, Barry. O'Brien, Mary. Shrader-Frechette and Westra (1997).
4. Scianta Predictive Modeling, An Overview and Assessment. www.scianta.com, Chapel Hill, NC, USA.
5. Autonomic computer systems, a necessary foundation for self-repair and real-time control, are beginning investigated by several computer companies. As an example, see **Autonomic Computing**—Creating self-Managing Computer System. http://www-306.ibm.com/autonomic/index.shtml
6. Little, J. D. (1961). A proof of the queuing formula $L = \lambda W$. *Operations Research, 9,* 383–387.
7. van Wyk, K. R., & Forno, R. (2001). *Incident response*. Sebastopol: O'Reilly Media.

Chapter 3
Event Management and Best Practice

In order to set up and configure an efficient Business Resilience System (BRS), we need to have a deep and broad understanding about event and event management with a focus on best practice. Such practice allows us to examine event filtering, duplicate detection, correlation, notification, and synchronization. In addition, it discusses trouble-ticket integration and how the trouble ticket and as part of BRS workflow can set the triggering point at BRS dashboard and set the maintenance modes and automation concerning event management. This chapter explains the importance of event correlation and automation. It defines relevant terminology and introduces basic concepts and issues. It also discusses general planning considerations for developing and implementing a robust event management system.

3.1 Introduction to Event Management

In order for us to introduce an event management, we turn our attention to IBM's International Technical Support Organization (ITSO) and tap into their Redbook under title of this chapter, which was published in 2004. Although the IBM ITSO team wrote this book in order to promote one of the IBM products around the event management, namely, Tivoli® Enterprise™. However, these authors are taking an agnostic approach, and at the end, we leave choice of the vendor, application tools, and product to the readers of this book for their particular BRS implementation for their organization and enterprise. Nevertheless, the intention of this chapter is for system and network administrators who are responsible for delivering and managing IT-related events through the use of systems and network management tools.

However prior to getting ready to set up your Business Resilience System (BRS), you should have a thorough understanding of the event management system in which you plan to implement these concepts. Thus, we go ahead and reflect the importance of event correlation and automation, defined by the IBM team and replicated here [1].

© Springer International Publishing AG 2017
B. Zohuri, M. Moghaddam, *Business Resilience System (BRS): Driven Through Boolean, Fuzzy Logics and Cloud Computation*,
DOI 10.1007/978-3-319-53417-6_3

We encourage readers to refer to this technical manual [1] for more details and information.

From the time of their inception, computer systems were designed to serve the needs of businesses. Therefore, it was necessary to know if they were operational. The critical need of the business function that was performed governed how quickly this information had to be obtained.

Early computers were installed to perform batch number-crunching tasks for such business functions as payroll and accounts receivable, in less time and with more efficiency than humans could perform them. Each day, the results of the batch processing were examined. If problems occurred, they were resolved and the batch jobs were executed again.

As their capabilities expanded, computers began to be used for functions such as order entry and inventory. These mission-critical applications needed to be online and operational during business hours required immediate responses. Companies questioned the reliability of computers and did not want to risk losing customers because of computer problems. Paper forms and manual backup procedures provided insurance to companies that they could still perform their primary business in the event of a computer failure.

Since these batch and online applications were vital to the business of the company and their business continuity management (BCM) and business process management (BPM), it became more important to ascertain in a timely fashion whether they were available and working properly. Software was enhanced to provide information and errors, which were displayed on one or more consoles. Computer operators watched the consoles, ignored the informational messages, and responded to the errors. Tools became available to automatically reply to messages that always required the same response [1].

With the many advances in technology, computers grew more sophisticated and were applied to more business functions. Personal computers and distributed systems flourished, adding to the complexity of the IT environment. Due to the increased reliability of the machines and software, it became impractical to run a business manually. Companies surrendered their paper forms and manual backup procedures to become completely dependent upon the functioning of the computer systems [1].

Managing the systems, now critical to the survival of a business, became the responsibility of separate staffs within an information technology (IT) organization. Each team used its own set of tools to do the necessary monitoring of its own resources for their implementation of BRS. Each viewed its own set of error messages and responded to them. Many received phone calls directly from users who experienced problems.

To increase the productivity of the support staffs and to offload some of their problem support responsibilities, help desks were formed. Help desks served as central contact points for users to report problems with their computers or applications. They provided initial problem determination and resolution services. The support staffs did not need to watch their tools for error messages, since software was installed to aggregate the messages at a central location. The help desk or an operations center monitored messages from various monitoring tools and notified the appropriate support staff when problems surfaced. However, the ideal BRS in place will eliminate some of these lengthy steps by being more automated with the right Service Level Agreement (SLA) in place for the monitoring dashboard to display early warning for the occurring adverse events for business to be handled, managed, and, finally, properly processed by taking right measures. Naturally, the automation for BRS and being an autonomous as much as possible a good knowledge-based (KB) database is required, and this KB database needs to constantly be up to date.

Today, changes in technology provide still more challenges. The widespread use of the Internet to perform mission-critical applications necessitates and 24 × 7 availability of systems. Organizations need to know immediately when there are failures, and recovery must be almost instantaneous. On-demand and grid computing allow businesses to run applications wherever cycles are available to ensure they can meet the demands of their customers. However, this increases the complexity of monitoring the applications, since it is now insufficient to know the status of one system without knowing how it relates to others. Operators cannot be expected to understand these relationships and account for them in handling problems, particularly in complex environments.

However, there are several problems with the traditional approach to managing systems as such, and they are pointed out here as:

- **Missed problems**
  Operators can overlook real problems while sifting through screens of informational messages. Users may call to report problems before they are noticed and acted upon by the operator.
- **False alarms**
  Messages can seem to indicate real problems, when in fact they are not. Sometimes additional data may be needed to validate the condition and, in distributed environments, that information may come from a different system than the one reporting the problem.
- **Inconsistency**
  Various operators can respond differently to the same type of events.
- **Duplication of effort**
  Multiple error messages may be produced for a single problem, possibly resulting in more than one support person handling the same problem.
- **Improper problem assignment**
  Manually routing problems to the support staffs sometimes result in support personnel being assigning problems that are not their responsibility.
- **Problems that cannot be diagnosed**

  Sometimes when an intermittent problem condition clears before someone has had the chance to respond to it, the diagnostic data required to determine the cause of the problem disappears.

Event correlation and automation address these issues by:

- Eliminating information messages from view to easily identify real problems
- Validating problems
- Responding consistently to events
- Suppressing extraneous indications of a problem
- Automatically assigning problems to support staffs
- Collecting diagnostic data

Event correlation and automation are the next logical steps in the evolution of event handling. They are critical to successfully managing today's ever-changing, fast-paced IT environments with the reduced staffs with which companies are forced to operate.

3.2 Event Management Terminology

Before we discuss the best ways to implement event correlation and automation, we need to establish the meaning of the terms we use. While several systems management terms are generally used to describe event management, these terms are sometimes used in different ways by different authors. In this section, we provide definitions of the terms as they are used throughout this Redbook [1].

3.2.1 Event

Since event management and correlation centers on the processing of events, it is important to clearly define what is meant by an event. In the context of this section, an event is a piece of data that provides information about one or more system resources.

Events can be triggered by incidents or problems affecting a system resource. Similarly, changes to the status or configuration of a resource, regardless of whether they are intentional, can generate events. Events may also be used as reminders to take action manually or as notification that an action has occurred.

3.2.2 Event Management

The way in which an organization deals with events is known as event management. It may include the organization's objectives for managing events, assigned roles and responsibilities, ownership of tools and processes, critical success factors, standards, and event handling procedures. The linkages between the various departments within the organization required handling events, and the flow of this information between them is the focus of event management. Tools are mentioned in reference to how they fit into the flow of event information through the organization and to which standards should be applied to that flow.

Since events are used to report problems, event management is sometimes considered a subdiscipline of problem management. However, it can really be considered a discipline of its own, for it interfaces directly with several other systems management disciplines. For example, system upgrades and new installations can result in new event types that must be handled. Maintaining systems both through regularly scheduled and emergency maintenance can result in temporary outages that trigger events. This clearly indicates a relationship between event management and change management.

In small organizations, it may be possible to handle events through informal means. However, as organizations grow both in size of the IT support staffs and the number of resources they manage, it becomes more crucial to have a formal, documented event management process. Formalizing the process ensures consistent responses to events, eliminates duplication of effort, and simplifies the configuration and maintenance of the tools used for event management.

3.2.3   Event Processing

While event management focuses on the high-level flow of events through an organization, event processing deals with tools. Specifically, the term event processing is used to indicate the actions taken upon events automatically by systems management software tools.

Event processing includes such actions as changing the status or severity of an event, dropping the event, generating problem tickets and notifications, and performing recovery actions. These actions are explained in more detail in Sect. 3.3 of this chapter.

3.2.4   Automation and Automated Actions

Automation is a type of actions that can be performed when processing events. For the purposes of this book, it refers to the process of taking actions on system resources, without human intervention in response to an event. The actual actions executed are referred to as automated actions.

Automated actions may include recovery commands performed on a failing resource to restore its service and failover processes to bring up backup resources. Changing the status or severity of an event, closing it, and similar functions are not considered automated actions. That is because they are performed on the event itself rather than on one or more system resources referred to or affected by the event.

The types of automated actions and their implications are covered in more detail in Sect. 3.3 of this chapter.

3.3   Concepts and Issues

This section presents the concepts and issues associated with event processing. Additional terminology is introduced as needed.

3.3.1   Event Flow

An event cannot provide value to an organization in managing its system resources unless the event is acted upon, either manually by a support person or automatically by software. The path an event takes from its source to the software or person who takes action on it is known as the *event flow*.

The event flow begins at the point of generation of the event, known as the event source. The source of an event may be the failing system itself, as in the case of a router that sends information about its health to an event processor. An agent that runs on the system to monitor for and report error conditions is another type of event source. Proxy systems that monitor devices other than itself, such as Simple Network Management Protocol (SNMP) manager that periodically checks the status of TCP/IP devices and reports a failure if it receives no response, are also considered an event source.

Event processors are devices that run software capable of recognizing and acting upon events. The functionality of the event processors can vary widely. Some are capable of merely forwarding or discarding events. Others can perform more, sophisticated functions such as reformatting the event, which is correlating it with other events received, displaying it on a console, and initiating recovery actions.

Most event processors have the capability to forward events to other event processors. This functionality is useful in consolidating events from various sources at a central site, for management by a help desk or operations center. The hierarchy of event processors used to handle events can be referred to as the event processing hierarchy. The first receiver of the event is the entry point into the hierarchy, and the collection of all the entry points is called the entry tier of the hierarchy. Similarly, all second receivers of events can be collectively referred to as the second tier in the hierarchy and so forth. For the purposes of this book, we refer to the top level of the hierarchy as the enterprise tier, because it typically consolidates events from sources across an entire enterprise.

Operators typically view events of significance from a console, which provides a Graphical User Interface (GUI) through which the operator can take action on events. Consoles can be proprietary, requiring special software for accessing the console. Alternatively, they can adhere to open standards, such as Web-based consoles that can be accessed from properly configured Web browsers. The collection of event sources, processors, and consoles or dashboard, which is sometimes, referred to as the *event management infrastructure*.

3.3.2 Filtering and Forwarding

Many devices generate informational messages that are not indicative of problems. Sending these messages as events through the event processing hierarchy is undesirable. The reason is because processing power and bandwidth are needed to handle them, and they clutter the operator consoles, possibly masking true problems. The process of suppressing these messages is called event filtering or filtering.

There are several ways to perform event filtering. Events can be prevented from ever entering the event processing hierarchy. This is referred to as filtering at the source. Event processors can discard or drop the unnecessary events. Likewise, consoles can be configured to hide them from view.

The event filtering methods that are available are product specific. Some SNMP devices, for example, can be configured to send all or none of their messages to an event processor or to block messages within specific categories such as security or configuration. Other devices allow blocking to be configured by message type.

When an event is allowed to enter the event processing hierarchy, it is said to be forwarded. Events can be forwarded from event sources to event processors and between event processors. We will discuss "Event Management Categories and Best Practices" later on in this chapter, where we explain the preferred methods of filtering and forwarding events.

3.3.3 Duplicate Detection and Throttling

Events that are deemed necessary must be forwarded to at least one event processor to ensure that they are handled by either manual or automated means. However, sometimes the event source generates the desired message more than once when a problem occurs. Usually, only one event is required for action. The process of determining which events are identical is referred to as duplicate detection.

The time frame in which a condition is responded to may vary, depending upon the nature of the problem being reported. Often, it should be addressed immediately when the first indication of a problem occurs. This is especially true in situations where a device or process is down. Subsequent events can then be discarded. Other times, a problem does not need to be investigated until it occurs several times. For example, a high CPU condition may not be a problem if a single process, such as a backup, uses many cycles for a minute or two. However, if the condition happens several times within a certain time interval, there most likely is a problem. In this case, the problem should be addressed after the necessary number of occurrences. Unless diagnostic data, such as the raw CPU busy values, is required from subsequent events, they can be dropped. The process of reporting events after a certain number of occurrences is known as throttling. Later on in later chapter, when we discuss about serve farming, this will be clear.

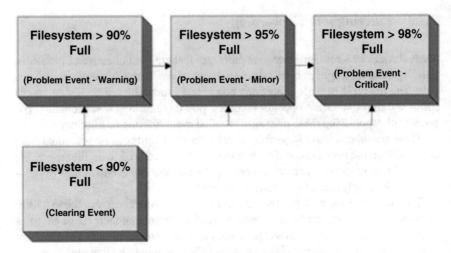

Fig. 3.1 Problem and clearing correlation sequence

3.3.4   Correlation

When multiple events are generated as a result of the same initial problem or provide information about the same system resource, there may be a relationship between the events. The process of defining this relationship in an event processor and implementing actions to deal with the related events is known as *event correlation*.

Correlated events may reference the same affected resource or different resources. They may be generated by the same event source or handled by the same event processor.

- **Problem and clearing event correlation**

  This section presents an example of events that are generated from the same event source and deal with the same system resource. An agent monitoring a system detects that a service has failed and sends an event to an event processor. The event describes an error condition, called a problem event. When the service is later restored, the agent sends another event to inform the event processor the service is again running, and the error condition has cleared. This event is known as a clearing event. When an event processor receives a clearing event, it normally closes the problem event to show that it is no longer an issue.

  The relationship between the problem and clearing event can be depicted graphically as shown in Fig. 3.1. The correlation sequence is described as follows:

  - Problem is reported when received (service down).
  - Event is closed when a recovery event is received (Service Recovered).

Taking this example further, assume that multiple agents are on the system. One reads the system log, extracts error messages, and sends them as events. The second agent

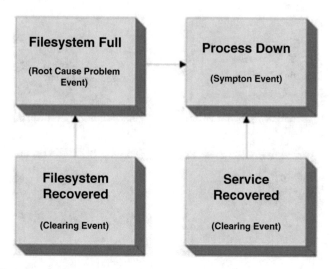

Fig. 3.2 Correlation of multiple events reporting the same problem

actively monitors system resources and generates events when it detects error conditions. A service running on the system writes an error message to the system log when it dies. The first agent reads the log, extracts the error messages, and sends it as an event to the event processor. The second agent, configured to monitor the status of the service, detects that it has stopped and sends an event as well. When the service is restored, the agent writes a message to the system log, which is sent as an event, and the monitor detects the recovery and sends its own event.

The event processor receives both problem events, but only needs to report the service failure once. The events can be correlated and one of them dropped. Likewise, only one of the clearing events is required. This correlation sequence is shown in Fig. 3.2 and follows this process:

– A problem event is reported if received from the log.
– The event is closed when the Service Recovered event is received from the log.
– If a service down event is received from a monitor, the service down event from the log takes precedence, and the service down event from a monitor becomes extraneous and is dropped.
– If a service down event is not received from the log, the service down event from a monitor is reported and closed when the Service Recovered event is received from the monitor.

  This scenario is different from duplicate detection. The events being correlated both report service down, but they are from different event sources and most likely have different formats. Duplicate detection implies that the events are of the same format and are usually, though not always, from the same event source. If the monitoring agent in this example detects a down service and repeatedly sends events reporting that the service is down, these events can be handled with duplicate detection.

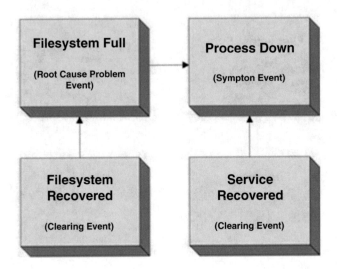

Fig. 3.3 Escalation sequence

- **Even escalation**

  Sometimes multiple events are sent to indicate a worsening error condition for a system resource. For example, an agent monitoring a file system may send a warning message to indicate the file system is greater than 90% full, a second, more severe event when greater than 95% full, and a critical event greater than 98% full. In this case, the event processor does not need to report the file system error multiple times. It can merely increase the severity of the initial event to indicate that the problem has become more critical and needs to be responded to more quickly.

  This type of correlation is sometimes called an escalation sequence. In Fig. 3.3, the escalation sequence is described as follows.

  For example, if file system >90% full is received, it is reported as a warning. When file system >95% full is received, it is dropped, and the reported event is escalated to a severe. Likewise, if file system >98% full is received, it is dropped and the reported event is escalated again to a critical.

  If file system >95% full is the first problem event received, it is reported. The same escalation logic applies. This type of correlation sequence assumes that severities are clearly defined and the allowable time to respond to events of those severities has been communicated within the organization. This is one of the purposes of the event management process described in Sect. 3.2.2 on this chapter.

- **Root cause correlation**

  A problem may sometimes trigger other problems, and each problem may be reported by events. The event reporting the initial problem is referred to as a root cause or primary event. Those that report the subsequent problems are called symptom or secondary events.

  At this point, it is important to note the difference between a root cause event and the root cause of a problem. The former is the event that provides information about the first of a series of related, reported problems. The latter is what caused the problem condition to happen.

Root cause events and root causes of problems may be closely related. For example, a root cause event reporting a faulty NIC card may be correlated with secondary events such as "interface down" from a Simple Network Management Protocol (SNMP) manager or "application unreachable" from a transaction-monitoring agent. The root cause of the problem is the broken card.

However, sometimes the two are not as closely associated. Consider an event that reports a file system full condition. The full file system may cause a process or service to die, producing a secondary event. The file system full event is the root cause event, but it is not the root cause of the problem. A looping application that is repeatedly logging information into the file system may be the root cause of the problem.

When situations such as these are encountered, you must set up monitoring to check for the root cause of the problem and produce an event for it. That event then becomes the root cause event in the sequence. In our example, a monitor that detects and reports looping application logging may be implemented. The resulting event can then be correlated with the others and becomes the root cause event.

Because of this ambiguity in terms, we prefer to use the term primary event rather than root cause event.

The action taken in response to a root cause event may automatically resolve the secondary problems. Sometimes, though, a symptom event may require a separate action, depending upon the nature of the problem it reports. Examples of each scenario follow.

– *Symptom event not requiring action*
  Assume that an agent on a UNIX® system is monitoring file systems for adequate space and critical processes for availability. One of the key processes is required to run at all times and is set up to automatically respond if it fails. The process depends upon adequate free space in the file system where it stores its temporary data files and cannot execute without it.

  The file system upon which the process depends fills up, and the agent detects the condition and sends an event. The process dies, and the operating system unsuccessfully attempts to restart it repeatedly. The agent detects the failure and generates a second event to report it.

  There are essentially two problems here. The primary problem is the full file system, and the process failure is the secondary problem. When appropriate action is taken on the first event to free space, within the file system, the process successfully responds, automatically. No action is required on the secondary event, so the event processor can discard it. In Fig. 3.4, the correlation sequence is described as follows:

  > The file system full event is reported if received.
  > The Process Down event is unnecessary and is dropped. Since the process is set to respond, it automatically starts when the file system is recovered.

Fig. 3.4 Correlation
sequence in which
secondary event does not
require action

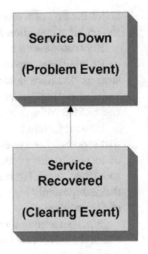

The file system full event is closed when the file system recovered
clearing event is received.

The Service Recovered clearing event is unnecessary and is dropped,
since it is superseded by the file system recovered clearing event.

– *Symptom events requiring action*

Now suppose that an application stores its data in a local database. An agent
runs on the application server to monitor the availability of both the application
and the database. A database table fills and cannot be extended, causing the
application to hang. The agent detects both conditions and sends events to
report them.

The full database table is the primary problem, and the hanging application is
the secondary problem. A database administrator corrects the primary problem.
However, the application is hung and cannot recover itself. It must be recycled.

Since restarting the application is outside the responsibility of the database
administrator, the secondary event is needed to report the application problem to
the appropriate support person.

In Fig. 3.5, the correlation sequence is as follows:

The file system full event is reported if received.

The Process Down event is reported as dependent upon the file system
being resolved.

The file system full event is closed when the file system recovered
clearing event is received.

The Process Down event is cleared when the Service Recovered clearing event is received.

An important implication of this scenario must be addressed. Handling the secondary event depends upon the resolution of the primary event. Until the database is

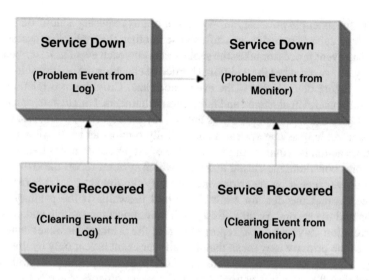

Fig. 3.5 Correlation sequence in which secondary event requires action

repaired, any attempts to restart the application fail. Implementation of correlation sequences of this sort can be challenging. More details can be found in Redbook by IBM [1].

- **Cross-platform correlation**

  In the previous application and database correlation scenario, the correlated events refer to different types of system resources. We refer to this as cross-platform correlation. Some examples of platforms include operating systems, databases, middleware, applications, and hardware.

  Often, cross-platform correlation sequences result in symptom events that require action. This is because the support person handling the first resource type does not usually have administrative responsibility for the second type. Also, many systems are not sophisticated enough to recognize the system resources affected by a failure and to automatically recover them when the failure is resolved. For these reasons, cross-platform correlation sequences provide an excellent opportunity for automated recovery actions.

- **Cross-host correlation**

  In distributed processing environments, there are countless situations in which conditions on one system affect the proper functioning of another system. Web applications, for example, often rely on a series of Web, application, and database servers to run a transaction. If a database is inaccessible, the transaction fails. Likewise, servers may share data through message queuing software, requiring the creation of the queue by one server before it is accessed from another.

  When problems arise in scenarios such as these, events can be generated by multiple hosts to report a problem. It may be necessary to correlate these events to determine which require action. The process of correlating events from different systems is known as cross-host correlation.

In the example presented in "Symptom events requiring action" on p. 12, the database can easily reside on a different server than the application accessing it.

The event processor takes the same actions on each event as described previously. However, it has the additional burden of checking the relationship between hosts before determining if the events correlate. Cross-host correlation is particularly useful in clustered and failover environments. For clusters, some conditions may not represent problems unless they are reported by all systems in the cluster. As long as one system is successfully running an application, for example, no action is required. In this case, the event processor needs to know which systems constitute the cluster and track which systems report the error.

In failover scenarios, an error condition may require action if it is reported by either host. Consider, for example, paired firewalls. If the primary firewall fails and the secondary takes over, each may report the switch, and cross-host correlation may be used to report failure of the primary. However, a hard failure of the primary may mean that the failover event is sent only by the secondary. This event should indicate the failure of the primary firewall as the condition that requires action. Again, the event processor needs to know the relationship between the firewalls before correlating failover events.

- **Topology-based correlation**
When such networking resources as routers fail, they may cause a large number of other systems to become inaccessible. In these situations, events may be reported that refer to several unreachable system resources. The events may be reported by SNMP managers that receive no answer to their status queries or by systems that can no longer reach resources with which they normally communicate. Correlating these events requires knowledge of the network topology and therefore is referred to as topology-based correlation.

This type of correlation, while similar to cross-host correlation, differs in that the systems have a hierarchical, rather than a peer, relationship. The placement of the systems within the network determines the hierarchy. The failure of one networking component affects the resources downstream from it.

Clearly, the event reporting the failing networking resource is the primary, or root, cause event and needs to be handled. Often, the secondary events refer to unreachable resources that become accessible once the networking resource is restored to service. In this case, these events may be unnecessary. Sometimes, however, a downstream resource may need to be recycled to resynchronize it with its peer resources. Secondary events dealing with these resources require corrective action.

Since SNMP managers typically discover network topology and understand the relationships between devices, they are often used to implement topology-based correlation.

- **Timing considerations**
An important consideration in performing event correlation is the timing of the events. It is not always the case that the primary event is received first. Network delays may prevent the primary event from arriving until after the secondary is

received. Likewise, in situations where monitoring agents are scheduled to check periodically for certain conditions, the monitor that checks for the secondary problem may run first and produce that event before the root cause condition is checked.

To properly perform event correlation in this scenario, configure the event processor to wait a certain amount of time to ensure that the primary condition does not exist before reporting that action is required for the secondary event. The interval chosen must be long enough to allow the associated events to be received, but short enough to minimize the delay in reporting the problem.

3.3.5 Event Synchronization

When events are forwarded through multiple tiers of the event management hierarchy, it is likely that different actions are performed on the event by different event processors. These actions may include correlating, dropping, or closing events.

Problems can arise when one event processor reports that an event is in a certain state and another reports that it is in a different state. For example, assume that the problem reported by an event is resolved and the event is closed at the central event processor but not at the event processors in the lower tiers in the hierarchy. The problem recurs, and a new event is generated. The lower-level event processor shows an outstanding event already reporting the condition and discards the event. The new problem is never reported or resolved.

To ensure that this situation does not happen, status changes made to events at one event processor can be propagated to the others through which the event has passed. This process is known as event synchronization.

Implementing event synchronization can be challenging, particularly in complex environments with several tiers of event processors. In addition, environments, designed for high availability, need some way to synchronize events between their primary and backup event processors. For further information, refer to Redbook [1].

3.3.6 Notification

Notification is the process of informing support personnel that an event has occurred. It is typically used to supplement use of the event processor's primary console, not to replace it. Notification is useful in situations when the assigned person does not have access to the primary console, such as after hours, or when software licensing or system resource constraints prevent its use. It can also be helpful in escalating events that are not handled in a timely manner (see Sect. 3.3.8 in this chapter).

Paging, e-mail, and pop-up windows are the most common means of notification. Usually, these functions exist outside the event processor's software and must be implemented using an interface. Sometimes that interface is built into the event processor. Often, the event processor provides the ability to execute scripts or BAT files that can be used to trigger the notification software. This is one of the simplest ways to interface with the notification system.

It is difficult to track the various types of notifications listed previously, and the methods are often unreliable. In environments where accountability is important, more robust means may be necessary to ensure that support personnel is informed about events requiring their action.

The acceptable notification methods and how they are used within an organization should be covered in the event management process, which is described in Sect. 3.2.2 in this chapter, previously.

3.3.7 Trouble Ticketing

Problems experienced by users can be tracked using *trouble tickets*. The tickets can be opened manually by the help desk or operations center in response to a user's phone call or automatically by an event processor.

Trouble ticketing is one of the actions that some event processors can take upon receipt of an event. It refers to the process of forwarding the event to a trouble-ticketing system in a format that system can understand. This can typically be implemented by executing a script or sending an e-mail to the trouble-ticketing system's interface or Application Programming Interface (API).

The trouble-ticketing system itself can be considered a special type of event processor. It can open trouble tickets for problem events and close them when their corresponding clearing events are received. As such, it needs to be synchronized with the other event processors in the event management hierarchy. The actions of opening and closing trouble tickets are also referred to as trouble ticketing.

In environments where accountability is important, robust trouble-ticketing systems may provide the tracking functions needed to ensure that problems are resolved by the right people in a timely manner.

3.3.8 Escalation

In Sect. 3.3.4, we discuss escalating the severity of events based on the receipt of related events. This escalation is handled by the event source, which sends increasingly more critical events as a problem worsens. A few kinds of event escalation require consideration and they are as follows:

1. **Escalation to ensure problems are addressed**
   An event is useless in managing IT resources if no action is taken to resolve the problem reported. A way to ensure that an event is handled is for an event processor

to escalate its severity if it has not been acknowledged or closed within an acceptable time frame. Timers can be set in some event processors to automatically increase the severity of an event if it remains in an unacknowledged state.

The higher severity event is generally highlighted in some fashion to draw greater attention to it on the operator console on which it is displayed. The operators viewing the events may inform management that the problem has not been handled, or this notification may be automated.

In addition to serving as a means of ensuring that events are not missed, escalation is useful in situations where the IT department must meet Service Level Agreements (SLAs). The timers may be set to values that force escalation of events, indicating to the support staff that the event needs to be handled quickly, or SLAs may be violated.

For escalation to be implemented, the allowable time frames to respond to events of particular severities and the chain of people to inform when the events are not handled must be clearly defined. This is another purpose of the event management process described in Sect. 3.2.2 of this chapter.

2. **Business impact escalation**

Events can also be escalated based upon business impact. Problems that affect a larger number of users should be resolved more quickly than those that impact only a few users. Likewise, failures of key business applications should be addressed faster than those of less important applications.

There are several ways to escalate events based upon their business significance:

- **Device type**
  An event may be escalated when it is issued for a certain device type. Router failures, for example, may affect large numbers of users because they are critical components in communication paths in the network. A server outage may affect only a handful of users who regularly access it as part of their daily jobs. When deploying this type of escalation, the event processor checks to see the type of device that failed and sets the severity of the event accordingly. In our example, events for router failures may be escalated to a higher severity, while events of servers remain unchanged.

- **Device priority**
  Some organizations perform asset classifications in which they evaluate the risk to the business of losing various systems. A switch supporting 50 users may be more critical than a switch used by five users. In this escalation type, the event processor checks the risk assigned to the device referenced in an event and increases the severity of those with a higher rating.

- **Other**
  It is also possible to perform escalation based on which resources a system fails, assigning different priorities to the various applications and services that run on a machine. Another hybrid approach combines device type and priority to determine event severity. For example, routers may take higher priority than

servers may. The routers are further categorized by core routers for the backbone network and distributed routers for the user rings, with the core routers receiving heavier weighting in determining event severity.

An organization should look at its support structure, network architecture, server functions, and SLAs to determine the best approach to use in handling event escalation.

3.3.9   Maintenance Mode

When administrative functions performed on a system disrupt its normal processing, the system is said to be in maintenance mode. Applying fixes, upgrading software, and reconfiguring system components are all examples of activities that can put a system into maintenance mode.

Unless an administrator stops the monitoring agents on the machine, events continue to flow while the system is maintained. These events may relate to components that are affected by the maintenance or to other system resources. In the former case, the events do not represent real problems, but, in the latter case, they may.

From an event management point of view, the difficulty is how to handle systems that are in maintenance mode. Often, it is awkward to reconfigure the monitoring agents to temporarily ignore only the resources affected by the maintenance. Shutting down monitoring completely may suppress the detection and reporting of a real problem that has nothing to do with the maintenance. Both of these approaches rely on the intervention of the administrator to stop and restart the monitoring, which may not happen, particularly during late night maintenance windows.

Another problem is that maintenance may cause a chain reaction of events generated by other devices. A server that is in maintenance mode may only affect a few machines with which it has contact during normal operations. A network device may affect large portions of the network when maintained, causing a flood of events to occur.

How to predict the effect of the maintenance and how to handle it are issues that need to be addressed. See Sect. 4.11 in the next chapter, for suggestions on how to handle events from machines in maintenance mode.

3.3.10   Automation

You can perform four basic types of automated actions upon receipt of an event and they are:

1. **Problem verification**

   It is not always possible to filter events that are not indicative of real problems. For example, an SNMP manager that queries a device for its status may not receive an answer due to network congestion rather than the failure of the device. In this case, the manager believes the device is down. Further processing is required to determine whether the device is really operational. This processing can be automated.

2. **Recovery**

   Some failure conditions lend themselves to automated recovery. For example, if a service or process dies, it can generally be restarted using a simple BAT (an executable) file or script.

3. **Diagnostics**

   If diagnostic information is typically obtained by the support person to resolve a certain type of problem, that information can be gathered automatically when the failure occurs and merely accessed when needed. This can help to reduce the mean time to repair for the problem. It is also particularly useful in cases where the diagnostic data, such as the list of processes running during periods of high CPU usage, may disappear before a support person has time to respond to the event.

4. **Repetitive command sequences**

   When operators frequently enter the same series of commands, automation can be built to perform those commands. The automated action can be triggered by an event indicating that it is time to run the command sequence. Environments where operators are informed by events to initiate the command sequences, such as starting or shutting down applications, lend themselves well to this type of automation.

Some events traverse different tiers of the event processing hierarchy. In these cases, you must decide at which place to initiate the automation. The capabilities of the tools to perform the necessary automated actions, security required initiating them, and bandwidth constraints are some considerations to remember when deciding from which event processor to launch the automation.

3.4   Planning Considerations

Depending upon the size and complexity of the IT environment, developing an event management process for it can be a daunting task. This section describes some points to consider when planning for event correlation and automation in support of the process.

3.4.1   IT Environment Assessment

A good starting point is to assess the current environment. Organizations should inventory their hardware and software to understand better the types of system resources managed and the tools used to manage them. This step is necessary to determine the event sources and system resources within scope of the correlation and automation effort. It is also necessary to identify the support personnel who can assist in deciding the actions needed for events related to those resources.

In addition, the event correlation architect should research the capabilities of the management tools in use and how the tools exchange information. Decisions about where to filter events or perform automated actions, for example, cannot be made until the potential options are known.

To see the greatest benefit from event management in the shortest time, organizations should target those event sources and system resources that cause the most pain. This information can be gathered by analyzing the volumes of events currently received at the various event processors, trouble-ticketing system reports, and database queries, and scripts can help to gain an idea about the current event volumes, most common types of errors, and possible opportunities for automated action.

BRS, through any vendor of your choice, such as IBM or other players in the game of Relationship Database Management System (RDBM) offers a service to analyze current event data. For example, IBM, offering, is called the Data Driven Event Management Design (DDEMD), uses a proprietary data mining tool to help organizations determine where to focus their efforts. The tool also provides statistical analysis to suggest possible event correlation sequences and can help uncover problems in the environment.

3.4.2   Organizational Considerations

Any event correlation and automation design needs to support the goals and structure of an organization. If event processing decisions are made without understanding the organization, the results may be disappointing. The event management tools may not be used, problems may be overlooked, or perhaps information needed to manage service levels may not be obtained.

To ensure that the event correlation project is successful, its design and processes should be developed with organizational considerations in mind.

1. **Centralized versus decentralized**

   An organization's approach to event management is key, to determine the best ways to implement correlation and automation. A centralized event management environment is one in which events are consolidated at a focal point and monitored from a central console. This provides the ability to control the entire enterprise from one place. It is necessary to view the business impact of failures.

   Since the operators and help desk personnel at the central site handle events from several platforms, they generally use tools that simplify event management by providing a common graphical interface to update events and perform basic corrective actions. When problems require more specialized support personnel to resolve, the central operators often are the ones to contact them.

   Decentralized event management does not require consolidating events at a focal point. Rather, it uses distributed support staffs and toolsets. It is concerned with ensuring that the events are routed to the proper place. This approach may be used in organizations with geographically dispersed support staffs or point solutions for managing various platforms.

When designing an event correlation and automation solution for a central-ized environment, the architect seeks commonality in the look and feel of the tools used and in the way events are handled. For decentralized solutions, this is less important.

2. **Skill levels**

The skill level of those responsible for responding to events influences the event correlation and automation implementation. Highly skilled help desk personnel may be responsible for providing first level support for problems. They may be given tools to debug and resolve basic problems. Less experienced staff may be charged with answering user calls and dispatching problems to the support groups within the IT organization.

Automation is key to both scenarios. Where first level support skills are strong, semiautomated tasks can be set up to provide users the ability to easily execute the repetitive steps necessary to resolve problems. In less experienced environ-ments, full automation may be used to gather diagnostic data for direct presenta-tion to the support staffs who will resolve them.

3. **Tool usage**

How an organization plans to use its systems management tools must be under-stood before event correlation can be successfully implemented. Who will use each tool and for what functions should be clearly defined. This ensures that the proper events are presented to the appropriate people for their action.

For example, if each support staff has direct access to the trouble-ticketing system, the event processor or processors may be configured to automatically open trouble tickets for all events requiring action. If the help desk is responsible for dispatching support personnel for problems, then the events need to be pre-sented to the consoles they use.

When planning an event management process, be sure that users have the technical aptitude and training to manage events with the tools provided to them. This is key to ensuring the success of the event processing implementation.

3.4.3   Policies

Organizations that have a documented event management process, as defined in Sect. 3.2, "Terminology," before in this chapter, may already have a set of event management poli-cies. Those that do not should develop one to support their event correlation efforts.

Policies are the guiding principles that govern the processing of events. They may include who in the organization is responsible for resolving problems; what tools and procedures they use; how problems are escalated; where filtering, correlation, and automation occur; and how quickly problems of various severities must be resolved.

When developing policies, the rationale behind them and the implications of imple-menting them should be clearly understood, documented, and distributed to affected parties within the organization. This ensures consistency in the implementation and use of the event management process.

Table 3.1 Sample policy

Policy	Rational	Implication
Filtering takes place as early as possible in the event life cycle. The optimal location is at the event source	This minimizes the effect of events in the network, reduces the processing required at the event processors, and prevents clutter on the operator consoles	Filtered events must be logged at the source to provide necessary audit trails

Table 3.1 below shows an example of policy, its rational, and implication.

It is expected that the policies need to be periodically updated as organizations change and grow, incorporating new technologies into their environments. Who is responsible for maintaining the policies and the procedure they should follow should also be a documented policy.

3.4.4 Standards

Standards are vital to every IT organization because they ensure consistency. There are many types of standards that can be defined. System and usernames, Internet Protocol (IP) address, workstation images, allowable software, system backup and maintenance, procurement, and security are a few examples.

Understanding these standards and how they affect event management is important in the successful design and implementation of the systems management infrastructure. For example, if a security standard states that only employees of the company can administer passwords and the help desk is outsourced, procedures should not be implemented to allow the help desk personnel to respond to password expired events.

For the purposes of event correlation and automation, one of the most important standards to consider is a naming convention. Trouble ticketing and notification actions need to specify the support people to inform for problems with system resources. If a meaningful naming convention is in place, this process can be easily automated. Positional characters within a resource name, for example, may be used to determine the resource's location and, therefore, the support staff that supports that location.

Likewise, automated actions rely on naming conventions for ease of implementation. They can use characters within a name to determine resource type, which may affect the type of automation performed on the resource. If naming conventions are not used, more elaborate coding may be required to automate the event handling processes.

Generally, the event management policies should include reference to any IT standards that directly affect the management of events. This information should also be documented in the event management policies.

Reference

1. IBM. (2004). *Event management and best practices*. International Technical Support Organization.

Chapter 4
Event Management Categories and Best Practices

Event management issues need to be addressed when an organization begins monitoring an IT environment for the first time, decides to implement a new set of systems management tools, or wants to rectify problems with its current implementation. Often it is the tool implementers who decide the approach to use in handling events. Where multiple tools are implemented by different administrators, inconsistent policies and procedures arise. The purpose of this chapter is to provide best practices for both the general implementation approach an organization uses to monitor its environment and the specific event management concepts defined in Chap. 2, "Introduction to Event Management."

4.1 Introduction

This chapter is a compilation of best practices in regard to event management and the rationale behind them. When reading this chapter, keep in mind the following rules:

- Place more weight on the reasons that affect organizations when determining which best practices to implement.
- Use of business impact software may affect which best practices to implement.

Specifically, do not blindly start using a best practice from this chapter. Keep the context of the best practice in mind and apply it to your specific needs. Be sure to weigh the rationales given for these best practices.

© Springer International Publishing AG 2017
B. Zohuri, M. Moghaddam, *Business Resilience System (BRS): Driven Through Boolean, Fuzzy Logics and Cloud Computation*,
DOI 10.1007/978-3-319-53417-6_4

4.2 Implementation Approaches

There are many approaches for implementing an event management design. Those that are used generally arise from the knowledge, experience, and creativity of the systems management product implementers. Each approach has its advantages and disadvantages, which organizations must weigh when deciding how to precede. Thus, we go ahead and reflect the importance of event management categories and best practice, defined by the IBM team and replicated here [1].

Regardless of the approach selected, effective communication within the organization, clear direction from management, a culture that does not assign blame for or finger-point over missed problems, and clear event management policies are critical to its success (see 3.2, "Policies and standards" in Chap. 3).

Five of the more common approaches and their pros and cons are discussed in the following sections.

4.2.1 Send All Possible Events

When implementing tools, sending events is enabled for all monitors. All error messages are extracted from logs of interest and forwarded through the event management hierarchy.

This approach, while generally easy to implement, has many drawbacks. The event management consoles trouble-ticketing systems, or both become cluttered with messages that may not represent actual problems. Operators and support staffs are left with the challenge of sifting through large amounts of information to determine which events or trouble tickets represent real problems and require action. Bandwidth and processing cycles are wasted in handling the extraneous events.

Of all the methods deployed, this one usually results in the most frustration to the tool users. These uses often ignore events, close large numbers of them at once, or cease using the tools to deal with the large event volumes.

4.2.2 Start with Out-of-the-Box Notifications and Analyze Reiteratively

The default settings provided with the systems management tools are used as a starting point to determine which events to handle. Received events are analyzed periodically to determine whether they are truly relevant to an organization.

This strategy relies on the belief that the vendor has chosen to report meaningful information and that the values chosen result in a reasonable event volume. If the tool vendor has done a good job in selecting defaults, this approach can be a good one. If not, the tool users may still have the challenge of finding the real problems among many extraneous events.

Another downside of choosing tool defaults as the starting point is that it does not consider conditions or applications that are unique to the organization. For example, if a homegrown application is used, no vendor-supplied tool will supply defaults as to which application errors to report to the event management processors.

4.2.3   Report Only Known Problems and Add Them to the List as They Are Identified

In this method, only events that indicate real problems are reported. When new problems occur, the technicians responsible for resolving them determine whether a log message can be extracted or a monitor deployed to check for the error condition. The new events are implemented. When the problem reoccurs, the events received are analyzed to ensure that they successfully reported the condition in a timely manner.

This approach is sometimes used when an organization enlists an application or systems support department to pilot use of the systems management tools. Presenting the support people with real problems that they know how to handle makes their jobs easier, helps reduce the mean time to repair, and hopefully makes them advocates of the tools to other support departments.

Support personnel like this approach. The events they receive are those that actually require action. This eliminates the situation where administrators spend time researching events only to find they do not require any corrective action. If this occurs too often, the support personnel eventually ignore the events presented to them.

Adding small numbers of new events at a time minimizes the possibility of event floods, and therefore, problem tickets or service dispatches. The events report conditions that the administrator has already resolved, so the resolution to the problems is usually known and easily handled. Finally, since those informed of the problems are already responsible for fixing them, they do not have the impression that the tool is giving them additional work.

The drawback of this approach is that the problem must occur and be identified as a real problem before it is reported. This relies on non-automated methods such as user phone calls to report the problem. Thus, when an error occurs for the first time, it is not automatically detected or reported.

4.2.4   Choose Top X Problems from Each Support Area

This is a variation of the previous approach. Representatives from each support area provide information about their top problems. The conditions can be situations on which they spend the most time handling or those that, while infrequent, can have the greatest impact on the systems for which they are responsible.

The approach differs from the previous one in that the problems can be conditions that have not yet happened but are so critical or pervasive in nature that they require immediate action if they occur. Also, monitors are implemented in an attempt to prevent them from occurring at all.

Again, administrators like this approach because they control which notifications they receive. Their most time-consuming and repetitive problem determination and recovery tasks can be automated, freeing them for more interesting challenges. Finally, they can stop manually monitoring for the situations that can potentially cause the most serious outages or problems.

The downside is that the condition must be already known to the support staff before it is reported. It does not catch problem conditions of which the administrator is not yet aware.

4.2.5 Perform Event Management and Monitoring Design

Using the Event Management and Monitoring Design (EMMD) methodology, all possible events from sources in the IT environment are analyzed, and event processing decisions are made for them.

This approach, while most time-consuming to implement, is the most thorough. It offers the advantages of the other approaches while addressing their drawbacks. Again, support personnel are only notified of real problems and have control over the notifications they receive. Reviewing all possible events may highlight previously unidentified potential problems, and monitors may be implemented to detect them. This makes this approach more proactive than the others do.

1. **Event Management and Monitoring Design**

   Since events are typically obtained from network and system monitoring agents, event management and monitoring are related topics. The proper monitors must be implemented to receive meaningful events into an event management hierarchy at a rate at which they can be handled.

   Therefore, another method of implementing event correlation and automation is to simultaneously analyze the monitors that are available or already implemented and the events they produce. IBM developed a proprietary methodology and patented set of tools to address both monitoring and event processing. The methodology and tools are available to clients as IBM Global Services offering *Event Management and Monitoring Design* (EMMD).

2. **EMMD approach**

   The EMMD approach systematically analyzes monitors and events based on either the resources that comprise a business function or the types of agents that produce events. The client's environment typically determines the manner in which the EMMD is used. In an organization with a high degree of separation by business application, it makes sense to perform a service decomposition to catalog which system resources and monitoring agents are applicable to critical business functions and analyze those using the methodology. The support personnel responsible for the application are also accountable for the server performance and can make decisions on the proper handling of both kinds of events.

Alternately, monitoring agent analysis may be done in organizations where support is by platform type or component rather than application. In these environments, a support group handles all the events of a certain type, regardless of the application or server that produces them. Therefore, the support group can take responsibility for all events produced by the monitoring agents they handle.

3. **Methodology**

Regardless of which way you use EMMD, you must follow these steps in the methodology:

(a) **Set scope:** Determine the scope of the project. Decide which services or monitoring agent event sources to analyze. Often, critical or high-visibility business functions are chosen when using the service decomposition approach. For component analysis, the monitoring sources that produce the most events, or those that report problems for the least stable platforms, are typically selected.

(b) **Determine event handling policies:** As discussed throughout the IBM Redbook [1], best practices dictate that a set of guidelines exists to determine how to handle events. This provides consistency across the organization and makes it easier to decide which action to take for a particular event. In this step, key members of the organization work together to develop these policies.

(c) **Document event repertoires:** The events that can be produced by event sources are compiled into worksheets used to document decisions about the events. These lists can include all possible events from a given source or those routinely received at the event processors. Typically, all events from a source are analyzed if there is a manageable number.

For sources that can produce a plethora of possible events, the events are usually limited in one of two ways. The limitation can be based on the event policies such as "Filter events related to configuration changes" or "Do not report information only events." Alternately, the events to be analyzed can be limited to those that are typically generated by the source. This list is comprised of each type of event produced by the source within a representative time frame such as two to 4 weeks.

(d) **Select monitors and thresholds:** Review existing monitors for relevance, and suggest new monitors to ensure necessary events are reported. Meaningful thresholds are set based on both best practices and on the baseline performance of the machines to be monitored.

(e) **Document event handling decisions:** Appropriate subject matter experts (SMEs) decide the filtering, forwarding, notification, and automation actions for each event in the event repertoire. These are documented in the event repertoire worksheets. The captured information is reported based on the level in the event processing hierarchy at which the processing should occur. It includes such things as event severities, trouble-ticket priorities, and automation script names.

(f) **Conduct event correlation analysis:** Determine which events correlate together, and assign primary, secondary, or clearing status to them. The SMEs can suggest possible correlation sequences based upon the meaning of the various events and upon their experience in solving past problems that may have produced the events. Help desk personnel are also invaluable in determining which events occur together since they frequently view all events and visually correlate them to determine which require trouble tickets and which should be dispatched to support staffs. In addition, you can use an IBM-patented data mining tool to determine statistically which events often occur together. This same tool can suggest possible correlation sequences. The event relationships are depicted diagrammatically using Visio.

(g) **Review the deliverables:** The project deliverables include the event handling policies, completed event repertoire worksheets, and correlation diagrams. Review these to ensure that they are understood both by those responsible for handling the events and the implementers of the monitoring agents and event processors.

(h) **Define an implementation plan:** Discuss ways to implement the design and develop an implementation plan. The plan includes, among other things, the order in which the event sources should be configured, the tasks required to complete the implementation, responsible parties, and testing and back out procedures.

4. **Tools**

To aid in the various steps of the methodology, IBM developed and patented a set of tools. These tools serve to automate the design steps wherever possible and produce configuration information that can be used in the implementation:

- **EMMD tool:** This is a Visual Basic tool that automates the creation of event repertoire worksheets. Blank worksheets may be produced that are used by the IT practitioner to document the events from a given source. The tool can also populate worksheets with event information from Simple Network Management Protocol (SNMP) Management Information Bases (MIBs), NetView trapped files, and, for example, IBM Tivoli Enterprise Console BAROC files. The worksheets include such information as the event name, description, filtering decisions, throttling parameters, forwarding, and automation commands or script names. They also include notification and trouble-ticketing targets and methods for each tier or event processor in the event management hierarchy.

  Additionally, the tool can help to generate Visio stencils that represent each event that requires correlation. These are used in the Visio diagrams to document the event correlation sequences.

- **EMMD workbooks:** Based on Microsoft® Excel, these workbooks contain macros that assist in the documentation of the event handling decisions. The functions may include shadowing events that are filtered, propagating information between the sheets that represent the various tiers of the event management hierarchy, and generating IBM Tivoli Enterprise Console™ classes and rules based on predefined templates to implement the design using tools such as IBM Tivoli Enterprise Console©.

- **EMMD diagrams:** The Visio diagrams depict the relationships among events, showing primary and secondary problem events and those that clear them. Using the stencils generated by the EMMD tool, the practitioner creates a multipage diagram that shows the event sequences. The diagram includes a table of contents for easy navigation among the pages. In addition, macros defined in a base diagram allow for such functions as generating IBM Tivoli Enterprise Console rules to implement the correlation sequences. These rules are a starting point for correlating events and should be modified to fit into a modular IBM Tivoli Enterprise Console rule base when implementing the EMMD design.
- **Data Driven Event Management Design (DDEMD) tool:** This data mining tool can help to process lists of commonly received events. The events to be processed are input via ASCII files. They can come from a wide variety of sources such as Microsoft Windows® Event Logs, IBM Tivoli Enterprise Console **wtdumprl** output, and log files.

  The events are parsed to determine event type and relevant information within the event. The various analysis functions of the tool, including reporting event frequency by type and host, event rates by time of day, and statistical correlation of events, can use the event details. There are also predictive functions within the tool that enable the practitioner to see the impact of implementing various filtering and correlation rules for the given list of events.

4.3   Policies and Standards

Critical to the success of event management is the process of creating event processing policies and procedures and tracking compliance with them. Without this, an organization lacks consistency and accountability. When different support groups implement their own event management, the tools used are not configured to standards, making them difficult to configure and maintain. Inconsistent tool use can affect measurements, such as mean time to repair, make accountability more difficult, or skew the problem counts that may be used to determine staffing in the various support groups.

Each event handling action—filtering, forwarding, duplicate detection, correlation, escalation, synchronization, notification, trouble ticketing, and automation—should be described in a documented policy. This makes it easier to make event processing decisions and implement systems management tools.

In this section, we discuss important policies and procedures to develop and document in addition to those that specifically describe the major event handling actions of filtering, duplicate detection, correlation, escalation, and automation. For each, we recommend that you list the policy and its implications.

Table 4.1 Policy and implications

Policy	Rational	Implication
Automatically page for all severity one events	Improves mean time to repair	To be meaningful, the notification should include the trouble-ticket number
	Minimizes assigning the wrong support person to the problem	An audit trail is required to ensure the process is working properly and to record who is assigned to the event

Note that some implications always follow from the policy, and others depend upon your systems management toolset or organizational structure. Table 4.1 shows an example in which the implications always follow from the policy.

If your problem management system is incapable of performing paging, you may add an implication stating that you need to develop an interface to send the trouble-ticket number to whichever event processor will trigger paging.

4.3.1   Reviewing the Event Management Process

You must first keep in mind the dynamic nature of the event management process. Organizations grow and change, developing new requirements and outgrowing existing ones. While the event management process is developed to address known issues, time, organizational changes, and experience often bring others to light, requiring changes to the event management process.

These changes can be made to the event handling guidelines documented in the policies and procedures or to the implementation of the event processors that filter, forward, correlate, automate, and notify trouble-ticket events. Periodically review these at time intervals related to the rate at which your organization changes.

(a) **Updating Policies and Procedures**

Assign the responsibility for the iterative review and update of policies and procedures to one person. This individual should understand your organizational structure, be privy to information about upcoming changes, know the roles of the various support groups, and be able to work with them to gather new requirements and integrate them into the existing policies and processes.

The reviews can be scheduled to occur periodically, such as once a year. Or they can be triggered by events such as reorganizations within the company, mergers and acquisitions, changes in upper management (and hence visions and goals), outsourcing, or additions of lines of business or applications.

(b) **Modifying Systems Management Tool Implementations**

There are two types of tool modifications. The first relates to how the tools address event management policies and procedures. The second relates to how they handle specific events.

Updates to the event management policies and processes often imply changes to the tools that implement them. Hence, these types of changes frequently coincide. Other factors that affect this type of change are implementation of new systems management tools or upgrades to existing ones. Typically, the new or upgraded software is capable of providing new function or more effectively implementing existing ones. Sometimes this means that the tool must be implemented or reconfigured to enforce the policies and procedures previously defined. Other times, the improved tools provide a better means of addressing event management and implying changes to the underlying policies and processes.

Changes to how the tools process individual events are affected by several factors and should be iterative to account for them. These changes are more frequent than the changes that address policies and processes and should be implemented more often:

- **New event sources**
  As new hardware and software are added to the environment, new events may be generated, or there may be changes to the use of existing ones. Review the new events through a methodology such as IBM EMMD, document the decisions for handling the events, and implement the changes.
- **Problem post mortems**
  These are ideal situations to help identify the causes and resolutions for problems. Use these sessions constructively to identify ways to monitor for the future failures, new events to forward, and correlation sequences, and implement them.
- **Experience of operators/staff using the events and their correlations**
  Those who monitor the consoles for events often have a good sense of which events flood, occur together, or can be ignored safely. Use their ongoing input to tweak your event management implementation.

  Often event processing decisions are initially made based on information in message manuals or on the educated guesses of SMEs. When an error condition occurs, it may behave differently than anticipated. For example, the message manual states the meaning of an error message that indicates an important failure for which notification is desired. However, it fails to mention that the subsystem will retry the failing process five times per second and generate the error message each time it fails. Those watching the console detect this and can provide the feedback necessary to suppress the duplicate messages or find another means of detecting the condition.

4.3.2   Defining Severities

The severity assigned to an event is an indication of how critical a problem is that it reports, which relates, in turn, to how quickly service must be restored to the failing system or resource. Event processors may use different terminologies, but most provide a range of severities that can be used to designate varying degrees of criticality, ranging between "immediate attention required" and "this can wait."

Severity levels may already be defined as part of documented problem or change management processes or Service Level Agreements. In this case, use the existing definitions. Otherwise, define severities and acceptable recovery times as part of the event management process. For each severity, choose notification methods appropriate for events of those critical levels.

Consider mapping severities to business impact. For example, define *fatal* to imply a key business process is down and affects many users. Designate *warning* to mean problems with noncritical resources or those that affect few users.

This is important for two major reasons. First, it is easier to decide the severity to assign to an event. When implementing event management, personnel from the support organization may be called upon to review events and identify those for which they desire notification. Part of this process is to assign severities to the events and choose a notification type. Knowing the meanings of the various severities simplifies this process.

For example, when the server that runs a key business application fails, it must be handled immediately. If a *fatal* severity is defined to mean, "respond immediately" and "inform by paging," the administrator can easily decide this is the proper severity to assign to events reporting outages involving that server.

Second, clear definitions facilitate tool setup and maintenance. When the same types of notifications are performed on all events of the same severity, the event processing tool can be configured to merely check an event's severity and take the appropriate action. If there is no consistency, the event processor must check additional parts of the event before making its determination, which complicates the implementation.

When event processors use different severities, define mappings to show how they relate. This is necessary to ensure events forwarded from one event processor to another are assigned the proper severity at the receiver. Show how the trouble-ticket severities map between all event processors in the hierarchy, from the event sources and monitoring agents, through the correlation engines, to the trouble-ticketing system. See "Event management products and best practices" in Redbook, on p. 173, for a sample severity mapping [1].

4.3.3 Implementing Consistent Standards

One of the primary goals of the event management process is to automate the filtering, duplicate detection, notification, correlation, and escalation of events and the recovery of systems. Automation of any kind is easier when the resources that are handled adhere to standards.

Organizations that already adhere to standards can more easily implement event management than those that have not yet defined or implemented them. If your organization is in the process of developing standards while implementing event management at the same time, you may have to initially set up two types of processing: one for machines that follow the convention and one for machines that do not follow the convention. This allows you to proceed with event management without waiting until your environment is completely converted to the new standards.

Use standards in your event management process, and keep them consistent across the enterprise. There are always special cases, which arise, but these should be kept to a minimum. This minimizes the processing cycles that are required and simplify configuring and maintaining tools.

The following sections document two examples of how commonly implemented standards can facilitate event management.

1. **Naming Conventions**

   Naming conventions is one of the most important standards to have. Machines, scripts, files, domains, and the like should all have consistent names across an organization because they are commonly used or referenced in automation.

   Depending upon your support structure, you may need to notify, page, or route a trouble ticket to people based upon the geographic location of the failing machine or by its device type. Having a standard naming convention simplifies this process.

   If the standard is to use a geographic location or device type either in the domain names or in positional characters within the host name, the event processor can easily determine whom to notify. When a failure occurs to RDUWWS01 (AMD PCNET Family Ethernet Adapter), for example, automation can determine from the first three characters that the device is in Raleigh and from the next two that it is a Windows Web server, and notify based on this information.

   It is much easier to determine the proper support group for a problem based on geography or device type than by individual host. If the trouble-ticket queue names or e-mail user IDs follow a convention similar to the machine names, it may be possible to create the notification target from values extracted from the host name, avoiding the need to maintain the information separately in spreadsheets or files for individual hosts. This is advantageous, particularly in large organizations with many systems located across many sites.

2. **System Configurations**

   Automation can be consistently applied to machines that have the same configuration. For example, when a performance problem occurs on a machine, it may be desirable to gather diagnostic data from the time that degradation occurs. Otherwise, the condition may clear and the information disappears, leaving the assigned support person unable to research the problem.

   The tools needed to gather the diagnostics may be installed on some systems but not others. Therefore, the automation can only be applied to a subset of the environment. The systems management tool implementer must then determine which systems have the required diagnostic tools installed and implement two types of event handling (one with diagnostics and one without) for the same event.

   Standard machine configurations eliminate this problem by ensuring that predefined groups of systems are all configured the same.

4.3.4   Assigning Responsibilities

Potentially many people may be involved in the event management process, including support staffs, tool implementers, help desk and operations center personnel, and managers. All have opinions about how to handle events based on their own experience in their jobs.

Definitely use their input when developing the event management process. Including their ideas helps to ensure a robust solution that meets the needs of its users. People who feel that they have influence are also more likely to embrace the resulting processes, facilitating their enforcement.

However, assign specific responsibilities to the involved parties and designate a final arbiter who will decide issues in dispute. Otherwise, the development process may stall as the involved people seek consensus (which is often difficult to obtain, especially in large groups).

Some of the roles to assign include:

- **Process owner:** Heads the development of the policies referenced in this section. Has ultimate responsibility for the success of the event management process
- **Systems management architect:** Designs the technical solution to meet the event processing requirements while adhering to appropriate policies and standards
- **Tool implementers:** Install, configure, and support the systems management tools to process events to the specifications of the architect's design
- **Subject matter experts:** Supply knowledge about a particular platform or system and determine the processing required for events within their areas of expertise
- **Support staff:** Handles problems with platforms and systems
- **Help desk:** Provides the first level of support for users and give feedback that is used by the SMEs to make event processing decisions
- **Managers:** Enforce adherence to policy by their staffs and ensure problems are addressed in a timely manner by responding to escalation procedures

4.3.5   Enforcing Policies

Given the importance of both the automated event handling policies and those defined in this section, it is crucial to see they are followed. Therefore, define, implement, and track compliance with policies. This ensures consistency of design and ease of tool implementation and maintenance, resulting in a successful event management endeavor.

The ramifications of not following policies and procedures vary with the policy itself. Data validity, for example, may be adversely affected by not following the policy requiring operators and administrators to close problems when they are resolved. Only closed events are recorded in the enterprise data warehouse database that is set up for this purpose. Leaving problems in an open state can prevent them from being recorded and reported within the warehouse, leading to incomplete or misleading service level reports.

One implication of enforcing policy is the necessity of a method of tracking adherence to it. Record that who takes which actions on an event. This lets you know whom to contact for the status of open problems and provides a means of determining who invoked wrong actions on an event so you can ensure it does not recur.

4.4 Filtering

Filtering is, without question, an essential part of work in each event management effort. This section discusses filtering, which in IT terms is described as the process of blocking information based on a defined set of rules and standards. In general, we define filtering as the part of the whole engagement, where we try to remove as much redundant and unnecessary data as possible.

The most difficult part is to determine what the right data is that we need to effectively implement an event management, which signals possible alert situations. The following sections discuss the aspects of filtering in a systems management environment. They also discuss best practices for the filtering task itself. They do not cover device-specific filtering methods and best practices.

4.4.1 Why Filter

Obviously, there is a need to restrict the data entering our event management system or filter them somewhere on their path toward the event management system. Otherwise, we would not dedicate a whole section to this topic.

There are several reasons why filtering of events is most recommended:

- **The pure amount of data produced by managed objects**
  In a complex IT environment, the amount of events produced by managed objects can reach a high amount of discrete events being sent to a single management instance. Many devices provide, by default, all events they are available to offer, which for some of those management agents can easily be a list of several hundred events. If this is, multiplied by the number of different managed devices in a managed environment, we see that this amount of possible events cannot seriously be managed.
- **Redundant information produced by various monitoring agents inside a single managed object**
  Often, various monitoring instances on a device provide the same information and send them in the form of events. For example, the *syslog* subsystem on a UNIX server provides critical information, while the SNMP agent running on that server provides trap information about the same event.

- **Network and bandwidth considerations, event server limitations**
  In a large and complex distributed environment, the event-related traffic is basically unwanted waste of resources. This applies to both the traffic produced from status polling of devices and the traffic generated by devices sending asynchronous unsolicited events over the network. Event-related traffic can occupy a reasonable amount of bandwidth. In most environments, network bandwidth is still a precious resource and is normally reserved for productive matters. An increased system management traffic can be treated itself as a degrading event.

  In addition, the receiving instance, whether a simple event console or a more sophisticated business management system, cannot accept an unlimited number of events per time frame. Often, the receiving management system itself polls managed objects in regular intervals to monitor critical resources on that object. If a threshold is exceeded, this information is translated into an event and enters the event stream. Obviously, if management stations have a limited capability to receive events, they have a limitation on the amount of polling operations per time frame. You can find a discussion about specific products and their capabilities in Chap. 6, "Event management products and best practices" of Redbook [1].

- **Manageability**
  As a first rule, keep the event management system as simple as possible. Too many events from a large number of different sources can lead to confusion. An operator can soon start to ignore incoming events if they alarm for duplicate events or, even worse, lead to wrong alarms and actions.

All of these points are reason enough to limit the amount of data arriving in the event management system. Together they make the need for filtering essential.

4.4.2   How to Filter

The main question we must ask is: "Which events do you need to do your job?" Or better, we should ask: "Because most IT organizations today are service units for the various other departments, which events do you need to fulfill the agreements you made with your customers?"

In general, if a piece of information arrives in our management system and does not indicate a loss or a degradation of services, it should not appear and should be blocked. If it does not affect your Service Level Agreements, remove it. Keep in mind that a particular event, in which you are not interested, may be of some importance to other departments. Therefore, preventing the event from being generated may not be the best idea.

Suppressing the delivery of an event without making it completely unavailable to other recipients makes the simple term filtering more difficult. Everyone may agree on filtering itself, but where the actual filter is applied can vary from one viewpoint to the other.

4.4.3 Where to Filter

Now we must ask: "Where do you need to filter unwanted events?" If we remember the discussion in "Why filter" on page 39, we considered the occupied network bandwidth as a good reason for filtering. The possible large amount of events was another reason.

This can only lead to one rule: Filter as close to the source as possible. Filtering as close to the source is, in most cases, the best method to block an event from being delivered to the rest of the world. It saves bandwidth, helps to keep event management systems manageable, and saves system resources needed for production.

Filtering as close as possible, preferably directly at the source, should be the first choice. But sometimes, you cannot achieve this goal for the following reasons:

- The event may be of some importance to other management instances. For example, network operations may be interested in a toggling Integrated Services Digital Network (ISDN) interface. The organization running the corporate-wide event console is, in most cases, not interested as long as the network performance is not degraded.
- The event cannot be filtered at the source, because the event generator itself is an all-or-nothing implementation. Either you buy all the events or you block all of the events.
- Events generated because of a status poll operation are normally not of a particular importance on the focal point level of the management system, in case it is an event console implementation such as the IBM Tivoli Enterprise Console. The status information is definitely needed for the actual status puller to maintain a list of the current states of the managed objects. Status information is also required if the focal point for the event management is a system dedicated to business impact management. Business impact management systems keep a record about the actual state of its managed object to monitor complete business environments.
- Trying to filter the event at the source can result in an effort which is more costly than just trying to catch the event on a higher level of event management. For example, after a rollout of a high number of devices, it turns out all the devices are, by default, configured to a positive forward all state. Remotely accessing these devices and blocking the unwanted event one by one at the source can be time-consuming.

4.4.4 What to Filter

Now we need to specify what to filter. This is by far the most time-consuming task related to filtering. Under normal circumstances, the various resources to be managed are well known. But regardless of whether these resources are capable of providing valuable information to a management system in form of events is not necessarily known.

After the set of event sources is specified, we need to address all events of each event source and analyze them for their value to the event management. Of course, we do not limit the analysis of the events to filter. Decisions, correlation candidates, and throttling parameters may be discussed and documented during this stage.

Speaking of best practices, two suggested approaches exist to make filter decisions and the correct correlation and escalation decisions. Refer to Sects. 4.6 and 4.8 in this chapter, for more information about correlation and escalation.

The first approach applies to environments, where little or no event processing takes place. It may also apply to environments where events are generated, but are treated as unwanted noise until a working systems management environment is set up. In this case, you must complete these tasks:

1. Identify and define the event sources, which are important to the management organization. Often it helps if element chains are documented and there is a given management view in place.
2. For each event source, build a list of all events offered by the event source.
3. Find an expert for the resource being analyzed and discuss the importance (or lack of importance) of each event.
4. Find an expert who is responsible for the management of that particular resource. Often this is the same person, who knows the events, which are needed. Discuss whether the events should be included in the event management.
5. Document these decisions.

The approach is somewhat static because it defines a set of event sources to be analyzed and the results being implemented. If no iterative process is set up after the initial work, the result is quickly outdated.

Define a process that, after the initial analysis, detects changes in the event flow or additions and deletions in the set of event sources. It should also ensure that the event management process is iterative.

This approach can be called and filtered by subject matter experts (SMEs). The analysis and the resulting filter decisions depend on the expertise of the people who are responsible for a given. Another approach to obtain fundamental information about the events appearing in a given environment is to analyze the events itself using a data mining approach: resource.

1. Obtain information about all events received in your organization over a time frame of at least three months to have a solid base for analysis.
2. Normalize the data by extracting only the relevant information, such as:

   – Time stamp
   – Event type
   – Event name

Try to limit the event information to a minimum. It is sufficient if the event can be uniquely identified.

3. With a small part of the whole data repository, typically the events of a two-week time frame run an initial analysis. Make sure that the data contains the information you expect.

   – Use the whole data and analyze it.
   – Are there any large amounts of a single event?
   – Is there another event from the same source having the same or a similar count?

   Such a pattern often signals a violation event and its corresponding clearing event.

   – Are there groups of events from different event sources appearing with similar counts?

   This can be an initial problem causing other secondary exceptions to occur. What makes them correlation candidates?

   – Are there more than two different events from one event source appearing with the same or a similar count?

   This can be a primary or secondary condition, too. For example, an interface down Simple Network Management Protocol (SNMP) trap sent by a network device is often followed by an interface down event produced by the SNMP network manager, generated by a status poll operation against this interface. An unsuccessful poll for status against an interface can result in a node down event being generated if the interface was the object's only interface.
   This type of group of events is a good filter candidate. You really need only one indication to signal a problem. The same applies to the associated clearing events. You often find an interface up trap, an interface up event, and a node up event.

4. Define filtering. After you run such an event analysis, you still need SMEs for the particular event source to finally define the filter conditions. Having solid data about event occurrence, and the amount of events, for a particular situation, helps to keep the discussion short.
   One last way to perform filter analysis is through *problem post mortems*. Analysis of situations, where a service degradation occurred and nobody was informed, may help to revise or find some filter decisions that were not made before. Regardless of the method used to determine events to filter, filtering is never implemented perfectly on the first try. You must continuously evaluate and redefine your filtering methodology for filtering to be most effective. As business needs and requirements change, you must also update your filtering methodology.

4.4.5 Filtering Best Practices

Up to now, we discussed the need to filter different methods to eliminate unwanted events. There are some best practices for which events not to filter and which events should never find their way into the management system.

Here are some best practices to implement for filtering:

- Do not filter or block events that have an exactly defined meaning, where an automatic action can be issued. Nor should you filter the corresponding clearing event.
- Do not filter clearing events for problem events you report. Administrators do not know when an event has been resolved if they do not receive the clearing event. In addition, during correlation, a problem event may not be closed if a clearing event is not received. Remember that clearing events are essential for de-escalation purposes.
- Report any exception from the normal process only once. For example, we mentioned the interface down trap, which causes an interface down and a node down event. Only one event should be passed to the management system. If possible, the primary event should be passed.
- There is an exception to this rule. Sometimes it is useful to take the double events to verify the situation. A node down may result from timing or network problems. The interface down trap always signals a real exception. When the interface down trap does not find its way into the management system, the interface down and node down events are the only indications of a failing interface.
- When using business impact software, do not filter status change events. This renders the business impact software useless for providing status of objects.
- Always pass actionable events and their clearing events. An *actionable event* must be acted upon by either automation or administrator intervention.
- Do not double monitor a resource. Having multiple monitors check for a single condition causes processing overhead and produces redundant events. Only one problem should be reported to prevent more than one support person from handling the same issue.
- A possible exception to this rule is when multiple agents are needed to validate that a problem is real and not a false positive. In this case, it is acceptable to double monitor the resource as long as the events produced by each monitor are correlated and only one reported.
- Filter false positives if possible to avoid unwanted and unneeded alerts and events. If you page someone at 3 a.m., you had better be sure it is a real problem.

4.5 Duplicate Detection and Suppression

Duplicate event detection is the process of determining which events represent the same instance of the same problem and summarizing or suppressing them in a way that simplifies the event management process. Its purpose is to save cycles and system resources on event processors and minimize bandwidth used to send unnecessary events.

4.5.1   Suppressing Duplicate Events

Duplicate detection and suppression can be done in more than one way. Depending on the case, the best practice can vary. Here are some common methods to perform duplicate detection (sometimes referred to as *de-duplication*):

- Send the first event and suppress the others. This approach is typically used with events that state a failure in a system or equipment. Immediately reporting the event minimizes the mean time to repair.
- Send an event only after a predefined number is received. This practice, commonly referred to as throttling, is often used for events that represent peak conditions.
- While some monitored variables, such as processor utilization, occasionally reach peaks, this is not a problem unless sustained for a period of time. For these types of events, do not send the first event. Summarize it with the subsequent events and send one event if they reach a threshold. After the event is sent, drop future occurrences until the time period expires or the condition clears. For example, if more than five peak events are received in 10 min, there may be a problem that requires notification. Count the events and send the fifth along with a count of the times it occurred.

Send an event only after a predefined number is received and send all future occurrences. While similar to the previous method, this differs in that all events are forwarded when the threshold is reached.

This approach may be used when it is necessary to know the actual values the variable reached. For these events, do not send the first event. Summarize it with the subsequent events and send one event if they reach a threshold. For example, if more than five peak events are received in 10 min, it may represent a problem and you need to be notified. When subsequent events are received, extract the relevant values from them and update the reported event.

This method is not generally used because it requires sending and processing all events generated after the predefined number. In general, if all the monitored values are required for problem determination or trending, this information should be provided by another type of tool and not by events.

4.5.2   Implications of Duplicate Detection and Suppression

Duplicate detection and suppression, when well done, are useful and can give fast results for the event management process. However, in some cases, it is important to ensure that you do not miss information, as illustrated in the following examples.

- **Time window considerations**
  The first two examples illustrate how time windows affect duplicate detection and suppression for peak events. In Fig. 4.1, a duplicate detection and suppression process was created for a processor utilization variable. In this case, when

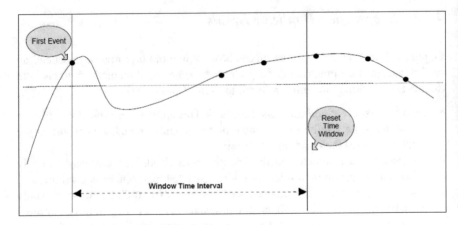

Fig. 4.1 Static time window for peak events

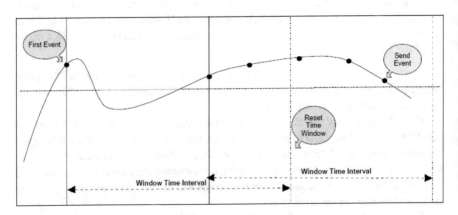

Fig. 4.2 Multiple time windows for peak event

one peak event arrives, it is buffered and a time window is created. If four more events occur inside the time window, one event is sent to the event management system. Note that no event is sent because only three events occurred during the time window.

The problem in this case is that the last five events occurred over a time period that is shorter than the defined time window, but no event is sent. This is because one time window was opened upon receipt of the first event and no others were started until the first one expired.

Figure 4.2 shows the same situation, but with another duplicate detection and suppression process. For this example, every time a new event arrives, a new time window is created. During the first time window, three events occur, the same as in the last scenario. However, during the second window, five events occur and one is sent.

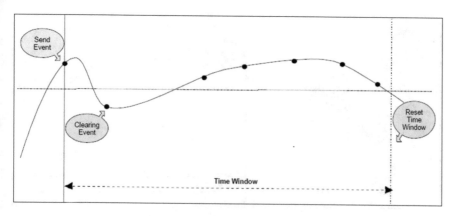

Fig. 4.3 Clearing event does not reset the time window

There are appropriate times to use each of these methods to create time windows:

- Use static time windows for situations that generate many (more than the threshold) consecutive events in a short time. It is also appropriate when you have severe performance constraints, since it does not create many time windows, reducing overhead in the event processors.
- Use multiple time windows when the performance constraints are not so rigid, fewer events arrive during the time windows, and the trigger number of occurrences must always send an event.

Obviously, the methods employed depend upon the capabilities of the event processing tool used.

- **Effect of clearing events on failure reporting**
  Make sure that resolved problems are closed or that duplicate detection suppresses new occurrences of same problem, considering an outstanding problem already reported. The next two examples discuss the effect of clearing events on failure reporting.

  Based on 4.5.1, "Suppressing duplicate events" on page 45, when a failure event arrives, the first event reporting should be sent to the event management system. Duplicate events should be suppressed during the time window. In the previous example, if a clearing event occurs within the time window and is not used to close the reported problem, subsequent failures within the time window are treated as duplicates and not reported. Figure 4.3 shows this example.

  For short time windows, operators viewing the events may notice the clearing event and manually close the original problem, which resets the window. However, this method is unreliable. The clearing event may become lost within large volumes of events, particularly for long event windows. The originally reported problem stays open, and subsequent failures are not reported.

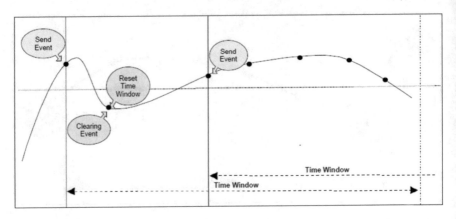

Fig. 4.4 Clearing event rests time window

The example in Fig. 4.4 illustrates how resetting the time window and opening a new one when the problem next occurs will resolve this problem.

A good practice is to correlate the failure event with its clearing event and close the problem. This results in resetting the time window, clearing the duplicate counter, and ensuring new occurrences of the problem are reported.

4.5.3 Duplicate Detection and Throttling Best Practices

We make several recommendations for duplicate detection and throttling. We also provide the rationale behind these recommendations:

- Perform duplicate detection as close to the source as possible.
  This practice saves cycles and system resources on event processors and minimizes bandwidth used for sending unnecessary events.

  When possible, configure the event source to detect duplicates and suppress them. If it is incapable of performing these actions or if implementing at the source causes undue, cumbersome tool configurations, use the closest event processor capable of performing the function.

  Uses throttling for intermittent problems that may clear themselves automatically and do not always require action. After a problem is reported, suppress or drop duplicates.

  It is frustrating for a support person to investigate a problem only to find that the problem has disappeared or requires no action. If this occurs too often, the support person loses faith in the systems management tools and begins to ignore its notifications.

For events that indicate problems always requiring action, inform when the first event is received, and suppress or drop duplicates.

This notifies the support person most quickly and minimizes the mean time to repair for the problem.

- Do not use duplicate events to reinform whether the original problem has been handled. Instead, use escalation.

Using duplicate events as reminders that a problem is still open is a bad practice. The extra events clutter consoles, possibly forcing the operator to sift through many events to find the meaningful ones. If there are too many events, the console user may begin to ignore them or close them in mass.

See "Unhandled events" on page 60 for a discussion about using escalation for events that have not been handled in a timely manner.

This bad practice typically arises in organizations that point fingers for unhandled problems and assign blame. Those who are responsible for resolving problems often need to justify why they miss problems and blame the tool for not informing them. The event management implementers, fearing these reproaches, configure the tools to send all occurrences of a problem rather than just one. Unfortunately, this compounds the problem because now the support person has to handle many more events and can still miss ones that require action.

Management needs to create an environment on which problem post mortems are used constructively, minimizing blame and enabling administrators to pursue and fix the causes of missed events rather than creating event floods to avoid blame.

- Use duplicate detection for open and acknowledged events.

Duplicate detection is often implemented for open events but not acknowledged ones. These are equally as important because they indicate that the problem is being addressed but is not yet corrected. Since someone is notified of the problem, there is no reason to reinform with subsequent events.

4.6 Correlation

Several sources can help to determine the relationships among events. Subject matter experts (SMEs) within the organization can provide input about events in their areas of expertise. Vendors can also furnish information about the events produced by their software. Sometimes the message manual for a product supplies event groupings by stating that a message is one of a series and listing the other messages that appear with it.

Data mining tools that employ statistical analysis, such as the proprietary software used in the IBM Data Driven Event Management Design (DDEMD) services offering, can suggest possible correlation sequences. Problem post mortems can help to determine events that frequently occur together or to validate proposed event sequences. Which sources of information an organization uses depends upon the skill level of its SMEs, the willingness or ability of its vendors to share information, and its access to statistical tools.

Often the event management implementer is left with the challenge of making sense of the information gathered and taking a best guess as to which of these potential correlation sequences to implement. Overzealousness may lead to correlating events that do not necessarily associate and dropping real problems as a result. An overly cautious approach may require more operator intervention to manually close or associate events. The best practices that follow are general guidelines to use in determining which sequences to implement.

4.6.1   Correlation Best Practice

The first step in determining what to correlate is to collect information from the sources identified previously. After the potential sequences are identified, apply these guidelines to choose which one to implement:

- Only correlate events whose relationship is clearly understood.
  Correlation should be implemented only when the association between the events is known. Implementing a best guess can result in correlation sequences that inadvertently drop real problems because they are erroneously thought to be symptom events. It is better to have an operator manually handle an event several times until its relationship to other events is known than to implement automatic associations that may not be valid.

  As discussed in 4.3, "Policies and standards" on page 32, the event management process should be iterative, and the event handling decisions should be updated as new information becomes available. Start with the sequences you know, and add to them based on the experience of you operations and support staffs.
- Automatically clear problem events when possible.
  Implement this type of correlation sequence whenever feasible. This ensures that the accuracy of the list of open events from which the operations and support staffs work. It also prevents duplicate detection from flagging new occurrences of a problem as duplicates of the resolved problem.

  It is usually easy to identify the clearing events associated with problem events. Monitoring agents can often be configured to generate them. The product documentation frequently describes the association between the problem and clearing events. Messages extracted from logs can be more difficult to clear. Sometimes only the problem is recorded and the recovery condition is not. In these cases, a policy is needed to ensure that operators manually close problems when they are resolved.

  Implementing clearing event sequences is also easy. Many monitoring products supply the rules necessary to do this type of correlation.

  Remember to close the clearing events as well as the problem events. This minimizes the number of events that display on operator consoles and reduces overhead at the event processor.

- Correlate events that show a worsening condition.
  Sometimes when a condition intensifies, multiple events are generated to show the progression of the problem. Correlate these types of events, and keep only the first in the series. Update it with the higher severities as the other events are received. If desired, include a threshold or other appropriate information, such as the time stamp from those events, and then drop them.

  This processing ensures that the original, reported event is available for event synchronization. When the problem is closed in the trouble-ticketing system or in another event processor, this event is then correctly closed. If another event is kept, the synchronization process does not know to look for it, and the problem may remain open indefinitely.

- Report all events requiring action, not just primary events.
  Part of understanding the relationship among events is knowing how action taken to resolve the problem referenced by one event affects the conditions reported by related events. This information must be obtained before determining whether a secondary event requires action.

  Sometimes resolving the primary problem automatically fixes the symptom problem. In this case, no further action is required. Other times, the symptom condition does not clear automatically when the primary problem is resolved, necessitating action for the secondary event. See "Root Cause Correlation" in Chap. 3 for examples of correlation sequences in which the secondary events require action and those in which they do not. To ensure all problems are resolved, report all events requiring action, even if they are symptom events.

- Drop secondary events that do not require action, unless they are status events required by a business impact manager.
  If an event does not require action, it is not needed at the event processor. To prevent clutter on the processor's console, drop the event. Some implications of this are discussed in 3.5.2, "Implementation considerations" in Chap. 3.

  The only exception to this rule is listed in the following practice for users of business impact software.

  Forward all status events to business impact software, even those that do not require action.

  Business impact managers are used to show the effects of system resource failures upon business functions. The proper functioning of business impact software relies upon it accurately reflecting the status of each of the system resources that constitute its business views.

  When a resource known by the business impact manager changes status, its new state needs to be reflected in both tool's database and business views. The effect the status change has on other resources is calculated, and the affected resources and views are modified to reflect their new states. Therefore, if using business impact software, forward all status events to it.

4.6.2   *Implementation Considerations*

While product-specific implementation is discussed in Chap. 6 of Redbook [1], "Event management products and best practices" on page 173, some general guidelines apply regardless of the tools that are used:

- If no clearing event is sent to report a resolved condition, require operators to close the appropriate event.

  Leaving open problems in an event processor can lead to incorrect error reporting. Namely, the duplicate detection process may think there is an outstanding problem and discard an event that reports a new problem. In addition, some data warehousing tools only load closed events into their databases.

  Keeping problems open prevents them from being recorded in the warehouse and subsequently being included in trending and Service Level Agreement reports.

  To prevent this from happening, implement clearing events. However, sometimes this is not possible. For these events, the operator should manually close the event through the trouble-ticketing system or event processor's console.

  This policy should be documented in the event management process guidelines, and compliance with it should be enforced and tracked.

- Link symptom events requiring action to the problem events upon which they depend.

  If a symptom event requires action to resolve the problem it reports, the action most likely cannot be executed until the primary problem is resolved. For example, a process that depends upon free space in a file system cannot be restarted until that space is available.

  To show dependency between events, link them in the event processor or trouble-ticketing systems. The events may be linked by updating fields in each to show its relationship to the other. Another approach is to copy the relevant information from the symptom event into the primary and then drop the secondary.

  The support person assigned to the problem can read the additional text to determine what actions may be required to handle the symptoms. Then they can either perform those actions if appropriate or requeue the event or events to a different group once the primary problem is resolved.

- Repeat lower-level correlations at higher levels.

  Correlation may fail for a couple of reasons:

  - The events may arrive too far apart. When it receives an event, the event processor often sets a timer to determine how long to wait before reporting the event as a problem. If an associated event or perhaps even its root cause event is received after the timer expires, no correlation occurs and both events are treated as primary problems that require action.

  - In a multitiered environment, the events may be forwarded to a higher-level event processor. They may arrive at the higher level closer together.

- – Defining the same correlation sequence at that processor allows for the chance that they arrive within the correlation window. The events can possibly be correlated and the appropriate event can be reported.
- – Memory constraints in the event processor may prevent the events from being correlating. If the processor relies on the presence of the events in cache to associate them, correlation fails if one of the events is dropped from cache due to memory shortages.
- – Again, the higher-level processor may not have the same constraints and may be able to perform the correlation.

• Allow sufficient time to receive events before you correlate them.
Setting timers in an event processor is an art. Waiting too little for events can result in missing the correlation between events and reporting them all as problems, even the symptom events not requiring action. Lengthier correlation windows eliminate this problem, but may introduce others. If the timer is set too long, there may be a delay in reporting the problem, resulting in a longer mean time to repair. In addition, events that are caused by different error conditions may be erroneously correlated.

Observe the rate at which associated events are generated by sources and received by event processors to choose meaningful timer values. This information can be obtained from logs at the event sources and processors and from problem post mortems.

• Correlate events from clusters, and escalate problems as more machines in the cluster experience them.
Clusters are groups of two or more systems typically used for load balancing or failover. If one machine in the cluster reports an event, there may not be a problem. For example, in a failover environment, only one machine in the cluster may need to run an application at a time. In this case, if the monitoring agents detect the application is not running on one server, this is a normal state and does not need to be reported. The problem exists when the application is not running on any clustered system. This concept also applies to grid computing.

In a load balancing cluster, the situation is slightly different. If one system experiences a problem, it should be addressed. However, it is less critical than if every system in the cluster has the same problem. Use differing severities to reflect the business impact of these two distinct conditions.

Implement cross-host correlations to handle the unique event reporting requirements for clusters and to ensure the proper error conditions are detected and reported.

• Perform topology-based correlation for network events using an SNMP manager.
Since the SNMP manager knows the network topology, it is capable of performing topology-based correlation. Many SNMP managers provide the ability to correlate network events out-of-the-box.

Performing topology-based correlation between network and server events requires supplying the network layout information to the event processor at which these events converge, typically not an SNMP manager. While it is possible to provide the topology to other types of event processors, the procedure is often complex and difficult to implement. Most networks are dynamic, implying frequent updates to the topology information supplied to the processors. This quickly becomes impractical, particularly in large networks.

If the SNMP manager can detect the status of nonnetworking resources, such as services, it can be used to perform topology-based correlation for events concerning those resources. You can find a description of how NetView implements this type of correlation in 6.3.1, "Correlation with NetView and IBM Tivoli Switch Analyzer" on page 218 [1].

4.7   Notification

Notification is a key step in the event management process. It is useless to detect an error condition unless action is taken to correct it. While automation is used increasingly to recover from problems, there are still many situations that require the intervention of an administrator to resolve. Notification is the means by which the appropriate person is informed to take action.

4.7.1   How to Notify

This section discusses the methods of notification from your event processing tool and the advantages and drawbacks of each. Figure 4.5 shows an overview of the types of notification. Depending on the structure of your business, you will handle your notifications in different ways.

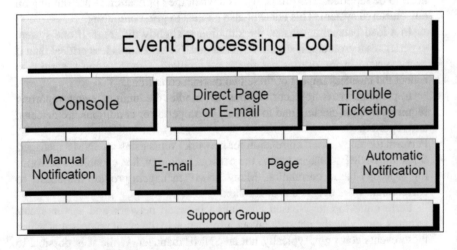

Fig. 4.5 Types of notification

With event processing tools, there are typically three ways to implement notifications:

- Console viewing by operators
  Operators watch the console looking for events that require action. When they see an event, they respond by taking action themselves or manually inform another person. Having operators view the console and then manually notify support teams gives you the advantage of having someone to verify events manually when they happen and give the notification to the right person.

    The disadvantages include human error, for example, missing an event on the console. The disadvantages also involve the costs of keeping and training a person to watch the console.
- Direct paging and e-mailing from the event processing tool
  Directly paging from the event processor, through scripts or executables that are triggered by criteria of an event, gives you the functionality of not having an operator continuously watch the console. However, it is difficult to maintain the proper lists of which groups to notify for which problems. This information, already kept in the trouble-ticketing system, needs to be duplicated in the event processor in different places such as rule bases or custom scripts. It is also difficult to track the notifications or ensure they are received.
- Integration with a trouble-ticketing system for automatic notifications
  The main advantage of integrating with a trouble-ticketing system is that you tie in with tools and processes that already exist within your organization at the help desk, operations, or command center. It is much easier to track on-call lists and the right support groups for notifications. Timetables for notifications are also easier to create and maintain within trouble-ticketing systems.

    It is also easier to assign each trouble ticket to the proper groups based on information within each event. For example, if you integrate your trouble-ticketing system with your asset management system, you can automatically assign tickets to the proper support group based on the host name slot from your event.

4.7.2   Notification Best Practices

This section discusses the best ways to design and implement problem notification:

- When possible, handle all notifications through integration with a problem tracking tool.
  Many trouble-ticketing systems provide all three notification methods: a console for viewing open problems, an e-mail, and paging capabilities. Performing these functions from a single place simplifies system setup and maintenance. See 4.7.1, "How to notify" on page 56, for more details.

- Notify someone about all problem or actionable events.
  Before you start thinking about sending notifications for your event process-
  ing tool, review the events you are receiving and classify them into two sepa-
  rate groups:
  – Informational events: Events that do not show a system as being down or
    having a problem such as clearing events.
  – Problem or actionable events: Events that indicate a system or applica-
    tion is unavailable or has a problem such as process or service down
    events.
  – Usually it is a good idea to notify only on events that require an individual
    or support group to take action, specifically problem events. You do not
    want to notify someone about informational events, especially if it is after
    hours.

While it is not a common practice to notify about clearing events, support
groups may want to know when their system is back up. If this is the case,
write your notification rules or scripts so that the notification comes from the
original down event being closed and not the clearing event.

Note

Next go through all of your problem events and decide which group or individual
receives the notification. It is a good idea to set up a database or spreadsheet if
you are not using a trouble-ticketing system.

- Consider the severity of the problem, the time of day, and the critical level of
  the failing system when determining what notification action to take. Page for
  higher severity problems and critical resources, and notify others by e-mail.
  After you decide which types of events require notification, go through each
  type and determine the severity at which you want to notify. Two automated
  methods are used for notification (console viewing is not automated):

  – Paging: Sending text or numeric messages to a paging device or cell phone
  – E-mail: Sending messages through the organization's e-mail system

- When trying to determine the severity of the event, keep in mind the importance of the host from which the event is coming. Try to relate the severity of
  the event to the importance of the failing system.
  Determining the severity for the event is directly related to the notification
  method chosen. When you page someone, you can reach him or her at all
  hours of the day. When you e-mail, you cannot reach someone unless they are
  logged in and checking their mail. Therefore, it is a good idea to have the
  higher severity or priority problems send a page and have the lower severity
  problems send an e-mail.
  While you determine this, keep in mind the time of day that these notifications are sent. Although it is not suggested, some trouble-ticketing systems

can send different types of notifications at different times of day such as sending an e-mail during the day and paging at night. Usually it is best to keep your notifications standard, either e-mail or page based on the problem severity. This is for easier maintainability. However, you must watch for sending unwanted or false pages, especially after hours.

- Ensure that after-hours paging does not occur for noncritical problems.
  After you set up your severities and methods for notifications, double-check to make sure that you are not sending notifications for noncritical problems.

  Also, remember that the notification process is ongoing and that when a mistake or false page is sent, you must take the proper steps to ensure that it does not happen again.

- Report problems with a notification system by some other means.
  Obviously, if the primary notification system is experiencing problems, it cannot be relied upon to report a problem with itself. Use another event processor or backup method to report problems with the notification system.

4.8   Escalation

Escalation is the process of increasing the severity of events to correct perceived discrepancies in their importance and to ensure problems receive appropriate and timely attention. Best practices always depend on an organization's environment and policies. It is no different for escalation.

In this section, some escalation recommendations are provided for different environments. Use those which most closely reflect your organization.

Also, keep in mind hardware and performance issues when creating escalation processes. The number of escalated events needs to be well defined and controlled, ensuring that perceived discrepancies are corrected while minimizing processing overhead.

4.8.1   Escalation Best Practices

As discussed in Chap. 3, "Introduction to Event Management," there are several different types of escalation. Each is important to the event management process and should have at least one policy that defines the standard way to implement it within an organization. As with all standards and policies, this facilitates tool implementation and maintenance.

This section covers the best practices for the three types of escalation.

- **Unhandled events**
  Without a clear, well-defined escalation policy for unhandled events, it is possible that problems will be reported, but never resolved. An assigned support person may be too busy to work on a problem or not know how to proceed.

Escalation ensures that events are handled in a timely manner by creating a greater sense of urgency through raising event severities, notifying additional people, or both.

Escalating problems is often used in help desks and network operations centers. It helps management by advising of situations in which those responsible for an event do not follow the procedures or cannot complete the job in time. With this information, managers can better coordinate the workforce. For example, if an operator has too much work and cannot handle an event in time, the problem is escalated to the manager, who can allocate an idle operator to help with its resolution. See the first best practice in the list that follows this discussion.

The question becomes how long to wait to see if an event is being handled. It is possible to set different durations based on event type, host priority, time of day, and other factors. However, this quickly becomes a maintenance nightmare. An easy, effective alternative is to base it on severity. See the second and third best practices in the list that follows this discussion.

Next, decide what needs to occur within the time interval. Two typical actions are acknowledging and closing the event. The person assigned to a problem can be given a limited time, based on the severity of the problem, to acknowledge the event or close it. This informs the event management system that the problem is being addressed or is successfully resolved. If the time interval expires without event acknowledgment or closure, the problem is escalated. See the fourth best practice in the list that follows this discussion.

Regardless of whether you choose to escalate based on acknowledging or closing events or both, you must define the escalation chain or hierarchy of people to inform for unhandled problems. See the fifth best practice in the list that follows this discussion.

The best practices to consider for unhandled events are:

- Increase the severity of outstanding events after an appropriate time interval.

  This is an effective way to draw attention to such events, which should result in corrective action. In addition, higher severities are usually defined to mean that the problems need to be resolved more quickly. Since time has already elapsed, there truly is less time to resolve the problems to meet Service Level Agreements.

- Set consistent time intervals for all events of the same severity.

  This method means that all events of severity *warning* should be escalated if they are not handled within the same time duration. Events of another severity, such as *critical*, may be, and usually are, assigned a different interval, but that interval still applies to *all* events of that severity. The exception is the severity assigned to clearing events. Since clearing events should be closed automatically, there should never be a need to escalate them.

  When severities are defined properly, they represent the urgency with which an event should be handled. They already account for Service Level Agreements (SLAs) and Operations Level Agreements

(OLAs). If they are developed considering the severities discussed in Sect. 4.3.2 of this chapter, the severity definitions already contain information about the acceptable time to resolve problems.

Moreover, it is generally easier to implement automated escalation based on severity than on a combination of other factors. When adding a new event that requires action, ensure that it has the right severity. Then little or no additional configuration or coding is required to integrate it into an existing, automated escalation process.

- Set escalation intervals to shorter than acceptable resolution times.

  The severity definitions tell how quickly to handle an event. Escalate before this time interval expires to allow the informed support person to take corrective action within an acceptable time frame.

  Avoid waiting to escalate until after the documented resolution time has passed. This is too late because the service level is already in violation, rendering it impossible to resolve the problem within SLA guidelines.

- Escalate when an event remains unacknowledged or unresolved.

  Checking for both of these conditions is the best way to ensure that SLAs are met. Set the time interval for acknowledgment to a shorter duration than for closure. That way, if the event is unacknowledged and the problem escalated, the support person notified has enough time to work on the problem to meet the SLAs.

- When escalating an unhandled event, inform both the originally assigned support person and others that the problem now has a higher severity.

  The responsible person may have accidentally forgotten about the event or may have been busy with other problems. The escalation serves as a reminder of the outstanding problem. It is also a courtesy to the administrator who may be measured on problem resolution time.

  Notifying others that the event has changed increases the chance that it will be handled. If the original support person, previously unable to respond to the event, is still not in a position to pursue it, someone else can take responsibility for it.

  Also, if the informed parties have authority, they can more readily ensure that the problem is handled by either reprioritizing the assigned support person's tasks or delegating the problem to someone else.

  Always notify the originally assigned support person when an event is escalated, because that individual is accountable. However, do not notify everyone for each escalation. Create levels of escalation, and choose to whom to notify for each escalation. For example, a manager may not care each time an event is escalated, but needs to know if a service level violation has occurred.

- **Business impact**

  This type of escalation is based on the premise that problems with a greater business impact should be handled more quickly than others. For example, suppose two servers fail. One affects only a few internal employees, while the other prevents customers from placing orders. Business impact escalation increases the severity of the second server down event to ensure it receives priority handling.

Escalating based on the criticality of resources implies knowledge of the business impact of failures. It is necessary to understand the components that comprise a business application to use this form of escalation. Often, organizations can easily determine which server provides the front end to their business functions. They may be less likely to know the back-end servers with which that system communicates for data and other processing.

When an organization determines the systems used for business applications and their relationships, it can perform a risk assessment. This term is generally used to denote the process of assigning priorities to resources based on their value to the business. Designating a system as high risk implies that its failure has detrimental effects on critical business functions.

– Increase severities of key events reporting problems with the greatest business impact.
Use the risk assessment value to determine what severity to assign to an event. Assign the same severity to events that reference resources of the same risk classification. For each risk classification, choose the severity defined with the most appropriate problem resolution time.

For example, routers may be classified as core and distribution. Core routers handle the traffic in the network backbone and are used by most business applications. *Distribution routers* connect remote locations to the backbone and serve smaller groups of users.

Core routers may be assessed as high risk, and distribution routers may be assessed as medium risk. Suppose that critical severity was defined to mean more than one user is affected and that fatal was defined to mean that most users are affected. The proper severity for a distribution router down event is critical. For a core router down, it is fatal. Since there are probably fewer core routers, set the severity of the router down event to critical, and escalate it to fatal if it is received for a core router.

– Perform business impact escalation for key events only.
Some events by their nature are not critical and should not be treated as such, even when reporting problems with a critical resource.

Consider again the server running the customer order application. If a server down event is received for this device, it requires immediate attention and the event severity should be adjusted to reflect this. However, if a backup power supply on the server fails, it may not need to be changed immediately. Do not perform escalation for the second event, even though it applies to the *high-risk* server.

– Escalate for business impact as early in the event life cycle as possible.
Ideally, an event is generated with the severity that best reflects both the nature of the problem reported and its business impact. This minimizes overhead in the event processors that handle it. In reality, many event sources are incapable of determining business impact or are not easily configured to do so. In these cases, an event processor must perform the business impact escalation.

Escalating as early in the event life cycle as possible minimizes the need for event synchronization later. It ensures that the event severity is accurately represented to users of intermediary event processors. Also, since the change occurs before initial notification of the problem, there is no need to re-notify for the severity change.

Do not duplicate business impact escalation at multiple levels of the event management hierarchy. Otherwise, events may be escalated several times, increasing their severities higher than originally intended.

- **Worsening condition**

This form of escalation differs from those previously mentioned in that it deals with multiple events. In the escalation types previously discussed, business impact and repair time trigger changing the severity of a single event. Here the driving factor is receipt of a new event indicating a worsening condition of the original problem.

  - When escalating a worsening condition, keep the first event and escalate its severity, adding information to it if necessary.

    There are several reasons to keep the first rather than subsequent events. A trouble ticket may have already been opened for the first event. When the ticket is closed, event synchronization procedures attempt to close the corresponding event in other event processors. If it does not exist, the synchronization fails. The problem event that still exists in the event processor remains open, leading to subsequent occurrences of the problem being discarded by duplicate detection.

    Also, keeping the first event ensures that the time at which the failure first occurred is recorded and available for problem determination.

    Update the original event with desired information from the new event such as the value of a monitored variable exceeding its threshold. After updating the original event, discard the others. This reduces overhead at the event processors.

  - When escalating worsening conditions, inform both the originally assigned support person and one or more others of the problem's higher severity.

    The same reasons apply here as for unhandled problems since the increased severity again implies less time to resolve the underlying issue (see "Unhandled events").

    In addition, if a monitored variable reported in the event is governed by SLAs, notify those responsible for the SLAs when the reported value is about to or has caused a violation.

  - Do not de-escalate for lessened conditions.

    Sometimes the term *de-escalation* is used to denote lowering the severity of an event. The new severity can indicate a lessened or resolved problem.

    De-escalate events only when they are resolved. There are several reasons for this. For example, you do not want to inform someone late at night about a critical problem only to give him or her a warning. In addition, a problem may oscillate between two severities. The event processors incur unnecessary overhead by repeatedly changing event severity.

    Most monitoring agents do not send events to indicate a lessened severity. They normally inform as a problem increases in severity and re-arm only when the condition is resolved.

4.8.2   Implementation Considerations

Escalation can be performed automatically by a capable event processor or manually by console operators. Therefore, consider the first best practice in the list that follows.

Any monitoring agent or tool capable of the function can escalate a problem. The best place depends on both the tools used and the type of escalation. Consider the last two best practices in the list that follows.

For implementation, consider the following best practices:

- Automate escalation whenever possible.

  When escalation is automated for unhandled problems, it occurs as soon as an acceptable, predefined time interval has expired. Similarly, a worsening condition and business impact escalation, when automated, occur immediately upon receipt of the relevant events. Operators perform escalation less precisely, only when they notice that a condition requires it. If you do not have a well-defined escalation process or it is too complicated to escalate using your toolset, allow operators to do it. Holding them accountable for ensuring the timely handling of problems gives them incentive to perform the required escalation.

- Use the trouble-ticketing system to escalate problems that do not receive timely action.

  This type of escalation typically requires modifying an open trouble ticket and notifying support personnel. These functions are best performed in the trouble-ticketing system itself. See 2.6, "Notification" on page 56, and 2.9, "Trouble ticketing" on page 68, for details.

  If trouble-ticketing software is not used, automate the escalation using a capable event processor, preferably the same one used to notify for problems.

- Perform worsening condition and business impact escalations at the first capable event processor that receives the relevant event or events.

  These types of escalation are event driven, rather than timer dependent. They can be performed most quickly when handled immediately by the first receiver of the relevant events. Escalating at the lowest level of the event processor hierarchy facilitates event synchronization because it is generally easier to synchronize events upward through the hierarchy than downward. See the following section for details.

  Escalating for business impact at the first capable processor ensures that the event has the correct severity when sent to subsequent event processors in the hierarchy. This minimizes the need to synchronize the events between the processors.

4.9   Event Synchronization

Changes made to events at one event processor can be propagated to others through which the event has passed. This is known as event synchronization.

Fig. 4.6 Upward and downward event synchronization

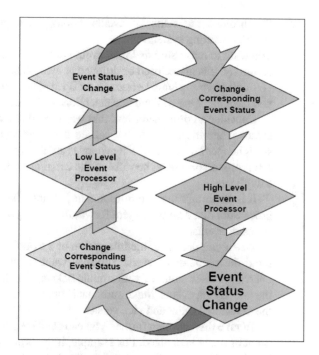

There are two main areas where event synchronization is key:

- Forwarding and receiving events through a hierarchy of event processors
- Integrating with a trouble-ticketing system

Any event processor or trouble-ticketing system can change an event. Depending on the event processor that performs the update, the event changes must be propagated upward through the hierarchy (see Fig. 4.6). Typically, the trouble-ticketing system notifies support personnel about problems and is at the top of the hierarchy. Therefore, changes made to trouble tickets are generally propagated downward to other event processors. If any event processor modifies an event, the synchronization is upward to the trouble-ticketing system and any event processors above it in the hierarchy and downward to lower event processors.

In general, it is easier to synchronize upward. Most event processors have the capability to automatically synchronize events upward, sometimes by merely re-forwarding the changed event through the event processor hierarchy. Downward synchronization is more difficult and often requires custom coding.

4.9.1   Event Synchronization Best Practices

When dealing with forwarding and receiving events through an event processor hierarchy, the most important aspect of the event is its status. By *status* we mean whether the event is open, closed, acknowledged, or dropped.

• Synchronize the status of events among event processors, including the trouble-ticketing system.

You want to make sure that if an event is closed or dropped at one level of the hierarchy, it is also closed or dropped at every other level. If this is not done, you will have *orphaned events* in an open state. This can cause problems if you have any type of duplicate detection on that event.

Consider an orphaned event that was closed at a different level event processor when the problem that caused the event was resolved. The problem starts happening again, which generates another event. The event is sent to the event processor where, because of the existing orphan event, it is dropped as a duplicate.

It is important that the rules and any scripts that you set up at your event processors that deal with event forwarding and synchronization deal with the status of events.

When you deal with the integration of a trouble-ticketing system, keep in mind the status of an event. You may want to start with synchronizing the open and closed status of your trouble tickets with the open or closed status of the underlying events. Make sure that if your trouble ticket is closed, it closes the associated event and vice versa.

To take this one step further, you can send events back and forth when the event or ticket is updated. For example, if the problem ticket that was opened is acknowledged by the support group that opened it, you can have a communication sent back to the event processor changing the status of the event that caused the ticket to be generated.

• At a minimum, propagate event severity changes upward to the trouble-ticketing system and higher-level event processors.

When a lower-level event processor escalates an event, this information should flow upward. Notification typically occurs either at the trouble-ticketing system or a higher-level event processor. As discussed in 4.8, "Escalation" on page 60, when events are escalated, someone needs to be informed.

Therefore, the event processor used for notification needs to be told that the event has been escalated. Upward event synchronization performs this function.

When consoles are used at different levels of the event processor hierarchy, severity and other event changes may need to propagate downward. Suppose an organization has a central event processor at its main site that is used by its after-hours help desk. It also has distributed event processors in various time zones for use by support personnel during normal business hours. The central help desk raises the severity of an event on its console based on a user call. When the distributed support personnel assume ownership of the call in the morning, they need to know that the problem has been escalated.

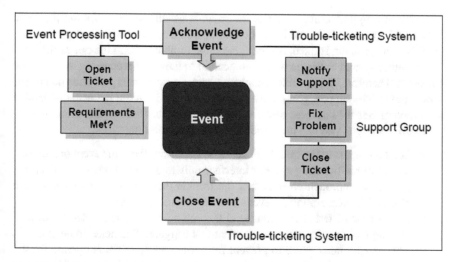

Fig. 4.7 Trouble-ticketing process flow

4.10   Trouble Ticketing

This section discusses integrating your event processing tool with a trouble-ticketing system. The focus is on event management, not problem management. This section presents some best practices for problem management. They are mentioned, as necessary, as they relate to the integration of a trouble-ticketing system with an event processor.

4.10.1   Trouble Ticketing

This section covers the typical integration of a trouble-ticketing system with an event processor by discussing the process flow, which is illustrated in Fig. 4.7.

We start with an event. When this event arrives at the event processing tool, it has already gone through filtering and correlation. If it is determined through these processes that this event requires action, then the event processing tool sends this event to the trouble-ticketing system. See the first best practice in the list that follows this example.

After the event reaches the trouble-ticketing system, that system cuts a ticket. The ticket is assigned to the proper support group based on criteria within the event. That support group is then notified based on which notification was set up for this problem's priority. See the second best practice in the list that follows this example.

After the support group receives the page, it usually starts to fix the problem. During this process, it should keep the trouble ticket updated with their progress.

When the problem is fixed, it should close the trouble ticket. This causes the trouble-ticketing system to trigger a communication to the event processing tool to close the event. Therefore, if the problem happens again, a new event is generated that opens a new ticket. The support group must then determine why it did not actually resolve the problem. See the last three best practices in the list that follows this example.

In this example, you can implement the following best practices:

- Send all events to the trouble-ticketing system from the same event processor. Most trouble-ticketing systems interface only to one event processor at a time. Also, this approach ensures that the trouble-ticketing system can initiate a close of the corresponding event when the ticket is closed.
- At the time of ticket creation, send a communication back to the event processing tool to acknowledge the event that triggered this ticket to be opened. This usually takes place to prevent further tickets from being opened for the same problem. Duplicate detection should take place with the event as long as it is in acknowledged status. See 4.4.4, "Duplicate detection and throttling best practices" on page 50, for more details.
- If a trouble-ticketing system is in use, use it to report all problems requiring action and only those problems.
  Now you can take the events from the previous section on notification that you decided were problem or actionable events. At this time, you can consider sending them to the trouble-ticketing system.

  If you are not careful in choosing events that are going to open a ticket, you may start having problems in a couple different areas. If you have too many events going back and forth between your event processing tool and your trouble-ticketing system, you start to use up resources. This can happen both in the network and on the machines that are running the software.

  The more events you send between your event processing tool and the trouble-ticketing system also take a toll on your event synchronization. If you send loads of unnecessary events to open problem tickets, there is a greater chance that the acknowledging or closing events may be missed or dropped.

  Another reason to choose your events carefully is to avoid misnotifying support teams. You should only really notify critical system down, or business affecting events that require action from support teams. If you start sending needless or unimportant pages to support groups, there is a big chance they may ignore the important ones.

  You must also be careful with a reporting standpoint, which is usually carried out from the trouble-ticketing system. You do not want to have unimportant problems skew the reports.

  In today's IT environment, it is essential to keep a good partnership between systems management and the various support teams. If the support teams have problems, with the way tickets are opened and notifications are handled, it is hazardous to your event management system.

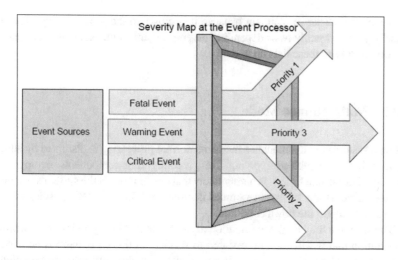

Fig. 4.8 Mapping severities at the processor

- Prioritize tickets based on event type, time of day, and criticality of source.
  After you have events opening trouble tickets, consider the priority of the
  tickets that are opened. Figure 4.8 displays a sample event severity mapping.
     Usually it is a good idea to set the severity of your events at the source.
  This means that when you send your event to the event processing tool, it
  should be sent with the severity that matches the priority of the ticket in the
  trouble-ticketing system.
     Be aware of your company's current service levels when determining
  severity or priority. The severity that you use to send the event in should match
  the service levels defined in your organization for the event type, time of day,
  and criticality of the source. When setting up an integration, with trouble ticketing, follow the processes and procedures already defined at your help desk.
  This makes it easier with the whole event management setup to tie into systems that are already in place.
- Implement on-call lists.
  There are two ways to notify support groups from your trouble-ticketing
  system:
  - **On-call lists**: Notify only the person on call. Maintain lists within the
    trouble-ticketing system of the current on-call personnel.
  - **Group paging**: Page the whole support group. The person who is on call
    takes the call and the rest ignore the notification.

   Usually it is wiser to go with the on-call list notification. This is because you send
only one notification for each problem. It is less likely that notifications are dropped,
or ignored by support personnel, who believe someone else is handling the problem.
Escalation policies implemented in the trouble-ticketing system ensure that the
problem is handled in a timely manner. See "Unhandled events" in this chapter for
more information.On-call lists are generally held within the trouble-ticketing system.

This makes it easier to maintain because it is closer to the group listings. The lists usually go by a weekly basis depending on your organization. Keep in mind that it is important to keep these lists up to date.

4.11   Maintenance Mode

When a system is in maintenance mode, its normal processing is disrupted by administrative functions such as applying fixes, reconfiguring components, or upgrading software. During that time, it is highly likely that one or more of the system's resources is unavailable. If any of these resources is monitored for availability, events may be generated to report the down condition.

The resource may be down intentionally because it is being maintained, with the expectation that it will soon be restored to service. In this case, there is no need to alert anyone about its condition. However, its unavailability may be completely unrelated to the system maintenance being performed, requiring that someone be informed. The difficulty is in differentiating between the two situations.

Ideally, whenever system maintenance is performed, the administrator knows which resources are impacted and temporarily stops monitoring only those resources for the duration of the maintenance. In reality, this is nearly impossible to implement.

Often, it is unclear as to exactly which resources are affected by the maintenance. Even if they are explicitly identified, it may be awkward to shut down monitoring for only those resources. Maintenance mode often entails rebooting a system one or more times. The monitoring agents will most likely restart automatically and need to be stopped again. In addition, errors are often reported by other systems that normally access the system being maintained. It is impractical to assume that the appropriate monitors are shut down on other systems as well.

Although the ideal is impossible to attain, a mechanism must be in place that accounts for events from systems in maintenance mode to ensure that extraneous events are not reported and real problems are not lost. A good approach is to inform the appropriate event processors that a system is in maintenance mode. Then have them take special action on the events from or about that system.

4.11.1   Maintenance Status Notification

An event processor needs to be informed when a system enters maintenance mode so it can take special action for events from the system. It must also know when to resume normal processing for those events. Therefore, a method is required to place a system into and remove it from maintenance mode.

- Only those processors that normally receive events concerning a system should be notified about its maintenance status.

If an event processor handles events from or about a system, it should be informed that the system is entering or leaving maintenance mode. This ensures that it can take the appropriate actions for events it may receive, as described in 4.11.2, "Handling events from a system in maintenance mode." Limiting the notification to only relevant event processors prevents the others from using cycles to check events for reference to the machine in maintenance mode.

- Use events to notify an event processor that a system is entering or leaving maintenance mode.

The event processors already have the ability to act upon events. The events from or about a machine in maintenance mode can be easily correlated with the maintenance notification event, and appropriate action taken.

Using events rather than some other means of notification eliminates the need for the event processor to access outside sources to determine which machines are in maintenance mode. If external files or queues are used to store the maintenance information, additional code may be required for the event processor to access that data.

- Automate the generation of the events as much as possible.

In sophisticated environments, the organization may use the change management system to automatically generate the maintenance notification events based upon its schedule. The shutdown and startup scripts or command files used during maintenance may also be modified to send the notifications.

The most common practice is to have the administrator send the events. While this method relies on the administrator's memory, it allows for emergency changes that may not be included in the change management system. It also allows for maintenance that does not require specific shutdown scripts to execute.

The administrator is provided with a desktop icon or script to run that automatically produces the required event. The change management procedures are updated to include generating the maintenance mode notification events as a required step in maintaining a system.

- Report extended maintenance mode situations as problems.

Experience has shown that administrators are more likely to notify when a system is entering maintenance mode than when it is leaving. They want to ensure that they are not informed of bogus problems about a machine that they are servicing. However, it is human nature to want to complete the maintenance as quickly as possible, particularly when it is performed after hours. As a result, many support people neglect to inform when a system is again operational. A method is needed to handle this situation.

First, the event processor needs to be given a time estimate for the maintenance window. This information can be sent in the maintenance mode notification event, stored on external media, or hardcoded into the event processor's rules. While any of these approaches work, sending the information in the event allows the greatest flexibility with the least effort. The administrator can easily differentiate between such maintenance (e.g., parameter reconfiguration) and lengthier changes (such as software upgrades and database reorganizations). Modifying files and rules is more complex and more prone to error.

At the start of maintenance, the event processor can set a timer. After the elapsed time exceeds the estimate, the processor can generate an event to inform the administrator that the machine is in maintenance mode longer than expected, or it can merely resume normal event processing for the system. Administrators generally prefer to be notified. This prevents a potential flood of problem events, should the system still be legitimately in maintenance mode.

4.11.2 Handling Events from a System in Maintenance Mode

When an event processor knows that a system is in maintenance mode, it can take appropriate action for events received from or about that system. The best practices to use for handling those events depend upon the organization and its policies for system maintenance.

- **In environments where the administrator maintaining the box has total control over the machine, *host-based maintenance* may be appropriate.**
  Giving an administrator total control over a machine in maintenance mode implies that it is acceptable to affect any system resource during the maintenance window. This approach also assumes that the administrator is fully responsible for restoring all processes when the maintenance is complete. Therefore, events received from or about the box during this time do not require action and may be discarded.
  We refer to the processing of dropping the events for systems in maintenance mode as *host-based maintenance*. This is a relatively simple method of handling the events and is easy to implement. However, it relies on the administrator to properly restore all functions on the machine. A condition may arise such as a file system filling that an administrator does not normally notice during maintenance mode. These problems may go unreported in host-based maintenance.
- **Where host-based maintenance is not appropriate, cache events and report them after maintenance mode is terminated.**
  Caching events received from or about a system in maintenance mode ensures that real problems unrelated to the maintenance are not lost. It also preserves the correlation relationship among events. This solution should be favored in organizations where production applications may continue to run on a system that is undergoing maintenance for other processes. It may also be used to validate that the system is completely restored afterward.
  When the event processor receives events for the system undergoing maintenance, it caches them. It can also apply correlation rules to them to drop extraneous events and to clear problems. When the machine comes out of maintenance mode, the event processor waits a short time to receive and process clearing events for the system resources affected by the maintenance. It then reports any outstanding problems.

Events for which no clearing event is available are always reported using this method, even if the problem they reference no longer exists. Whenever possible, configure monitoring agents to send clearing events. This minimizes the number of these events that are inadvertently reported.

4.11.3 Prolonged Maintenance Mode

Sometimes the resolution to a problem is known but cannot be immediately implemented. For example, a short-on-storage condition may arise because a machine does not have adequate memory. Adding memory to the system is planned, but the hardware is scheduled to ship in a few days.

In these situations, it is undesirable to report the error every time it occurs. The solution is known, but cannot yet be implemented. There are several ways to handle this situation. The first is to reconfigure the monitoring agent so it does not report the error. This effectively stops the event from flowing. However, it relies upon the memory of an administrator to restart monitoring after the fix is implemented. In the case of hardware, the solution may not be implemented for several weeks. It is highly likely that the support person will forget to re-enable the monitor at that time.

A second approach allows the event to remain open until it is resolved and to discard duplicates in the meantime. This method also has some problems. The problem may occur intermittently and be cleared automatically by an event. Leaving the event open requires a reconfiguration of the event processing rules, which have the same drawbacks as reconfiguring a monitor. Also, some event processors perform duplicate detection only on events in cache, which is cleared when the processor is recycled. To address the shortcomings of the other two solutions, we propose that you temporarily ignore events whose resolution is known, but cannot yet be implemented.

To implement this solution, the event processor needs to be told which event to ignore, from which host, and for how long. This information may need to be stored in an external file or queue so it can be accessed by the event processor upon restart. If the event processor supports this, it may be loaded into cache at startup for more efficient runtime access.

Note that normal maintenance mode event processing does not require storing the list of nodes is being maintained on external media. Maintenance windows are expected to be relatively short. They may be scheduled so that they do not occur when the event processor is scheduled for reboot.

When the event processor receives an event, it checks to see if the event should be temporarily ignored and suppresses it if appropriate. If the event is not reported, a support person does not waste time pursuing a known problem.

There are several advantages of this solution. The monitors and event processing rules can remain intact. This implies that an administrator does not need to remember to restore monitoring of the system resource. As soon as the estimated time has passed, the resource is monitored again. During this prolonged maintenance mode, the system is still monitored for other conditions that are unrelated to the known, but not yet fixable problem. Finally, if desired, the event can be correlated before it is ignored. This may prevent the investigation of symptom events for which no action can be taken.

4.11.4   Network Topology Considerations

When a network device is placed in maintenance mode, a potentially large number of systems can be affected. If the component is a single point of failure, any network traffic that traverses a path containing it may be disrupted. In this case, it is necessary to know the network topology before determining whether an event is the result of the maintenance.

In this case, we propose the best practice to use the topology-based correlation capabilities of SNMP-based managers to handle events resulting from network maintenance.

While it is possible to provide network topology information to other types of event processors, the procedure is often complex and difficult to implement. Most networks are dynamic, implying frequent updates to the topology information supplied to the processors. This quickly becomes impractical, particularly in large networks.

Allowing the SNMP-based manager to correlate the events means that only the root cause event is presented to the other event processors. This event references the network component is being maintained. Since the other event processors were informed the network device is under maintenance, they can handle the event as described in Sect. 4.11.2 in this chapter.

When a network component is maintained, some events caused by the maintenance may be reported. For example, a Web server that is trying to access its database server across the network path may report a communication failure. Since neither the Web server nor the database server is in maintenance mode, the event is reported.

While it is possible, theoretically, to handle these cases, it is usually not worth the effort involved. The communication between any two devices may be affected by a large number of networking components. Moreover, in today's grid and on-demand environments, it is not unusual for an application to run on different hosts when required. Not knowing on which server the application runs at any given time makes it difficult to determine which applications are affected by a network failure.

Consider redundant network paths to minimize communication outages due to both network component failures and maintenance.

4.12 Automation

There are several ways to plan for and implement automation. Some organizations choose to implement correlation first and then to analyze their common problems, identifying events that may be used as triggers for automated action. Others decide automated actions at the same time as filtering, correlation, and notification. Still others use problem post mortem investigations to uncover automation opportunities.

The approach that an organization chooses depends upon the current monitoring environment, event management policies, and the employees' skill levels. If little or no monitoring is in place, a company may decide to analyze all possible events from a monitoring tool before implementing it and making filtering, correlation, notification, and automation decisions concurrently. In environments with robust monitoring already in place, automation may be added to it. Where staffs are highly skilled and can provide insight into which events should trigger automation, the decisions can be made reliably before problems happen. Novice support staffs may wait until problems occur before they identify automation opportunities, working with vendors' support centers to gain insight into the reasons, the problems occurred, and how to prevent them.

Regardless of how an organization decides to handle planning and implementing automation, there are several best practices to observe as explained in the following sections.

4.12.1 Automation Best Practice

The first step in implementing automation is deciding which events should trigger automated actions. Here are several guidelines to use in making this determination:

- **Do not over automate.**
  Sometimes organizations are overzealous in their correlation and automation efforts. In a desire to improve the efficiency of their monitoring environments, they quickly set up automation for everything they can.

  Some pitfalls to this approach arise from not understanding the ramifications of potential automated actions before implementing them. For example, a locked account resulting from a user mistyping a password several times needs to be reset, but if it was caused by hacking attempts, a security investigation may be required. Automatically resetting accounts based upon receipt of the account locked event is not a good idea.

  Similarly, an event may be used to report more than one problem, necessitating different actions depending upon which problem caused the event. Perhaps actions may be required when an event references a certain host but not others.

These are a few examples of things to consider when determining whether automated action is appropriate for an event. Remember, it is better to be judicious in choosing automated actions and implement fewer than to implement automation whose consequences are unknown.

- **Automate problem verification whenever possible.**

Sometimes it is not possible to filter all events that do not indicate real problems. For example, as discussed in 3.3.10, of Chap. 3 "Automation" on page 19, an SNMP manager that queries a device for its status may not receive an answer back due to network congestion rather than the failure of the device. However, the manager believes the device is down and sends an event. Before assigning this event to someone, it is advantageous to determine if it is truly a problem.

SMEs who understand various event sources well may be able to identify events such as this one that may sometimes be false alarms. Likewise, help desk personnel learn from experience which events do not always represent problems. Use the expertise of these people within the organization to list events that require verification.

After the events are identified, work with the SMEs to determine if the problem verification procedures lend themselves to automation and automate whenever possible. This minimizes the amount of time that the support staffs spend investigating false events.

- **Automate gathering diagnostic data if the data may disappear before the problem is investigated, multistep or long running procedures are required to capture it, or support staff lacks the skills to acquire it themselves.**

In cases where diagnostic data may disappear before a support person has time to respond to the event (such as the list of processes running during periods of high CPU usage), automating the gathering of diagnostic data may be the only way to determine the cause of the problem. All events reporting these types of intermittent problems should have automated diagnostic data collection associated with them.

The situation is less clear for events whose diagnostic information remains available in the system until requested by an operator. In these cases, automate the diagnostic data gathering if multiple steps are required to obtain the data. This way, the user does not have to remember the steps, and user errors, such as typos, are eliminated.

Also, if the diagnostic gathering procedures take a long time to run, automating the data collection ensures that the data is available to the support person sooner, reducing the mean time to repair for the problem. Automating diagnostic data gathering in circumstances where the user can type a single command to produce the data may not make sense, since the user has to run at least one command to view the collected data as well. In this case, unless the command takes a while to run or the support staff is not highly skilled, do not automate the data collection. This saves on cycles in the event processors handling the event.

- **Automate recovery only for real problems that always require the same sequence of actions.**

  Obviously, if an event does not represent a real problem, it does not require a recovery action, automated or otherwise. For real problems, be sure that the same sequence of actions is always required to resolve the problem before automating. This sequence can contain conditional logic, as long as all the possible conditions and the actions to take for them are known. Also, ensure that the steps can be automated. A sequence of actions that requires operator interaction with a Graphical User Interface (GUI), for example, may not be automatable. See the first best practice, "Do not over automate," in this list for additional considerations in determining whether to automate the recovery action.

- **Consider cross-platform correlation sequences for possible automation opportunities.**

  Cross-platform correlation refers to the process of associating events about different system resources. These event sequences are often excellent sources for automated recovery actions.

  Often, cross-platform correlation sequences result in symptom events that require action. This is because the support person handling the first resource type does not usually have administrative responsibility for the second. In addition, many systems are not sophisticated enough to recognize the system resources affected by a failure and to automatically recover them when the failure is resolved.

  In "Cross-Platform Correlation" in Sect. 3.3.4, we provide an example of a process that dies as a result of a full file system. The corresponding events can be associated, and the process can be restarted automatically when the *file system* problem clears.

4.12.2   Automation Implementation Considerations

We discuss several types of automation that can be executed upon receipt of an event. You must perform these in a specific order to ensure that the desired results are achieved.

- **Automate as close to the source as possible.**

  There are several factors that determine where to automate.

  - The selected tool must be capable of performing the automation. For example, when implementing automated action for a network device using SNMP commands, the tool used to perform the automation must be capable of communicating via SNMP.
  - The automation should be easily maintainable. Sometimes it is possible to perform the action from more than one tool. If it is difficult to implement and maintain the automation using the tool closest to the source, use the next closest one instead.

- Performance considerations may dictate which tool to use for automation. If a tool requires many processing cycles or uses much bandwidth to automate a particular function, it should be overlooked in favor of one that performs better.
- If correlation is required before executing an automated action, the closest point from which automation can be triggered is the event processor at which the events in the associated events converge.

- **Check to see if a device is in maintenance mode before performing any automated actions.**
  As discussed in 2.11, "Maintenance mode" in this chapter, if a device is in maintenance mode, any problems reported for that device are suspect. No automated recovery actions should be taken for events received about devices in maintenance mode. Diagnostic automation can be performed, if desired. This ensures that the diagnostic data is available should the problem still exist after the device comes out of maintenance mode.

- **Perform *problem verification* automation first.**
  If this type of automation is done, it should always precede all other types of event processing such as correlation, recovery and diagnostic automation, notification, and trouble ticketing. None of these actions are necessary if there is not really a problem.

- **Perform automated diagnostic data collection after verifying that the problem really exists and prior to attempting recovery.**
  Obviously, if there is not really a problem, no diagnostic data is required. However, you must perform this step before you perform the recovery actions, since the diagnostic data sometimes disappears after the problem is resolved. An example of this is logs that are overwritten whenever an application starts.

- **For real problems for which recovery action is desired, try the recovery action prior to reporting the problem. If it fails, report the problem. If it succeeds, report the situation in an *incident report*.**
  For the reasons stated earlier, this type of automation should succeed both problem verification and diagnostic data gathering automation. However, running recovery actions should be performed prior to reporting problems. This ensures that only events requiring operator intervention are reported through notification and trouble ticketing. If the recovery action succeeds, the problem should not be reported to prevent unnecessary events from cluttering the operator console.

  Sometimes it may be useful to know a problem occurred, even if it recovered automatically. Consider, for example, an automated action to remove core files from a full UNIX file system. When the file system fills, an event is sent, which triggers the removal of core files from the file system, freeing adequate space. Since the file system now has an acceptable amount of free space in it, the problem is, closed and not, reported. An application may be producing core files repeatedly and filling the file system. It is useful to know about this condition to identify and resolve the application problem.

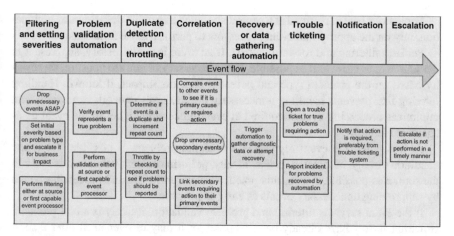

Fig. 4.9 Event processing flowchart

Opening incident reports for conditions that are automatically recovered is the preferred method to track them. Incident reports make the information available to a support person when they choose to examine it. Reviewing the incident reports highlights *flapping* or *fluttering* error conditions, such as the one described earlier, and may lead to their resolution.

4.13 Best Practices Flowchart

In this chapter, we discuss the best practices for the various types of processing to perform for events. The purpose of this section is to recommend the order in which to perform the processing. The flowchart in Fig. 4.9 shows a recommended sequence for handling events.

The exact event processing sequence that you implement depends upon the capabilities of the various monitoring agents and event processors in your environment. This flowchart was developed with the Tivoli product suite in mind.

Note

In general, if best practices dictate performing the event processing as close to the source as possible and a monitoring agent is capable, consider configuring the agent to perform the function. For example, IBM Tivoli® Monitoring or Microsoft SharePoint® can be configured to perform such tasks as filtering unnecessary events, throttling based on number of times a condition occurs, correlating multiple error conditions before reporting, and executing automation.

For actions best performed at a central site, such as trouble ticketing and notification, rely on the appropriate event processors to perform those functions.

Perform filtering and forwarding first. If an event is not necessary, suppress it in the source or as close to it as possible. If it is necessary or you are not sure, set severity based on the problem type and potential business impact, if known. Handling filtering first prevents handling unnecessary events, saves network bandwidth, and minimizes processing cycles required at event processors higher in the event management hierarchy.

At this point, the event represents a potential problem. If possible, verify the condition through automated means. Ideally, perform problem validation as close to the source as possible. Again, this practice saves bandwidth and processing power by suppressing unnecessary events as early as possible.

If the event survives filtering and problem validation, it likely is a real problem. However, it may have already been reported, or it may recover itself after a short time. Perform duplicate detection and throttling to block the event from traveling further if it is already reported. Ensure that it is forwarded if it happened a predefined number of times within a given time frame. Increment a counter of occurrences of an event within a time frame. This provides data that the support person can use to determine the extent of the problem and the load the event processor is handling.

Next, compare the problem event to other received events and classify them as a primary or secondary event. If it is a secondary event that requires action, link it to the primary and forward it. If it does not require action, drop it to prevent further processing load on this and other event processors. If necessary, delay the comparison long enough to allow other related primary events to be received.

At this point, the event represents a problem requiring action. Before informing a support person, perform automation to gather diagnostic data or attempt recovery, if desired. Use incident reports to record success recovery and trouble tickets for unsuccessful.

The appropriate event processor or trouble-ticketing system can then inform the support person there is a problem requiring action. If the problem is not acknowledged or resolved within a preset time interval, escalate the problem by raising its severity or notifying additional support personnel.

Reference

1. IBM. (2004). *Event management and best practices*. International Technical Support Organization.

Chapter 5
Dynamic and Static Content Publication Workflow

Dynamic content publishing is a method of designing publications in which layout templates are created which can contain different contents in different publications. Using this method, page designers do not work on finished pages but rather on various layout templates and pieces of content, which can then be combined to create a number of finished pages. In cases where the same content is being used in multiple layouts, the same layout is being used for several different sets of content, or both dynamic page publishing can offer significant advantages of efficiency over a traditional system of page-by-page design. This technology is often leveraged in Web-to-print solutions for corporate intranets to enable customization and ordering of printed materials, advertising automation workflows inside of advertising agencies, catalog generation solutions for retailers, and variable digital print-on-demand solutions for highly personalized one-to-one marketing. A digital printing press often prints the output from these solutions. The dynamic content publishing is a tool that can enhance Business Resilience System (BRS) to publish all the warning and events within ABS through the enterprise content management (ECM) at enterprises or organizations for stakeholder and decision-maker as well as build a database for knowledge base (KB) for its infrastructure. KB is a technology that is used to store complex structured and unstructured information used by a computer system.

5.1 Introduction

Dynamic content publishing is, often considered to be, a subset of automated publishing that enables generation of long documents, such as manuals or books from a content repository. Dynamic content publishing is often used to build high-value documents, such as ads, collateral, brochures, catalogs, direct mail, eBooks, and

© Springer International Publishing AG 2017
B. Zohuri, M. Moghaddam, *Business Resilience System (BRS): Driven Through Boolean, Fuzzy Logics and Cloud Computation*,
DOI 10.1007/978-3-319-53417-6_5

other documents on demand or in variable data printing workflows. Dynamic content publishing is often associated with XML authoring and technologies related to the semantic Web initiatives.

Having a powerful, dynamic enterprise content management (ECM) and document management solution that solves the problems of managing, finding, and tracking information in companies of all size is one of the important components of Business Resilience System (BRS). Availability of such a tool in multi-languages is additional gravy for BRS system to be able to publish all the events for stakeholder observation and should be easy to deploy. It allows to learn and use and has enabled thousands of businesses in their business process management (BPM) and business continuity management (BCM) as part of their risk atom, which we learned in Chap. 1. Implementation such tool as component of BRS to make dramatic gains in efficiency and productivity by improving the way they organize and coupe with events in real-time and manage their documents, information and processes as part of knowledge base (KB), to store complex structured and unstructured information used by a computer system.

A metadata-driven dynamic content publishing tool along with its navigation solves and intractable shortcoming of traditional folder-based approaches is limited to allowing a file to exit in only one location or having copies of the file or link to the file reside in other folders. The fact that we are in quest of at least near-real-time BRS if not real time demands a tool that is driven by metadata navigation capability that makes more desirable system, to publish all occurring events in dynamic mode.

Traditional ECM and document management systems are expensive, complicated products, which require major changes in business processes and extensive IT services and support. However, a good dynamic content publication workflow and its file management change this paradigm by providing a powerful yet easy-to-use solution that helps businesses organize, manage, and track events, documents, and information, more dynamically searchable and more efficient for specific data point within knowledge base master data management or big data. With the versatile and powerful file management workflow features in support of risk atom as core of BRS, routine company tasks can be automated, and task assignments can be given to the right decision-maker persons at the right time with no programming required. Workflow within file management of this tool can be instituted for non-interrupted BCM and BPM and other day-to-day continuous operation of an organization. This may include routine operation such as contracts, circulation of purchase invoices, and processing job applications and for myriad of other use cases that require review, edit, and/or approval by several entities as additional bounce added to Business Resilience System.

Leveraging the metadata-driven workflow capabilities in managing file of such dynamic publishing content will improve business process efficiencies, eliminate bottlenecks, and maintain consistency and quality in documentation. It assures employees do not accidently skip a step in important procedures that may interrupt daily routine operation of enterprise or for that matter any upcoming natural disaster such as hurricane or others that may be a threat to BCM and BPM organization. BCM or BPM may fall into a supply chain, inventory control that has tremendous impact on a manufacturing corporation with production line or e-commerce with their portal site or multitier grocery stores that need to replenish their sale and delivery of their goods to the front store.

Archiving content for future use in the file manager of the tool can be built based on metadata, instead of simply archiving certain folders to an external storage location, for purpose of disaster recovery. It also allows users to create dynamic archiving rules to control what documents and information are archived, such as archiving only documents that are classified as customer data information for purpose of customer relationship management (CRM) as part of BCM within a banking industry or healthcare records that are more than several years old.

Unlike traditional ECM systems, which are more restrictive and make use of antiquated security models based on folders, permissions control the management files way in the mentioned system means that a document's final access control settings are derived from its metadata and it is done so in a highly dynamic way. With changes to the metadata, driving changes in document permissions automatically allows real-time accesses more efficient for BRS to trigger warning point at the dashboard level of BRS. This one-of-a-kind architecture provides a revolutionary way for management file to manage access to any content, including confidential content, if such BRS system with its dynamic content publication workflow is in use by organization such as homeland security administration or military.

5.2 Business Workflow

Business workflow means building a better business with a comprehensive business process management (BPM). Faced with pressures to produce better products and services while holding down operational costs, organizations are striving to improve their business workflows. For many major companies in financial services, insurance, healthcare, communications, manufacturing, and other sectors, having an intelligent business workflow in place is a valued partner in these efforts. There are a lot of commercial and off-the-shelves software and software industry, and offering business process management (BPM) platform helps organizations to design and implement intelligent, agile business workflows [1].

A good business process management should deliver what your organization needs to design, implement, and evaluate business workflows. It should support all dimensions of business *workflow management,* including:

- **Intelligent routing and queuing**: With this capability, you can create and deploy an intelligently automated business workflow that uses defined business rules to route work to the right system or the right person and to prioritize work within queues.
- **Case management**: It should be able to reach across geographic, organizational, and IT system divisions to present workers with a comprehensive, unified view of all the information the organization possesses that is pertinent to a particular

work case. For each case passing through a BPM-based business workflow, the system also automatically attaches relevant service level conditions, specifying clear deadlines, escalation paths, and corrective actions in the event that targets are missed.

- **Guided task execution**: It should guide workers through task execution, providing needed information at just the right time.

Now the question remains, how Agile is your business workflow?

Your choice of a right BPM for your organization should fit in today's BPM marketplace for your application and line of business, specifically. Your BPM business should be growing at twice the pace of the market as a whole, and industry analysts should consistently rate it as the segment leader, when they evaluate your corporation performance in the market that you are operating and selling. It should lead the BPM market because it empowers organizations to change—efficiently, intelligently, and rapidly. With today's fast-paced technology, businesses need to find the agility and the rapid ROI that they are seeking in today's marketplace. Their BMP should have the following functionalities and capabilities as well [1]:

- Business users can quickly convert new business goals into new business workflows and processes, with minimal reliance on IT staff. It should offer a technology that automates the programming required to turn *business rules* and workflow models into running applications, greatly accelerating your rollout of new processes and services in conjunction with BRS in place within your organization.
- Business rules are sophisticated, flexible, and powered by a high-performance *business rules engine*, enabling you to fully capture your organization's objectives and insights into rule-driven enterprise applications.
- Industry-specific solution frameworks speed your progress, providing data models, business rules, business workflows, and user interface templates rooted in industry best practices.

A good BMP, with a good business user-driven change process and associated powerful *business rules platform*, the selected BPM model, should be able to support the risk atom, defined in Chap. 1, and is the ideal foundation for continuous process-improvement programs, *business transformation* initiative, or targeted solutions that make specific operations more effective, efficient, and agile.

5.2.1   Workflow Management

A good workflow management should be able to deliver agile, end-to-end workflow. In an effort to compete more effectively in today's demanding marketplace, many organizations are striving to optimize their internal workflows. Looking to deliver high-quality products and services while holding down costs and leading companies in industries including financial services, insurance, healthcare, communications, and manufacturing are turning to business process system for workflow management software solutions. The solution for such business workflow management should support comprehensive enterprise workflow management with a much greater degree of agility than what competing solutions allow [1].

An end-to-end workflow management on its business process management (BPM) platform should support all facets of enterprise workflow management including:

- **Getting the work to the right person**: This process management software should be able to deliver a robust functionality for routing management and to match the work item with the most suitable resources. It should queue management, to prioritize and move items into and out of queues, and role management, to catalog the organization's inventory of roles and skills.
- **Providing context for the work**: The BPM's case management functionality breaks down information silos and ensures the seamless integration and presentation of all information relevant to a work item. The BPM-based workflow management also automatically attaches relevant service level agreement conditions to work items, setting clear deadlines, escalation paths, and corrective actions in the event that service level targets are missed.
- **Guiding the execution of the work**: With its BPM, workers are going to be guided through task execution and provided the right information at the right time, within the business workflow.

Furthermore, workflow management agility is the clear BPM market component, with sales growing at twice the overall industry rate and consistent top rankings from major analysts. The BPM should be top-tier leader in large part because more than any other solution provider, it should empower organizations to change intelligently and rapidly, such as:

- **Business users drive business change**: With an intelligent BPM in place, business stakeholders use an intuitive interface to define business rules and workflow management models, and system-innovative build for change technology then automatically completes the programming necessary to turn rules and models into executable applications. New business goals become new processes and procedures more quickly than ever before.
- **Business rules as smart as your business**: The BPM features should be the most sophisticated and powerful business rules platform on the market, so your business rules and rule-driven processes can capture all the intelligence that your organization can deliver.
- **Best practices frameworks**: The right BPM helps a jump start to your workflow management initiative with industry-specific solution frameworks for financial services, communications, insurance, healthcare, and more. Incorporating industry best practices solution frameworks features industry-appropriate data models, role definitions, workflows, business rules, and user interface templates.

Innovative, agile, and driven by the industry's most powerful business rules engine, this choice of BPM is the ideal platform for business transformation initiatives and continuous process improvement programs or solutions to improve the effectiveness and efficiency of specific operations [1].

5.2.2 Business Rules Engine

A business rules engine (BRE) should be as dynamic as your business. A business rules engine separates business logic from your mission-critical applications in order to gain agility and improve operational performance. To get the most benefit from this application architecture, you need a business rules engine that:

- Empowers business users to create and manage business rules with minimal involvement from IT staff
- Supports sophisticated, powerful rules that can capture your business workflow and your policies as well as procedures in their entire dynamic complexity format
- Integrates seamlessly with your existing IT assets and scales for enterprise-class performance

The BPM in place should be knowledgeable in enterprise business rules and business process management solutions. It has to be innovative in business rules logic and automated decision-making functionality built into it.

- **Usability**
  The BPM business rules engine should be able to put business users firmly in charge of creating and managing business rules, for maximum agility. Usability features should include:

  - Intuitive HTML rule forms to easily configure and manage business rules
  - An MS Visio graphical front end to link rules with applications
  - Built-in review and approval processes
  - A version-controlled rules inventory that supports the efficient reuse of existing rules
  - A secure audit trail for all rule changes
  - Automatic documentation generation

- **Power**
  The BPM business rules engine should be unmatched in its support for a wide range of rule types, including:

  - Process rules that automate workflow management
  - Decisioning rules of all types, including decision tables, decision trees, and decision maps
  - Declarative rules that compute values based on detected changes in other related values
  - Transformation rules that appropriately transform data as it passes across heterogeneous systems
  - Integration rules that determine the right system to invoke in each situation

- **Scalability**

  With the right BPM tool in place within an organization and utilization of its business rules engine, distributed application nodes can share a common rule database, for optimum scalability. The system employs an open Java and XML architecture and runs on any major operating system, in conjunction with any major application server. Some of the largest corporations in the world run enterprise-scale business transformation programs that utilize the selected BPM business rules engine as well.

5.2.3 Business Rules Platform

The organization should be able to drive business agility with the right business rules platform installed. A business rules platform, by separating business logic from your business applications and giving you centralized control over that logic, can increase your business process agility as well as improving coordination across systems. For true agility, it is essential that the business rules platform enable business users to define and apply business rules with minimal involvement from overstretched IT staff. It is also imperative that the business rules platform be capable of fully capturing the complexity and nuance of your processes, policies, and procedures.

When the world's leading corporations need business rules solutions to drive their business transformation or continuous process improvement programs, they should turn to software, either off the shelves or customized one that meets their need. Consistently identified by industry analysts as the leader in process management software, the right software should have the right business rules platform provider for eight of the top ten global banks, seven of the top ten insurance companies, and four of the top five health insurance payers. Our high-performance business rules engine is available in a stand-alone business rules platform or as a core component in our enterprise business process management platform. This should consider the two following rules as consideration [1]:

1. **Business users take charge of business rules**

   The business rules platform drives business agility by enabling business users to quickly translate shifting business objectives into dynamic rule-based process applications. Usability is a defining quality of the business rule platform, with features such as:

   - Intuitive HTML forms for configuring and managing business rules
   - A Microsoft Visio front end to graphically visualize and manage business workflow as well as links between rules and applications
   - Built-in mechanisms for rule review and approval
   - Audit trail creation for all changes
   - Automatic generation of documentation

2. **Business rules as dynamic as your business**

The business rules platform delivers the most robust and flexible business rules support on the market, including:

- Process rules that automate all facets of workflow management
- A wide range of decisioning rule types, including decision trees, decision tables, and decision maps
- Declarative rules that compute values based on changes in other values
- Data transformation rules that enable data to be passed across dissimilar IT systems
- Integration rules that determine the appropriate system to invoke in each circumstance
- Forward and backward chaining

The business rules platform is built on open standards and easily integrates with your existing IT infrastructure. Built for enterprise scalability, the business rules platform should be able to drive process applications in some of the world's largest and most dynamic organizations for their day-to-day business activities.

5.2.4 Business Transformation

Power your business transformation with a right software tool, which matches your organization need in order to sustain continuous BPM and BCM with interruptions.

In an intensely competitive global marketplace rife with change, business transformation is the only path forward for many organizations. Choosing that path forward and leading companies in the financial services, insurance, healthcare, communications, and other dynamic sectors have to turn to the right software tool for transformation-driving technologies. The pacesetter in business process management (BPM) solutions and the right software help organizations achieve business transformation that better aligns operations with business goals and dramatically increases agility.

The right software tools for the right enterprise and organization help you build for change with your entity. The BPM should have capability to, at least, grow your BPM need at twice the pace of the overall BPM market against any competition in the market.

With the right process management software tools, you should be able to drive your business transformation with following points in mind:

- Integration is not a sticking point. Business transformation does not mean throwing out the baby with the bath water. With the right BPM in place, you can leverage your existing systems that are worth leveraging. This BPM should be built on open standards and is platform agnostic, to facilitate fast and effective integration with your diverse IT assets.

- Solution frameworks get you jump-started. Business transformation also does not require reinventing the wheel. Your choice of tool should extend your BPM system with industry-specific solution frameworks built on proven best practices in sectors including the financial services, insurance, communications, healthcare, and more. Chocked with industry-appropriate data models, business rules, role definitions, business workflow models, and user interface templates, the software tool solution frameworks should provide a robust starting foundation from which you can quickly build customized applications that capture your organization's unique insights, principles, and goals.

- Business users are in the driver's seat. More than any other solution on the market, the BPM empowers business users to be the drivers of business transformation. With an intuitive interface featuring familiar tools such as MS Excel and Visio, business users can easily define business rules and workflow management processes. The build for change technology then automates the programming required to turn business users' ideas into running applications. New business goals become new business processes more quickly and easily than ever before.

- Success breeds success. Build for change technology also drives agility by facilitating easy reuse of common processes and rules across your organization. Processes or campaigns proven successful in one corner of the company can be efficiently leveraged for more broad-based business transformation. At the same time, the business rules platform supports the layering of segmented, customized rules on top of baseline, enterprise-wide rules, to easily accommodate differences by region, channel, product, or customer.

Innovative, agile, and motored by the industry's most powerful business rules engine, the right BPM is the ideal foundation for transforming your business.

5.3   Dynamic Publishing and Processes

Optimizing the publishing processes enables organizations to generate new revenue while cutting costs, utilizing a good software tool in place for their dynamic publishing, and processing the information cross enterprise. Ninety percent of published information is stale; customers are overloaded with information, but usually cannot find what they need; the field force is dispatched to service calls with incorrect instructions; help desks desperately shuffle through a myriad of papers to find obsolete answers; organizations frequently publish incomplete or outdated information. It is no wonder that customer satisfaction is low, service costs are high, and legal exposure is high.

Creating and using redundant, inconsistent, and unstructured information for formal publishing requirements are a recipe for failure. Just as you would not tolerate uncontrolled data in your organization's financial management systems, you can no longer afford to handle your intellectual content in an uncontrolled, unconstrained, error-prone, and highly labor-intensive manner.

By using traditional publishing software, authors typically waste 30–50% of their time formatting documents instead of focusing solely on content creation and improvement. Often, authors must recreate content that already exists because that content cannot be located. Inconsistencies in the sequence and structure of information across similar documents make the information difficult for readers to understand. Lastly, lack of an automated publishing process forces authors to manually update multiple documents whenever a product or service changes, which is a time-consuming and error-prone process.

By optimizing your publishing process, your organization can gain a significant competitive advantage and achieve lasting differentiation from your competition.

The following paper will discuss how you can improve the quality of your publications and why a dynamic publishing system is optimal to solving your information quality problems.

- What if you could speed time to market by automating publication development and ensuring that new publications are released simultaneously in multiple languages and multiple media formats?
- What if you could increase employee productivity by 30% by improving employee access to current, relevant, and higher-quality information?
- What if you could deliver the most current, personalized, training materials automatically to all your training locations while eliminating 75% of your publication costs?
- What if you could maximize your ability to win new business by cutting proposal response time by 50%?
- What if you could automatically deliver personalized documentation to all your customers while quadrupling the frequency of documentation updates?
- What if you could eliminate the need to create 1/3 of all new documents by facilitating reuse of existing information? What if you could eliminate up to 80% of your localization costs?

Challenges with traditional publishing processes include the following:

1. **Poor information quality impacts organizational performance and corporate profitability**
   Organization's inabilities to reuse information and to automate the publishing process are two of the largest areas of inefficiency in today's corporate environments. Taken together, these inefficiencies have a dramatic, yet typically unrecognized, impact on internal costs, revenue recognition, productivity, time to market, and customer satisfaction.

2. **Typical publishing processes in organization**
   In today's competitive environment, customers demand more relevant information, tailored to their needs, while delivered in a timely manner and in the format and media of their choice. An inefficient publishing process can have a significant effect on the financial success of a product or service. Organizations typically have multiple publishing processes that involve multiple corporate functions and roles. When totaled, these processes have a significant impact on the overall

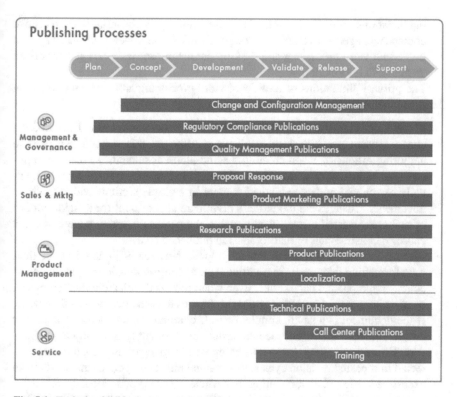

Fig. 5.1 Typical publishing processes steps [2]

performance of the organization. Organizations that employ traditional publishing software suffer from slow linear processes, lengthy revision cycles, and costly redundancies in their information release cycles. Manual publishing processes force compromises in the quality of documentation and place undue pressure on the publishing team, resulting in overtime and inaccurate information. The net impact of this approach is inevitably a loss of customer satisfaction, higher service costs, and possible liability exposure. A typical publishing process is depicted in Fig. 5.1 here as [1]:

Note that workforce inefficiencies related to publishing will cost organizations across the globe approximately "$800 billion" (A. T. Kearney).

3. **Traditional publishing systems limit the benefits of enterprise content management (ECM) software**

Organizations with limited content management experience believe that deploying a content management system alone will solve their publishing problems. Content management systems offer organizations the ability to impose controls over the process related to content creation and publishing. Users benefit by having a formal, predictable, and secure way of assigning tasks for creating new information or updating existing information. However, traditional publishing systems, such as word processing and desktop publishing software, waste organizations' time and resources while limiting the benefits that enterprise content management (ECM) can deliver. This waste arises from the way that traditional word processing and desktop publishing software lets authors create and publish information.

The primary limitations of traditional word processing and desktop publishing software include:

- Manual updates—Copying/pasting content is a common method for making changes when using desktop publishing. Although this method may add to authoring efficiency when creating a stand-alone document, it greatly complicates the maintenance of documents. Changes to content require authors to manually search and update redundant content in multiple sections within multiple documents. Changes are the writer's nightmare and account for a major portion of authors' "wasted effort." Additionally, the process is error-prone and often causes inconsistencies or inaccuracies in published content.

- Manual formatting—Authors typically spend between 30% and 50% of their time formatting documents in traditional publishing applications, applying character, paragraph, and page styles. Even when provided with authoring templates, most authors make formatting and style changes to each document as they write. If the document is a small, simple, one-off communication such as a memo or e-mail, the wasted effort of manual design and formatting is usually not significant. However, for long documents composed from legacy information and subjected to repeated revision cycles, this manual effort compounds and leads to an excessive waste of labor, resulting in inefficiency and production delays.

- Recreating existing content—Authors will recreate content that already exists if they cannot find it. When they find the content, authors typically copy and paste it into the new document. Both approaches are wasteful because making subsequent improvements to the information requires finding, updating, and reviewing all documents containing the repeated passage, caption, or phrase. Where localized versions of the content are required, translation fees can add exponential cost to a publishing operation. Furthermore, rewriting content instead of reusing it not only increases the cost and time to develop the content but also raises the risk that redundant information is inconsistent with other documents, making it even more difficult to update.

- Lack of structure—Inconsistencies in the sequence and structure of information across similar documents make the content more difficult for readers to understand and more difficult for your authors to find and update. Unstructured information is impossible to reuse or automatically format. A computer program cannot process inconsistent input—the classic "garbage in, garbage out" problem. Lack of structure adds redundant cycles of review and edit to the publishing process, as multiple updates must be circulated.

So, if desktop publishing and word processing are so problematic, why are they so popular? For organizations using these methods, traditional publishing software actually provides many benefits compared to the publishing tools it replaced. For example, desktop publishing provides a quick and easy way to make changes and previews the result before printing. However, while desktop publishing may be appropriate for some types of documents, many organizations are unaware that there is an alternative, which lets you publish better information more efficiently.

The traditional publishing process requires authors to create, format, and manage their content individually. They usually begin with a standard template for a specific type of publication, which they modify as they create the document. Authors add text, create or import graphics and tables, include boilerplate text (such as labels, disclaimers, etc.), and import content from other documents or databases from within or outside the organization.

Authors format the document as they write and then send it out for review and approval. Any change requires additional manual update and formatting. The process typically requires several iterations before the document is complete. On average, authors spend 30–60% of their time in low-value tasks such as decorating documents and updating multiple instances of the same information that proliferated through copying and pasting. Approved content is sent to the design team for re-formatting and configuration for publishing to additional media such as the Web. Often, content will be sent to a third party for multiple translations. Translated content typically varies in length and requires additional formatting before the publications are finalized. While not very efficient, this process might be tolerable were it not for the inevitability of content changes. Any update requires publications to undergo the same serial labor-intensive processes and requires all downstream functions to redo massive amounts of work. Furthermore, many organizations need to simultaneously manage multiple publications with overlapping release cycles.

These issues are further complicated by the typically chaotic environment already in place. Disconnected silos of information, lack of data version control, inadequate security/information access, and inefficient means of internal/external communication, all serve to exacerbate the above problems. As a result, most publications are stale, outdated, or inaccurate. This results in lost/delayed revenue, dissatisfied customers, and substantial legal exposure. A typical example of traditional publishing process is illustrated in Fig. 5.2.

Considering the tip of the iceberg of dynamic content publication workflow, which was mentioned in this chapter, a question comes up:

When should an organization consider adopting an automated approach to publishing?

Thus, you should consider evaluating the benefits of replacing traditional desktop publishing or word processing software if your content has one or more of the following characteristics:

- Multichannel delivery
- Multiple embedded diagrams
- Large volume
- Repeatable processes
- Personalized content
- Configurable documents
- Dynamic content
- Interactive content

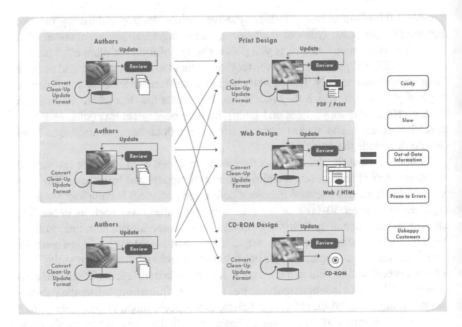

Fig. 5.2 Example of traditional publishing process [2]

As we have learned so far, dynamic content publishing is a strong function of business process management (BPM) and business continuity management (BCM) cross enterprise, and both variables of BPM and BCM are part of the nucleus of risk atom as was illustrated in Chap. 1 of this book. For further information on dynamic publication, the reader should refer to white paper in Reference 2 and for the business workflow to the web site in Reference 1 in this chapter.

5.4   E-commerce World and Dynamic Content Publication Workflow

Today's fast-paced and agile economy forces us to turn into online shopping and handling of our day-to-day need via dynamic Web pages online. Thus, it is worth to mention few important means of communications, with end users and customers of your organization who are using your products presented online. It does not matter if your business is online store, banking, or anything else that has established dynamic pages online for marketing and selling purposes; the content management requirements stay the same. If anything happens to your organization BPM or BCM that affects your customers and end users, it needs to be communicated to them via dynamic publications, so they benefit from it by protecting their assets online such as credit card information, personal identities, etc.

Table 5.1 Operation group

Access to the system	This class of users will access BRS.com site database and application for operation purpose like deployment, managing batch programs and external interfaces, database backup, tuning, exception management, etc.
User affiliations	BRS.com employees or other users designated by BRS
User skill level	Expert DBA and other operational staff
User subclasses	N/a
Access rights	Full access to back-end systems and batch programs that will be located at the co-location site
Usage pattern	Incident-based and pre-defined processes
Number of users	Low, may be less than five
Security constraints	Functionality to be accessible from BRS or other authorized locations (people support) only
Availability requirement	Continuous availability is not expected
GUI	N/a
Remarks	Most of the processes will be manual

As part of this requirement, Table 5.1 summarizes few of these requirements, as part of dynamic publishing and what you need to do in order to establish such means of online communications. This falls into your customer relationship management (CRM) tool that retains your loyal customers to be always loyal to you.

As part of dynamic content publication workflow, product catalog information including category grouping, associated images, prices, and other related information needs to be edited, approved, and tested before it is deployed on the live site. This should be done without disrupting the availability of the store to the end users.

Static HTML information including site policies, help pages, FAQs, display templates, and other related information needs to be edited, approved, and tested before it gets deployed in the live site.

This should be done without disrupting the availability of the store to the end users. The few steps that should be considered are:

- **Online interface requirements**
- **iChat rooms chat server**
- **iChat Pager instant messenger**
- **iChat discussion boards**
- **iSyndicate syndicated news**
- **CommTouch Web-based e-mail**
- **SMTP mail interface**
- **Google World Wide Web search**
- **Oracle interMedia Text search**
- **Anything missing from this section**
- **Batch interface requirements**
- **Blind e-mail management**
- **iSyndicate content updates**

- **Locator data loads**
- **Newsletter distribution**
- **Search index building**

The above are few that us as authors can suggest; however, they are not written into stone as mandatory requirements. Your organization and your need should dictate them as well.

The content should be organized and based on the analysis of audience needs. The desired information content should be compiled, cataloged, and presented with a highly effective design taking into consideration not only ease of access but also the way the readers process the presented information.

The approach to content management and information cataloging is based on subdividing large body of information to accommodate the limitations of the human brain in holding and remembering presented data. The overall approach is designed to combine effective graphic design, layout, and editorial division of information to devise effective organizational schemes to divide and organize data into functional discrete units for ease of navigation and processing. Therefore, the prepared content should be divided into smaller chunks of related information to be organized into modular units of information with consistent organization scheme to form the basis of hypertext links within the web site.

Few technical steps that we can suggest to you, from design point of view of a dynamic publishing site, are listed below, and you have to make up your need based on your organization requirements and end users of the site.

The approach to content management consists of the following steps:

1. **Logical division**
   The content should be divided into smaller, uniformly organized chunks of information units that can be located quickly and accessed nonsequentially. The discrete chunks of information lend themselves to Web links, enabling readers to find specific unit of information without having to read and filter large volume of information. However, too much subdivision of information is equally frustrating to most readers, and a highly balanced approach is the key to organizing information into effective units to avoid either extreme.

2. **Designing hierarchy**
   Hierarchy provides effective means for ranking information units in terms of their relative importance, relevance, and interrelationship. Organization hierarchy is designed for ease of navigation among information units, from the most important or most general concepts down to the most specific or optional topics. Hierarchical organizations are virtually a necessity on the Web, because most home page and link schemes depend on hierarchies, moving from the most general overview of the site (home page) down to submenus and content pages that become increasingly more specific as shown below.

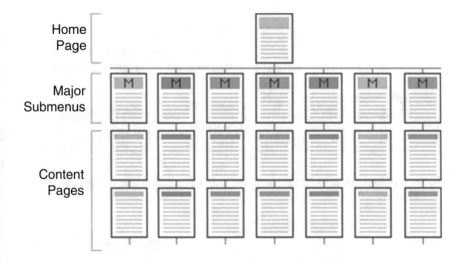

Home Page

Major Submenus

Content Pages

3. **Establishing relationships**

A critical step in content management is to provide clear, well-thought-out, and consistent logical organization of information units with consistent methods of grouping, ordering, labeling, and graphically arranging information. When confronted with a new and complex information, system users begin to build mental models and then use these models to assess relationships among topics and to make guesses about where to find things they haven't seen before. The success of the web site as an organization of information will largely be determined by how well the actual organization system matches the user's expectations.

4. **Functional refinement**

During this step, the created site (or prototype) should be analyzed in terms of its aesthetics and the practicality and efficiency of its organizational scheme to refine and balance the structure and relationship of menu or "home" pages and individual content pages or other linked graphics and documents. The goal is to build a hierarchy of menus and pages that feel natural to the user and doesn't interfere with their use of the web site. The web sites tend to grow almost organically and often overwhelm what was originally a reasonable menu scheme. WWW sites with too shallow a link hierarchy depend on massive menu pages that over time devolve into confusing "laundry lists" of unrelated information, listed in no particular order:

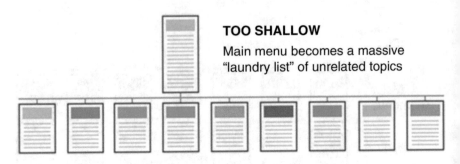

TOO SHALLOW

Main menu becomes a massive "laundry list" of unrelated topics

Menu schemes can also be too deep, burying information beneath too many layers of menus:

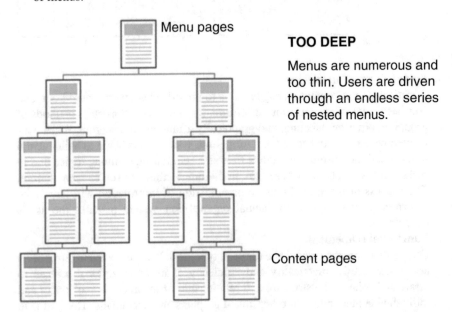

Menu pages

TOO DEEP

Menus are numerous and too thin. Users are driven through an endless series of nested menus.

Content pages

For actively growing, the proper balance of menus and pages is a moving target. User feedback and site analysis can help decide if the menu scheme has outlived its usefulness or has poorly designed areas. Complex document structures require deep menu hierarchies, but users should never be forced into page after page of menus if direct access is possible. The goal is to produce a well-balanced hierarchical tree that facilitates quick access to information and helps users understand how you have organized things.

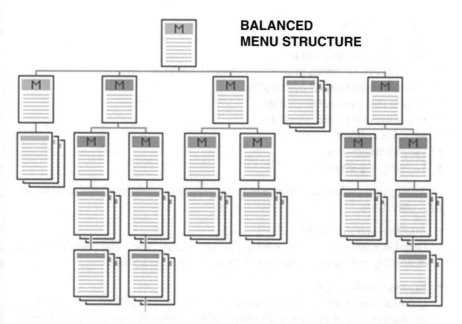

BALANCED MENU STRUCTURE

Scalability   Due to the nature of the Web and e-commerce, load on the BRS.com site will vary widely. The system architecture should be scalable and allow dynamic addition of more machines to support growth and special sales events.

Consistent Performance   During the uneven load requirements of the Web, the Web user is expected to get consistent response times.

Site Access   The BRS.com site should be accessible to all Web users. Part of the application may be restricted to members who have signed up. The physical site will be protected by firewall permitting only SSL and non-SSL http communications.

Site Security   A third party will host the public site. All machines constituting the public site need to be protected via firewall, single sign-on (SSO) technology should be in place for the identification to sign to the site, and SLA (service level agreement) should be implemented.

Operational Requirements   These requirements need to be studied along with other production procedure requirements as a separate effort. Some areas that need to be covered include:

- Backup/restore process and requirements
- Addition of servers for load balancing
- Software update procedures
- Application support requirements and procedures
- Recovery procedures
- Daily operation procedures
- Site auditing for security

- Load monitoring
- Exception handling procedures
- **Transaction management**
- **Content management process database transaction**
- **Contact management address book**
- **Article and product searching**
- **Error and exception handling**
- **Security and access control**
- **User communities**
- **Logging requirements**

Appropriate authentication needs to be provided for user, administrative, and CSR operations.

- **Auditing requirements**
  Detailed auditing would be provided in subsequent phases. The complete details of auditing requirements need to be captured for future phases.

- **Encryption/decryption**
  Database and machines are expected to be in a secure location, and only authorized users will be accessing database and host files (such as shell scripts, application parameters, etc.) directly. There will not be any encryption of the data in the database.

- **Network and Web security**
  Network and Web security requirements should be weighed against the degree of risk associated with doing business on the Internet. The sensitivity of information, productivity of users, and impact on revenue should all be assessed to determine the safeguards required to protect the network and infrastructure from hacker attacks.

  Security is achieved by any combination of the following five layers, physical layer, authentication layer, access and privilege layer, encryption layer, and audit trail layer. In addition to these five layers, three keys are used: (1) something you know, (2) something you have, and (3) something you are.

Prior and after implementing security measures, conduct an audit and risk assessment of the network. The effort should include an internal network security audit and an external penetration test.

The following recommendations are various means of providing protection layers or keys required to minimize security breaches:

- Setting the executables and the content to "read-only" status, it becomes very difficult.
- Implement proper firewall configuration to offer the best possible protection for the enterprise. Firewalls are deployed to prevent unauthorized "external" users from accessing the enterprise network while permitting "internal" users to communicate with "external" users and systems.
- Firewall logs provide a central point for logging and auditing Internet traffic.

- Server-side antiviral protection (e.g., from Trend Micro or Finjan) provides protection against incoming viruses and hostile ActiveX or Java applet.
- Deploy products such as Tripwire or Intercept to detect and prevent hacker attacks.
- Implement virtual private networks (VPNs) for secure remote access over the Internet.
- Often, the Web server or scripts run with privileges beyond what is required to execute the services the user needs and to include some that can be exploited by attackers. Although it may be possible for CGI and Web server designers to avoid design and configuration errors, the records of accomplishment show that secure design and configuration only rarely occurs. To prevent these attacks, we recommend syntax checking for forms and URLs delivered to Web servers as well as filtering inputs to the scripts to limit attacks via this mechanism.
- We recommend the use of Java as the mechanism for creating active content on Web pages from back-end data systems over all other techniques because of Java's capability to support a security policy manager that will uniformly apply a policy across all Java servlets running on the Web page.
- Utilize strong encryption and SSL to provide required encryption layer protection.
- For effective authentication, the user must first be strongly identified. For the most mission-critical, only two key methods of authentications are acceptable. Examples of two key authentication products include token cards such as SecureID or SafeWord.
- Coupled with the need for authentication is the need to control a user's access to specific content. We recommend use of a centralized authentication/authorization database. Use of this centralized database will enable a Webmaster to centrally administer identity and access rights. Vendors such as Netegrity and Blockade offer tools to implement centralized authentication and authorization.
- Although encryption will protect the communications channel between the Web browser and the Web server, information can still be corrupted if the Web server is not secure. Good systems administration is important, but it is not sufficient to protect the Web server. The server must be configured with only the tools needed to enable it to function as a Web server; nonessential functions need to be removed.
- Data must never be stored on the Web server, but must be accessed on back-end systems on behalf of the browser user.
- Encryption must be used between the Web server and the back-end database systems. This method will prevent external and internal attackers from posing as the authentic Web server and illicitly manipulating the authoritative database.
- In addition to implementing the aforementioned tools and techniques, organizations must have a security framework. The framework consists of security practices and incorporates the policy statement, roles and responsibilities, and risk management and analysis. It applies equally to the Internet, intranet, and web site security.

In addition, the following administrative procedures will enable enterprises to reduce web site vulnerabilities:

1. Creating secure password policies.
2. Creating an http server configured for least privilege.
3. Separating developments and production systems via a firewall or packet filtering router.
4. Maintaining active and accurate log files at all times and by authorized users only. Logs should be retained in a secure but retrievable format. Web usage/tracking tools are helpful in obtaining and presenting this data in a usable format.
5. Following proper anonymous FTP setup.
6. Having a valid password file accessible via anonymous FTP.
7. Disabling and filtering chargen, echo, and other unused UDP features.
8. Securing the firewall.

- **Authentication requirements**
  Basic login and identification information for Web users will be managed using Microsoft membership server.

  Site administrator and other support users will be maintained in the same membership server under individual authentication roots. Access rights to individual modules will be maintained in a security table.

Data analysis, data retention, and archival mechanisms need to be planned as well. The complete details of these requirements need to be captured for next few chapters.

5.5 Overall Concepts/Philosophy

The following basic choices drive the architecture:

- The published content data will be optimized for fast read.
- The order and other transaction data should be optimized for writes.
- Content creation process should not interfere with published content.
- Batch processes should not interfere with online transaction data.
- Data backup process should not interfere with Web usage.
- Support the very distinct security needs for each class of users.
- Allow "instantaneous" publication of all approved content without stopping the application.
- Allow for growth in site traffic.
- Allow for easy integration with future phases.

For justification, we can assume that a simple application has four classes of users with very different load, security, and user interface and availability requirements. The most practical option is to build different applications for each of the four user classes. These are:

1. **Partition**
   The four partitions are shown in the following diagram (Fig. 5.3).

Fig. 5.3 Application
partitioning

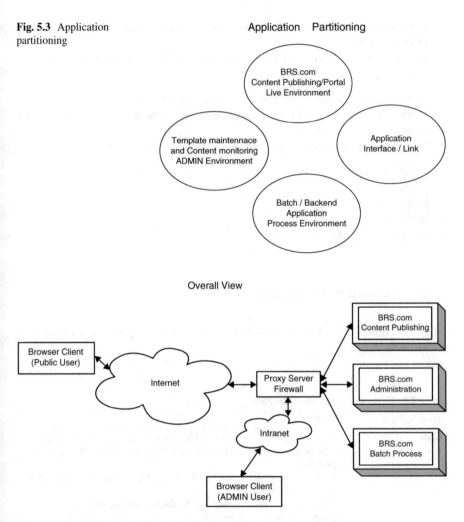

Application   Partitioning

BRS.com
Content Publishing/Portal
Live Environment

Template maintennace
and Content monitoring
ADMIN Environment

Application
Interface / Link

Batch / Backend
Application
Process Environment

Overall View

BRS.com
Content Publishing

Browser Client
(Public User)

Internet

Proxy Server
Firewall

BRS.com
Administration

Intranet

BRS.com
Batch Process

Browser Client
(ADMIN User)

Fig. 5.4 Overall view of physical layout

5.6   Business Resilience System Technology Topology Design

We assume a simple example and suggest some technology topology design as very holistic approach and not necessary will fit your organization needs and requirements.

The overall view of the users against the partitions (Fig. 5.4).

Story server-based solution that is going to be selected for the development of the portal for BRS.com is based on the following:

- Enables different members of the team to work collaboratively and productively
- Provides workflow for template and component development and content management

- Provides a unique method of delivering pages using cache to increase response time manyfold
- Personalization to make it attractive for users to come back and visit site more often

Suppose if we use Microsoft SharePoint software as an application for this infrastructure; then we need to know if SharePoint will provide such solution that we will implement in this phase.

In addition to the selected story server, the following also will be used:

- iSyndicate service
- iChat server
- iPager service
- Message boards service
- Google WWW search service

5.6.1 Terms and Definitions

This section defines all terms introduced to explain and implement the architecture (Table 5.2).

Table 5.2 Terms and definition of terminology description

Terminology	Description
CAS	Content application server (CAS) generates the requested page and if specified caches them and uses the templates and content, to generate the page Can have more than one CAS, but each of them should be associated with a Web server Pages can be cached depending on the volatility of the contents Cache can be cleared periodically
Content and membership database	Contents of all dynamic pages are stored in this database along with profiles of members and nonmembers
Locator database	Member personal database that BRS.com has or might purchase from various sources is stored in the locator database
BRS.com Web farm	A collection of Web servers with the same virtual Internet address managed with a load balancing service. The BRS.com portal is accessed via these machines
iChat discussion message boards server	Provides chat, instant message, and message board's service to the user
Batch process server	The site and dynamic content administration will be done from this Web server. Access is limited to site administrative staff. In addition, CSR administrators would be able to log in through this server

Fig. 5.5 Logical
architecture

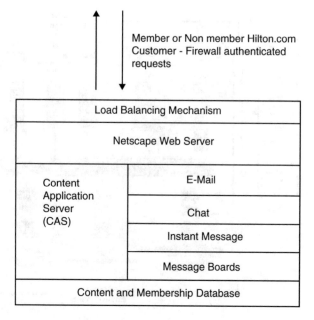

Member or Non member Hilton.com
Customer - Firewall authenticated
requests

| Load Balancing Mechanism |
| Netscape Web Server |

Content Application Server (CAS)	E-Mail
	Chat
	Instant Message
	Message Boards
Content and Membership Database	

5.6.2   Architecture Diagram and Logical Architecture

For this simple example, the architectural diagram and logical architecture are depicted in Fig. 5.5.

Moreover, physical architecture, including production environment, is depicted in Fig. 5.6.

5.6.3   Publication Process and Actors and Workflow

For the actors and workflow, these can be provided using project management tools (i.e., Microsoft Project).

For publication process, we can describe the publication process for dynamic content as (Fig. 5.7):

- Editing for a publication is entered as an editing group.
- An editing group has a scheduled publishing date and other status indicators.
- Content editors edit data—catalog, product information, etc. into that batch.
- Multiple batches can be active at a time.
- After the consolidation of the data on the working database, the content will be programmatically moved to the staging content database.
- The standby content database is deleted and recreated using database script.

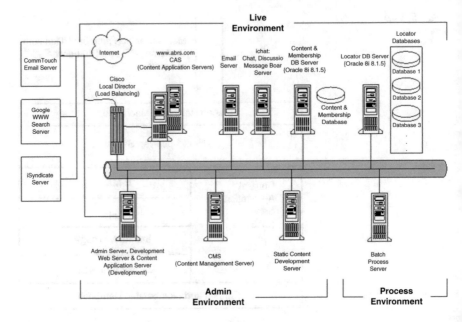

Fig. 5.6 Physical architecture—production environment

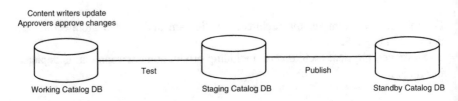

Fig. 5.7 Dynamic content update servers' layout

- The content data on the staging database will be loaded to the standby content database.
- When the standby database has to be made active, Global.asa file will be modified to swap active content database with standby database.
- The process is repeated for the next publication day.

5.6.4   Static Content Publication Workflow

For static content publication workflow, we can state the following: Static content publication will follow the publication process of the site server. Any dynamic content publication and all the associated content like the images will also be published (Fig. 5.8).

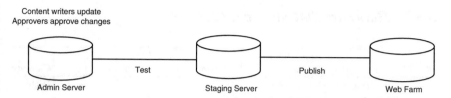

Fig. 5.8 Static content update servers' layout

5.6.5   Product Image Data Storage

All products related to and other site images will be kept as operating system files in Web-accessible directories. There are six images for each product and two images at SKU level. Each type of image will be kept in a separate directory on the Web server.

5.6.6   Product Image Data Publishing and Editing

Off-the-shelf products will be used to edit image files. It is ideal that new images to be published are kept in a separate directory for each edit batch with individual subdirectories for each image type. The subdirectory names should be same as the published directory name. This will facilitate publishing images effortlessly when product information is published programmatically. The image publishing process is expected to be manual.

5.6.7   Transaction Data Management Process

Much of the main transaction data created by the online shop consists of online orders and notification requests of customer events. The following list shows how this data will flow in the system:

- Transaction data will be created in the transaction database. The transaction database will not maintain any data versioning.
- Transaction data will be programmatically moved to process/history database in small batches.
- All subsequent batch programs will use the process/history database.
- Batch programs will be used to create text files to send data to Organization Corporation or Internet Service Provider where BRS.com is hosted.
- Batch programs will be used to update process/history database with data received from organization or any other sources within Organization Corporation.

5.6.8 Hardware/Software Specifications

#	Machine name	Application servers	Software	Databases	Capacity
1.	www.BRS.com Web farm	HTML, BRS. com application objects	IIS SSL Transaction server Site Server Microsoft Office SharePoint Server 2007 (MOSS)	–	High
2.	Membership server	–	Site Server (membership server)	–	High
3.	Catalog DB server	–	SQL server	Tax files Standby catalog database Active catalog database	High
4.	Transaction DB server	–	SQL server	Credit card database Transaction database Membership database	High
5.	CSR, site admin Web server	CSR and admin ASPs, BRS.com admin applications, BRS.com admin objects	IIS MS Transaction Server Site Server (Commerce server)	–	Medium
6.	Membership trial server	LDAP/AUO	MS Site Server (Membership Server)	–	Low
7.	Working catalog DB server	–	MS SQL Server	Trial transaction database Working catalog database Trial membership database	Medium
8.	Staging server	–	Site server (Content Server) SQL Server	Staging catalog database	Low
9.	Batch process and FTP server	Batch process applications	ftp server scheduler	Process DB (order history and management data)	Medium
10.	SMTP mail server	–	Exchange	–	Low

5.6.9   Webmail and Discussion Board Interface

- The webmail application will reside on a server separate from the organization application. The organization application will provide a link to that server through a URL.
- The webmail program will pick up the user id and password from the organization database.

5.6.10   Member Address Book

The user will be asked to export their Google, Netscape, FoxPro, or Microsoft Outlook address book into a comma-delimited file. The user will then upload the file onto the organization server using a graphical interface within the site (similar to attachments in online e-mail programs). This information will be inserted into the organization address book database and shown in the site.

5.6.11   Site and Database Search

BRS.com content site requires a tool, which can combine text and database search together. BRS recommends Oracle as the database, and Oracle interMedia is proposed as the search engine. It works as part of the database and can search across database text columns and URLs referenced in columns. It is possible to define multiple indexes to offer search to part of the site or entire site. Search commands are parts of SQL and combine with normal SQL joins.

5.6.12   Load Balancing and Web Farm Management

The load balancing will be done using load balancing routers to distribute Web traffic to Web servers. The router provides single virtual IP for all the Web server machines. Each Web server machine will use selected content delivery server. Content delivery servers will interact with the content management database.

This will be done at the hosting site, which will be decided by BRS. CISCO Local Director or F5 routers will be used. Two routers will be required to provide fail over.

User Session Management The client browser needs cookies to be enabled for shopping in the site.

Database Database design information will be provided as part of the physical data model document during phase I of design/development and production.

Access Security   See the *User Communities* section.

Data Security   The database is expected to be accessed outside the application only by the authorized database administrator. The system will not implement any data encryption in the database.

Usage Statistics and Reporting   This feature is out of scope of phase I. Initial usage data may be collected by using Web log analyzers.

Off-the-Shelf Components   Please see the following for any off-the-self components that are required for this site.

Credit Card Authentication Site Server COM Object   A COM object will be used to send authorization request to credit card authorizing company and receive pre-authorization code from it.

Microsoft Site Server Order Pipeline Components   The commerce application will be built with Microsoft Site Server (Commerce Edition) as the basis. Order processing will be handled by customizing Microsoft Order Processing pipeline components.

5.7   System Interfaces

The interfaces will be decided during the development process.

Limitations   The client browser needs cookies to be enabled for supply chain in the site. Browsers that do not support cookies or where cookies have been disabled cannot place orders in BRS.com site.

A standard "cookie policy" addressing the needs of the same may be added in the privacy policy of the site.

Assumptions   The following technical/architecture assumptions are made:

1. Organization and the site-hosting agency identified by the organization will be responsible for network configuration, firewall, and setting up of all the servers. However, BRS will assist their personnel to accomplish this task.
2. As is the case with all software-based projects, the third-party products used for the project may exhibit limitations or bugs that can impede the development process. To reduce the impact of such problems, alternative approaches may need to be evaluated during the course of the project, including possible modifications to the system's design and/or functional requirements.
3. If the BRS development team cannot code a particular functionality for any bugs identified with the development software, then the team will report such bugs to the appropriate vendor's technical support in order to acquire any patches/alternate suggestions to implement the same. In case a solution is not forwarded by the vendor's technical support within a short period, it will affect

the development schedule. In such situations, BRS will propose an alternate implementation strategy around the application development software features available at that time.

4. BRS will not be responsible for performance and technical limitations, based on products being integrated. BRS will work with the organization to alleviate the issues and develop work-arounds.

5. BRS will not be responsible for performance limitations based on the network response time.

6. BRS's estimated development time and development cost is based on the proposed software architecture. Any changes in the software architecture may have an impact on delivery date and project cost and should be treated as out of scope.

7. The system will be developed using specific versions of software development tools and operating systems.

8. The BRS.com site will be developed in US-English only as an example of language to be picked for US organizations. Any internationalization issues are out of scope of this project, i.e., if BRS wants to support other languages in BRS.com, they should be treated as independent sites.

9. Any delays due to an upgrade will require a change request and may affect the schedule and cost of the project.

10. Any additional time required in support of an upgrade will require a change request and may influence the schedule and cost of the project.

5.8 Technical Approach

This section identifies the technical requirements to support the baseline business/functional requirements. The technical requirements provide the basis for the technical approach presented in this section.

5.8.1 Technical Requirements for Phase I

The technical requirements for the solution are noted below:

- Support 24×7 operation for the Military Advantage web site.
- Solution should support 1.5 million (approximately) unique monthly visitors for the year 2000. Solution should be scalable to support growth in enrollment rates and associated site traffic.
- Provide support for cookies and auto-login based on cookies.
- Support personalization of home page and standard organization content components (news, information, community, etc.) based on user profile that we can administer.

- Ability to log page access and accrue "organization.com" points for defined access events.
- Support for anonymous users and registered users.
- Support common branding based on the branch and user preference-based personalization.
- Support Microsoft Internet Explorer 6.× and Netscape Communicator 7.× browser clients.
- Support online e-mail notifications for specific user actions.
- Support scalability and load balancing. It should be possible to add additional hardware depending on the system load.
- Support browser-based dynamic content editing, category creation, keyword creation, and content categorization.
- Support dynamic and static content publication workflow.
- Support dynamic content publication without any downtime to the web site.
- Provide interface to chat, discussion groups, and bulletin board service software.
- Provide interface to Web-based e-mail with one mail domain and instant messaging.
- Support site search and directed search on bases, associations, and home pages.
- Provide World Wide Web search.
- Support generation of newsletters and electronic distribution of these letters.
- Support manual publication and approval of user content.
- Support the loading of locator databases.
- Support people search (locator) within BRS member database and locator database.
- Support batch processing of personnel search through the BRS locator database.
- Provide online interface to three identified people search services such as Yahoo's People Search.
- Provide online interface to other services such as 411 and 1-800-USSEARCH (link to the co-branded URL sites).
- Provide interface to capture syndicated news feeds.
- Support creation of personal home pages by registered members.
- Provide interface to ad server.
- Accommodate links to commerce partners (e.g., third parties that charge for personnel search or book vendors).
- Integrated home page, contact management, chat/discussion, and event planning.
- Searchable databases of military text/graphics content on unit insignias, histories, etc. that will populate affinity group home pages as well as the home page for individual registered users.
- Can we include the options to make a given page the start page on your browser?

5.8.2 Technical Requirements Beyond Scope for Phase I

- Internationalization and multiple language support.
- Any mechanism to monitor appropriateness of language, content, and graphics of user content and any means of blocking such content are also beyond scope.
- Integration with spatial or map databases (such as those for database locations).
- Any validation of user claims on military involvement, credentials, stories, and any other profile data.
- Online inquiry/maintenance or redemption of BRS points.
- Creation and management of user accounts at machine level to support file transfer (ftp) and any disk management.
- Any kind of reporting beyond what is provided out of the box by selected tools—online or offline, decision support, auditing, or tracking.
- Any metering of disk usage or facility for charged premium services.
- Any data purging or archival policies/processes (we need the ability to locate and delete inactive pages).
- Securing databases beyond user id/password control. Any data level encryption support is beyond scope.
- Validation and cleanup of locator data obtained from external sources.
- Maintenance of pre-populated content such as base information, ships unit, etc.
- Any effort toward developing content/screens for the corporate information requirement (baseline requirements detail).
- Web server security beyond what is provided by the Web server.

5.8.3 Proposed Hardware and Network Architecture (Fig. 5.9)

The above hardware configuration is provided for reference only. Actual configuration will be decided during Work Segment I phase.

The following text describes the different components used in the proposed solution:

- Content application servers/Web servers: MOSS 2007 or higher.
- Web browser: Microsoft Internet Explorer 6 or higher and Netscape Navigator 7.x or higher.
- SMTP mail, chat, and bulletin board server: Compatible with the identified network architecture.

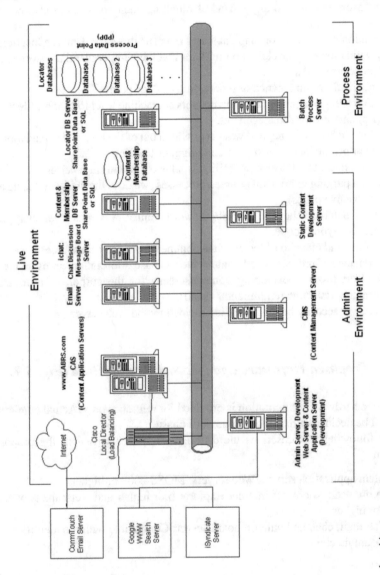

Fig. 5.9 Network architecture

- Database: SQL Server Version MOSS 2007 or higher. The following databases have been identified thus far:

  - BRS membership and profile database.
  - BRS content database:

    Feature articles and other editorial content BRS.com generates
    Libraries of searchable/browsable graphics and text, such as insignias, unit histories, how-to guides, reference information, etc.
    Home pages organized in a searchable/browsable manner along several dimensions (unit name, class of ship/aircraft)
  - Licensed locator database(s)—this will store the personnel information that Military Advantage will procure from different sources. This may be stored in one or more databases, depending on the format and data information provided by the different sources.

- Static content development server: This will be the machine that the BRS editorial team will use for categorizing content.
- Batch process server: This will serve all batch programs, generation of newsletters, off-line search for personnel, distribution of newsletters, etc.

  Profile server
  Ad server

  [Note: Version numbers of the software will be finalized during the analysis phase of the project. The information presented in this section is indicative only.]
  Some of the terms used above are explained below:

Terminology	Description
Content database	All dynamic content including enterprise content management page templates, data required to build the content pages
Content delivery server	The component of the ECM system which sits on the Web server machine to build and deliver dynamic pages
Locator databases	These are databases BRS will license to provide a more complete personal locator service. The structure, size, and number of these databases are unknown
Membership and profiling database	This database stores the profiling information and membership database for users of the site
Batch process server	This is the dedicated machine to run batch process without affecting the performance of the web site

5.8.4   *Logical Architecture for Live Environment*

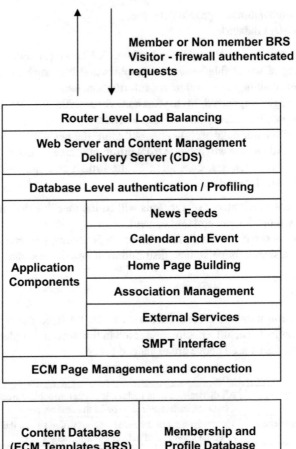

5.8.5   *Technical Risks*

The technical risk are listed here

- Given the e-mail and home page creation requirements, BRS will have to plan for significant storage requirements. This will be further evaluated through the hosting strategy. A validation of the above or more accurate data will be required to perform the capacity planning.
- Number of data sources and the formats for locator database, base information, or other pre-populated information are unknown at this point. Multiple data sources may lead to inconsistent data.
- Integration with multiple third-party packages.

5.8.5.1 Phase I Items Not Estimated

The following items are part of phase I but have not been estimated now. This will be, handled through the change management process once more details become known.

- Loading of locator databases
- Importing of organization information content from external sources (information on bases, ships, units, etc.)

5.8.5.2 Recommended Solution

Please provide ballpark:

- Development equipment/environment costs
- Other applications which are required at scale
- Estimated hosting costs

5.8.5.3 User Home Pages

Disk Space Requirement (First Year) The following is an initial estimate of user home pages to the estimated 1.5 million members of BRS. This is based on if become and ISP to all of BRS clients at enterprise and corporate level;
   Most likely scenario is the middle scenario.

% of members creating home pages	# of user home pages	Allocated disk space estimate at 10 MB per home page	Allocated disk space estimate at 5 MB per home page
1%	12,000	120 GB	60 GB
5%	60,000	600 GB	300 GB
10%	120,000	12,000 GB = 1.2 TB	600 GB

Disk space requirement varies from 60 GB to 1.2 TB to fulfill the home page creation requirement. Please note that this estimate does not include association membership and related data.

Development Environment Hardware for Development Environment
   The hardware for the development environment will be supplied by Military Advantage. The following information is indicative of what will be required. Exact requirements will be detailed during the analysis phase.

Development Server
The development server configuration requirements are as follows.

- PC box with MOSS 2007 or above machine to host Web server, content delivery server, content management server, content database, and other third-party software

- Single CPU 4.0 GHz
- 2 GB or higher RAM
- 300 GB or higher disk capacity

Test Server

The test server configuration requirements are as follows.

- PC box with MOSS 2007 or above machine to host Web server, content delivery server, content management server, content database, and other third-party softwares
- Single CPU 4.0 GHz
- 2 GB or higher RAM
- 300 GB or higher disk capacity

Software for Development Environment

Development Server

The following software licenses are required for the development server:

- Microsoft Office SharePoint Server
- SQL Server
- Other third-party softwares as listed in the section

PCs

The following workstation software licenses are required:

- Windows XP/7/8 or later operating system
- Microsoft Office Professional 2007/2010 or later
- Visio 6.0 or later

References

1. https://www.pega.com/business-workflow?&utm_source=google&utm_medium=cpc&utm_campaign=900.US.Active-Awareness&utm_term=%2Bdynamic%20%2Bworkflow&gloc=9031921&utm_content=cXFDtlTwlpcridl106047114849lpkwl%2Bdynamic%20%2Bworkflowlpmtlblpdvlcl
2. *Dynamic publishing*. White paper. Courtesy of PTC.com. Single-Sourcing Solution Inc.

Chapter 6
What Is Boolean Logic and How It Works

If you want to understand the answer to this question down at the very core, the first thing you need to understand is something called Boolean logic. Boolean logic, originally developed by George Boole in the mid-1800s, allows quite a few unexpected things to be mapped into bits and bytes. The great thing about Boolean logic is that, once you get the hang of things, Boolean logic (or at least the parts you need in order to understand the operations of computers) is outrageously simple. In this chapter, we will first discuss simple logic "gates" and then see how to combine them into something useful. A contemporary of Charles Babbage, whom he briefly met, Boole is, these days, credited as being the "forefather of the information age." An Englishman by birth, in 1849, he became the first professor of mathematics in Ireland new Queen's College (now University College) Cork. He died at the age of 49 in 1846, and his work might never have had an impact on computer science without somebody like Claude Shannon, who 70 years later recognized the relevance for engineering of Boole's symbolic logic. As a result, Boole's thinking has become the practical foundation of digital circuit design and the theoretical grounding of the digital age.

6.1 Introduction

Boolean logic, originally developed by George Boole in the mid-1800s, allows quite a few unexpected things to be mapped into bits and bytes. What George Boole did to be recognized as the father of modern information technology was to come up with an idea that was at the same time revolutionary and simple. Boole's work certainly started modern logic off on the right road, but it certainly was not anything to do with the *laws of thought*. The fact of the matter is that even today, we have no clear idea what laws govern thought, and if we did, the whole subject of artificial intelligence would be a closed one. How Boolean logic works, and what it is, can be

© Springer International Publishing AG 2017   183
B. Zohuri, M. Moghaddam, *Business Resilience System (BRS): Driven Through Boolean, Fuzzy Logics and Cloud Computation*,
DOI 10.1007/978-3-319-53417-6_6

simply defined by asking yourself, how a computer can do some logical things and tasks like balancing a checkbook, or play chess, or spell-check a document? These are things that, just a few decades ago, only humans could do.

Now computers do them with apparent ease. How can a "chip" made up of silicon and wires do something that seems like it requires human thought?

Computers and logic have an inseparable relationship. They are now, but at the start, things were much hazier. The first computers were conceived as automatic arithmetic engines, and while their creators were aware that logic had something to do with it all, they were not 100% clear as to the how or why. Even today, we tend to be over simplistic about logic, and its role in computation and understanding the world and George Boole the man who started it all off was a bit over the top with the title of this chapter on the subject.

Boolean logic is very easy to explain and to understand:

- You start off with the idea that some statement P is either true or false, it can't be anything in between (this called the law of the excluded middle).
- Then, you can form other statements, which are true or false, by combining these initial statements together using the fundamental operators **AND**, **OR**, and **NOT**.

Exactly what a *fundamental* operator is forms an interesting question in its own right—something we will return to later when we ask how few logical operators do we actually need?

The way that all this works more or less fits in with the way that we used these terms in English.

For example, if P is true, then NOT(P) is false, so if "today is Monday" is true, then "Not (today is Monday)" is false. We often translate the logical expression into English as "today is Not Monday," and this makes it easier to see that it is false if today is indeed Monday.

Well, this is the problem with this sort of discussion. It very quickly becomes convoluted and difficult to follow, and this is part of the power of Boolean logic. You can write down arguments clearly in symbolic form.

The great thing about Boolean logic is that, once you get the hang of things, Boolean logic (or at least the parts you need in order to understand the operations of computers) is outrageously simple.

In this chapter, we will first discuss what bits and bytes are and what we mean by these values, and then we talk about simple logic "gates" and then see how to combine them into something useful.

6.2   How Bits and Bytes Work

If you have used a computer for more than 5 min, then you have heard the words bits and bytes. Both RAM and hard disk capacities are measured in bytes, as are file sizes when you examine them in a file viewer. Figure 6.1 shows the logic for computer that works based on binary of 0s and 1s.

Fig. 6.1 Fundamental of
computer logic

You might hear an advertisement that says, "This computer has a 32-bit Pentium processor with 64 MG of RAM and 2.1 GB of hard disk space"; these are the information that we need to be a bit knowledgeable about them. In this section, we will discuss bits and bytes so that you have a complete understanding.

In order to have a better grasp and handle on computer logic of bits and bytes as well as understanding of how they work, we refer to the work by Marshall Brain technical site, where he is describing "How Bits and Bytes Work" [1].

The easiest way to understand bits is to compare them to something you know: digits and decimal numbers.

A digit is a single place that can hold numerical values between zero and nine. Digits are normally combined together in groups to create larger numbers. For example, 6357 have four digits. It is understood that in the number 6357, the 7 is filling the "1s place," while the 5 is filling the 10s place, the 3 is filling the 100s place, and the 6 is filling the 1000s place. Therefore, you could express things this way if you wanted to be explicit:

$$\left(6 \times 10^3\right) + \left(3 \times 10^2\right) + \left(5 \times 10^1\right) + \left(7 \times 10^0\right) = 6000 + 300 + 50 + 7 = 6357$$

What you can see from this expression is that each digit is a **placeholder** for the next higher power of 10, starting in the first digit with 10 raised to the power of 0.

That should all feel comfortable—we work with decimal digits every day. The neat thing about number systems is that there is nothing that forces you to have ten different values in a digit. Our **base-10** number system likely grew up because we have ten fingers, but if we happened to evolve to have eight fingers instead, we would probably have a base-8 number system. You can have base-anything number systems. In fact, there are many good reasons to use different bases in different situations.

Computers happen to operate using the base-2 number system, also known as the **binary number system** (just like the base-10 number system, it is known as the decimal number system). Find out why and how that works in the next subsection.

6.2.1   The Base-2 System and the 8-Bit Byte

The reason computers use the base-2 system is because it makes it a lot easier to implement them with the current electronic technology. You could wire up and build computers that operate in base-10, but they would be fiendishly expensive right now. On the other hand, base-2 computers are relatively cheap.

Therefore, computers use binary numbers and thus use **binary digits** in place of decimal digits. The word **bit** is a shortening of the words *binary digit*. Whereas decimal digits have ten possible values ranging from 0 to 9, bits have only two possible values: 0 and 1. Therefore, a binary number is composed of only 0s and 1s, like this: 1011. How do you figure out what the value of the binary number 1011 is? You do it in the same way we did it above for 6357, but you use a base of 2 instead of a base of 10. So:

$$\left(1\times2^{3}\right)+\left(0\times2^{2}\right)+\left(1\times2^{1}\right)+\left(1\times2^{0}\right)=8+0+2+1=11$$

You can see that in binary numbers, each bit holds the value of increasing powers of 2. That makes counting in binary pretty easy. Starting at 0 and going through 20, counting in decimal and binary looks like the presentation in Table 6.1.

When you look at this sequence, 0 and 1 are the same for decimal and binary number systems. At the number 2, you see carrying first take place in the binary

Table 6.1 0–20

0 = 0
1 = 1
2 = 10
3 = 11
4 = 100
5 = 101
6 = 110
7 = 111
8 = 1000
9 = 1001
10 = 1010
11 = 1011
12 = 1100
13 = 1101
14 = 1110
15 = 1111
16 = 10000
17 = 10001
18 = 10010
19 = 10011
20 = 10100

Table 6.2 8 bits in byte

0 = 00000000
1 = 00000001
2 = 00000010
...
254 = 11111110
255 = 11111111

system. If a bit is 1 and you add 1 to it, the bit becomes 0 and the next bit becomes 1. In the transition from 15 to 16, this effect rolls over through 4 bits, turning 1111 into 10000.

Bits are rarely seen alone in computers. They are usually bundled together into 8-bit collections, and these collections are called bytes. Why are there 8 bits in a byte? A similar question is, "Why are there 12 eggs in a dozen?" The 8-bit byte is something that people settled on through trial and error over the past 50 years. With 8 **bits** in a **byte**, one can represent 256 values ranging from 0 to 255, as it is shown in Table 6.2.

Next, we will look at one way that bytes are used.

6.2.2   The Standard ASCII Character Set

Bytes are frequently used to hold individual characters in a text document. In the ASCII character set, each binary value between 0 and 127 is given a specific character. Most computers extend the ASCII character set to use the full range of 256 characters available in a byte. The upper 128 characters handle special things like accented characters from common foreign languages.

You can see the 127 standard ASCII codes below. Computers store text documents, both on disk and in memory, using these codes. For example, if you use Notepad in Windows 95/98 or higher version to create a text file containing the words, "Four score and seven years ago," Notepad would use 1 byte of memory per character (including 1 byte for each space character between the words—ASCII character 32). When Notepad stores the sentence in a file on disk, the file will also contain 1 byte per character and per space.

Try this experiment: Open up a new file in Notepad and insert the sentence "Four score and seven years ago" in it. Save the file to disk under the name getty.txt. Then use the explorer and look at the size of the file. You will find that the file has a size of 30 bytes on disk: 1 byte for each character. If you add another word to the end of the sentence and re-save it, the file size will jump to the appropriate number of bytes. Each character consumes a byte.

If you were to look at the file as a computer looks at it, you would find that each byte contains not a letter but a number—the number is the ASCII code corresponding to the character (see Table 6.3). So on disk, the numbers for the file look like this:

Table 6.3 ASCII code corresponding to binary code

F	o	u	r		a	n	d		s	e	v	e	n
70	111	117	114	32	97	110	100	32	115	101	118	101	110

Table 6.4 Byte prefixes and binary mathematic units

Kilo (K)	$2^{10} = 1024$
Mega (M)	$2^{20} = 1,048,576$
Giga (G)	$2^{30} = 1,073,741,824$
Tera (T)	$2^{40} = 1,099,511,627,776$
Peta (P)	$2^{50} = 1,125,899,906,842,624$
Exa (E)	$2^{60} = 1,152,921,504,606,846,976$
Zetta (Z)	$2^{70} = 1,180,591,620,717,411,303,424$
Yotta (Y)	$2^{80} = 1,208,925,819,614,629,174,706,176$

By looking in the ASCII table, you can see a one-to-one correspondence between each character and the ASCII code used. Note the use of 32 for a space—32 is the ASCII code for a space. We could expand these decimal numbers out to binary numbers (so 32 = 00100000) if we wanted to be technically correct—that is how the computer really deals with things.

The first 32 values (0 through 31) are codes for things like carriage return and line feed. The space character is the 33rd value, followed by punctuation, digits, uppercase characters, and lowercase characters. To see all 127 values, check out Unicode.org's chart.

We will learn about byte prefixes and binary math next.

6.2.3   Byte Prefixes and Binary Math

When you start talking about lots of bytes, you get into prefixes like kilo, mega, and giga, as in kilobyte, megabyte, and gigabyte (also shortened to K, M, and G, as in Kbytes, Mbytes, and Gbytes, or KB, MB, and GB). The following table shows the binary multipliers. See Table 6.4.

You can see in this chart that kilo is about a thousand, mega is about a million, giga is about a billion, and so on. So when someone says, "This computer has a 2 gig hard drive," what he or she means is that the hard drive stores 2 GB, or approximately 2 billion bytes, or exactly 2,147,483,648 bytes.

Terabyte databases are fairly common these days, and there are probably a few petabyte databases floating around the Pentagon or Homeland Security and intelligent communities by now. Our futuristic Business Resilience System (BRS) needs to be structured and designed based on these sizes of data for handing all types of events, from e-commerce to banking, from power plants to renewable energy and sustaining demand on grids for energy consumptions, and so on.

Binary math works just like decimal math, except that the value of each bit can be only 0 or 1. To get a feel for binary math, let us start with decimal addition and see how it works. Assume that we want to add 452 and 751:

$$
\begin{array}{r}
452 \\
+751 \\
\hline
\cdots \\
\hline
1203
\end{array}
$$

To add these two numbers together, you start at the right: $2 + 1 = 3$. No problem. Next, $5 + 5 = 10$, so you save the zero and carry the 1 over to the next place. Next, $4 + 7 + 1$ (because of the carry) $= 12$, so you save the 2 and carry the 1. Finally, $0 + 0 + 1 = 1$. So the answer is 1203.

Binary addition works exactly the same way:

$$
\begin{array}{r}
010 \\
+111 \\
\hline
\cdots \\
\hline
1001
\end{array}
$$

Starting at the right, $0 + 1 = 1$ for the first digit. No carrying there. You have $1 + 1 = 10$ for the second digit, so save the 0 and carry the 1. For the third digit, $0 + 1 + 1 = 10$, so save the zero and carry the 1. For the last digit, $0 + 0 + 1 = 1$. So the answer is 1001. If you translate everything over to decimal, you can see it is correct: $2 + 7 = 9$.

To sum up, here is what we've learned about bits and bytes:

- Bits are binary digits. A bit can hold the value 0 or 1.
- Bytes are made up of 8 bits each.
- Binary math works just like decimal math, but each bit can have a value of only 0 or 1.

There really is nothing more to it—bits and bytes are that simple.

For more information on bits, bytes, and related topics, check out Marshall Brain and the links in his web site [1].

6.3   Logical Gates

Boolean logic affects how computers operate, and it is illustrated holistically here as Fig. 6.2, which uses bunch of gate arrays.

Several arrays of gates do exist in any computer that you need to know about and learn their operational functions. The simplest possible one is called an "inverter," or a NOT gate. This gate takes 1 bit as input and produces output as its opposite. The logic table is Table 6.5.

Fig. 6.2 Computer logical
gate arrays

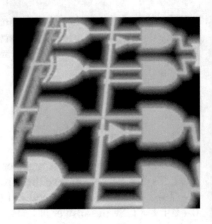

6.4   Simple Adders

In the article on bits and bytes, you learned about **binary addition**. In this section,
you will learn how you can create a circuit capable of binary addition using the
gates described in the previous section.

Let us start with a **single-bit adder**. Let us say that you have a project where you
need to add single bits together and get the answer. The way you would start designing a circuit for that is to first look at all of the logical combinations. You might do
that by looking at the following four sums:

$0 + 0 = 0$
$0 + 1 = 1$
$1 + 0 = 1$
$1 + 1 = 10$

That looks fine until you get to $1 + 1$. In that case, you have that pesky **carry bit**
to worry about. If you don't care about carrying (because this is, after all, a 1-bit
addition problem), then you can see that you can solve this problem with an XOR
gate. But if you do care, then you might rewrite your equations to always include **2
bits of output** like this:

$0 + 0 = 00$
$0 + 1 = 01$
$1 + 0 = 01$
$1 + 1 = 10$

From these equations, you can form the logic table:

1-bit adder with carry-out

A	B	Q	CO
0	0	0	0
0	1	1	0
1	0	1	0
1	1	0	1

Table 6.5 The logic table

NOT gate	The NOT gate has one input called **A** and one output called **Q** ("Q" is used for the output because if you used "O," you would easily confuse it with zero). The table shows how the gate behaves. When you apply a 0 to A, Q produces a 1. When you apply a 1 to A, Q produces a 0. Simple

A	Q
0	1
1	0

AND gate	The AND gate performs a logical "and" operation on two inputs, A and B:

A	B	Q
0	0	0
0	1	0
1	0	0
1	1	1

The idea behind an AND gate is "If A AND B are both 1, then Q should be 1." You can see the behavior in the logic table for the gate. You read this table row by row like this:

A	B	Q	
0	0	0	*If A is 0 AND B is 0, Q is 0*
0	1	0	*If A is 0 AND B is 1, Q is 0*
1	0	0	*If A is 1 AND B is 0, Q is 0*
1	1	1	*If A is 1 AND B is 1, Q is 1*

OR gate	The next gate is an OR gate. Its basic idea is, "If A is 1 **OR** B is 1 (or both are 1), then Q is 1"

A	B	Q
0	0	0
0	1	1
1	0	1
1	1	1

 **NAND gate**	Those are the three basic gates (that's one way to count them). It is quite common to recognize two others as well: the *NAND* and the *NOR* gate. These two gates are simply combinations of an AND or an OR gate with a NOT gate. If you include these two gates, then the count rises to five. Here's the basic operation of NAND and NOR gates—you can see they are simply inversions of AND and OR gates:

NAND gate		
A	B	Q
0	0	1
0	1	1
1	0	1
1	1	0

(continued)

Table 6.5 (continued)

	NOR gate		
NOR gate	**A**	**B**	**Q**
	0	0	1
	0	1	0
	1	0	0
	1	1	0

The final two gates that are sometimes added to the list are the *XOR* and *XNOR* gates, also known as "exclusive or" and "exclusive nor" gates, respectively. Here are their tables

XOR gate		
A	**B**	**Q**
0	0	0
0	1	1
1	0	1
1	1	0

XNOR gate		
A	**B**	**Q**
0	0	1
0	1	0
1	0	0
1	1	1

XOR gate

The idea behind an XOR gate is "If either A **OR** B is 1, but **NOT** both, Q is 1." The reason why XOR might not be included in a list of gates is that you can implement it easily using the original three gates listed

XOR gate		
A	**B**	**Q**
0	0	0
0	1	1
1	0	1
1	1	0

XNOR gate

If you try all four different patterns for A and B and trace them through the circuit, you will find that Q behaves like an XOR gate. Since there is a well-understood symbol for XOR gates, it is generally easier to think of XOR as a "standard gate" and use it in the same way as AND and OR in circuit diagrams

XNOR gate		
A	**B**	**Q**
0	0	1
0	1	0
1	0	0
1	1	1

By looking at this table, you can see that you can implement Q with an XOR gate and CO (carry-out) with an AND gate. Simple.

What if you want to add two 8-bit bytes together? This becomes slightly harder. The easiest solution is to modularize the problem into **reusable components** and then replicate components. In this case, we need to create only one component: a **full binary adder**.

The difference between a full adder and the previous adder we looked at is that a full adder accepts an A and a B input plus a **carry-in** (CI) input. Once we have a full adder, then we can string eight of them together to create a byte-wide adder and cascade the carry bit from one adder to the next.

In the next section, we will look at how a full adder is implemented into a circuit [1].

6.5 Full Adders

The logic table for a full adder is slightly more complicated than the tables we have used before, because now we have **3 input bits**. It looks like this:

1-bit full adder with carry-in and carry-out

CI	A	B	Q	CO
0	0	0	0	0
0	0	1	1	0
0	1	0	1	0
0	1	1	0	1
1	0	0	1	0
1	0	1	0	1
1	1	0	0	1
1	1	1	1	1

There are many different ways that you might implement this table. I am going to present one method here that has the benefit of being easy to understand. If you look at the Q bit, you can see that the top 4 bits are behaving like an XOR gate with respect to A and B, while the bottom 4 bits are behaving like an XNOR gate with respect to A and B. Similarly, the top 4 bits of CO are behaving like an AND gate with respect to A and B, and the bottom 4 bits behave like an OR gate. Taking these facts, the following circuit implements a full adder: see Fig. 6.3.

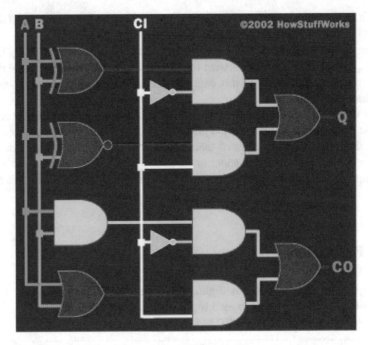

Fig. 6.3 Full adders can be implemented in a wide variety of ways [1]

This definitely is not the most efficient way to implement a full adder, but it is extremely easy to understand and trace through the logic using this method. If you are so inclined, see what you can do to implement this logic with fewer gates.

Now we have a piece of functionality called a "full adder." What a computer engineer then does is "black-box" it so that he or she can stop worrying about the details of the component. A black box for a full adder would look like this:

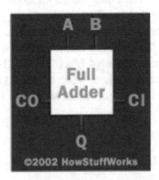

With that black box, it is now easy to draw a **4-bit full adder**.

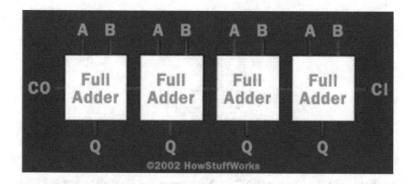

In this diagram, the carry-out from each bit feeds directly into the carry-in of the next bit over. A 0 is hardwired into the initial carry-in bit. If you input two 4-bit numbers on the A and B lines, you will get the 4-bit sum out on the Q lines, plus 1 additional bit for the final carry-out. You can see that this chain can extend as far as you like, through 8, 16, or 32 bits if desired.

The 4-bit adder we just created is called a **ripple-carry** adder. It gets that name because the carry bits "ripple" from one adder to the next. This implementation has the advantage of simplicity but the disadvantage of speed problems. In a real circuit, gates take time to switch states (the time is in the order of nanoseconds, but in high-speed computers, nanoseconds matter). Therefore, 32-bit or 64-bit ripple-carry adders might take 100–200 ns to settle into their final sum, because of carry ripple.

For this reason, engineers have created more advanced adders called **carry-lookahead** adders. The number of gates required to implement carry-lookahead is large, but the settling time for the adder is much better.

There are more to logical gate, and they can be found on the web site of Marshall Brain [1].

6.6   Truth Tables

The rules for combining expressions are usually written down as tables listing all of the possible outcomes. These are called truth tables, and for the three fundamental operators, these are:

P	Q	P AND Q
F	F	F
F	T	F
T	F	F
T	T	T

P	Q	P OR Q
F	F	F
F	T	T
T	F	T
T	T	T

P	NOT P
F	T
T	F

Notice that while the Boolean AND is the same as the English use of the term, the Boolean OR is a little different.

When you are asked would you like "coffee OR tea," you are not expected to say yes to both!

In the Boolean case, however, "OR" most certainly includes both. When P is true and Q is true, the combined expression (P OR Q) is also true.

There is a Boolean operator that corresponds to the English use of the term "OR," and it is called the "exclusive or" written as EOR or XOR. Its truth table is:

P	Q	P XOR Q
F	F	F
F	T	T
T	F	T
T	T	F

This one really would stop you having both the tea and the coffee at the same time (notice the last line is true XOR true = false).

6.6.1   Practical Truth Tables

All this seems very easy, but what value has it?

It most certainly isn't a model for everyday reasoning except at the most trivial "coffee or tea" level.

We do use Boolean logic in our thinking; well politicians probably don't but that's another story, but only at the most trivially obvious level.

However, if you start to design machines that have to respond to the outside world in even a reasonably complex way, then you quickly discover that Boolean logic is a great help.

For example, suppose you want to build a security system which only works at night and responds to a door being opened. If you have a light sensor, you can treat this as giving off a signal that indicates the truth of the statement:

P = It is daytime.

Clearly NOT(P) is true when it is nighttime, and we have our first practical use for Boolean logic!

What we really want is something that works out the truth of the statement:

R = Burglary in progress

from P and

Q = Window open

A little raw thought soon gives the solution that

R = NOT(P) AND Q

That is, the truth of "Burglary in progress" is given by the following truth table:

P	Q	NOT(P)	NOT(P) AND Q
F	F	T	F
F	T	T	T
T	F	F	F
T	T	F	F

From this, you should be able to see that the alarm only goes off when it is nighttime and a window opens.

6.7 Summary

The term "Boolean," often encountered when doing searches on the Web (and sometimes spelled "boolean"), refers to a system of logical thought developed by the English mathematician and computer pioneer George Boole (1815–1864). In Boolean searching, an "AND" operator between two words or other values (e.g., "pear AND apple") means one is searching for documents containing both of the words or values, not just one of them. An "OR" operator between two words or other values (e.g., "pear OR apple") means one is searching for documents containing either of the words.

In computer operation with binary values, Boolean logic can be used to describe electromagnetically charged memory locations or circuit states that are either charged (1 or true) or not charged (0 or false). The computer can use an AND gate or an OR gate operation to obtain a result that can be used for further processing. The following table shows the results from applying AND and OR operations to two compared states:

0 AND 0 = 0	1 AND 0 = 0	1 AND 1 = 1
0 OR 0 = 0	0 OR 1 = 1	1 OR 1 = 1

For a summary of logic operations in computers, see logic gate in above sections of this chapter.

Reference

1. Brain, M. *How bits and bytes work*. Retrieved from http://computer.howstuffworks.com/bytes. htm.

Chapter 7
What Is Fuzzy Logic and How It Works

The idea of fuzzy logic was first advanced by Dr. Lotfi Zadeh of the University of California at Berkeley in the 1960s. Dr. Zadeh was working on the problem of computer understanding of natural language. Natural language (like most other activities in life and indeed the universe) is not easily translated into the absolute terms of 0 and 1. (Whether everything is ultimately describable in binary terms is a philosophical question worth pursuing, but in practice much data we might want to feed a computer is in some state in between and so, frequently, are the results of computing.) It may help to see fuzzy logic as the way reasoning really works, and binary or Boolean logic is simply a special case of it.

7.1 Introduction

Fuzzy logic (FL) is an approach to computing based on "degrees of truth" rather than the usual "true or false" (1 or 0) Boolean logic on which the modern computer is based.

Fuzzy logic includes 0 and 1 as extreme cases of truth (or "the state of matters" or "fact") and also includes the various states of truth in between so that, for example, the result of a comparison between two things could be not "tall" or "short" but "0.38 of tallness."

Fuzzy logic seems closer to the way our brains work. We aggregate data and form a number of partial truths, which we aggregate further into higher truths, which in turn, when certain thresholds are exceeded, cause certain further results such as motor reaction. A similar kind of process is used in neural networks, expert systems, and other artificial intelligence (AI) applications. In one short statement, *fuzzy logic is a logic that is centered on multitier defaming and looks at aggregation of data and information, which is partially true or partially false.*

Each of these cases is described as follows:

© Springer International Publishing AG 2017
B. Zohuri, M. Moghaddam, *Business Resilience System (BRS): Driven Through
Boolean, Fuzzy Logics and Cloud Computation*,
DOI 10.1007/978-3-319-53417-6_7

1. **Neural Networks**

   A neural network usually involves a large number of processors operating in parallel and arranged in tiers. The first tier receives the raw input information—analogous to optic nerves in human visual processing. Each successive tier receives the output from the tier preceding it, rather than from the raw input—in the same way neurons further from the optic nerve receive signals from those closer to it. The last tier produces the output of the system.

   Each processing node has its own small sphere of knowledge, including what it has seen and any rules it was originally programmed with or developed for itself. The tiers are highly interconnected, which means each node in tier n will be connected to many nodes in tier $n - 1$—its inputs—and in tier $n + 1$, which provides input for those nodes. There may be one or multiple nodes in the output layer, from which the answer it produces can be read.

   Neural networks are notable for being adaptive, which means they modify themselves as they learn from initial training, and subsequent runs provide more information about the world. The most basic learning model is centered on weighting the input streams, which is how each node weights the importance of input from each of its predecessors. Inputs that contribute to getting right answers are weighted higher.

2. **Expert Systems**

   Typically, an expert system incorporates a knowledge base containing accumulated experience and an inference or rules engine—a set of rules for applying the knowledge base to each particular situation that is described to the program. The system's capabilities can be enhanced with additions to the knowledge base or to the set of rules. Current systems may include machine learning capabilities that allow them to improve their performance based on experience, just as humans do.

   The concept of expert systems was first developed in the 1970s by Edward Feigenbaum, professor and founder of the Knowledge Systems Laboratory at Stanford University. Feigenbaum explained that the world was moving from data processing to "knowledge processing," a transition, which was being enabled by new processor technology and computer architectures.

   Expert systems have played a large role in many industries including in financial services, telecommunications, healthcare, customer service, transportation, video games, manufacturing, aviation, and written communication. Two early expert systems broke ground in the healthcare space for medical diagnoses: Dendral, which helped chemists identify organic molecules, and MYCIN, which helped to identify bacteria such as bacteremia and meningitis and to recommend antibiotics and dosages.

3. **Artificial Intelligence**

   AI (pronounced AYE-EYE) or artificial intelligence is the simulation of human intelligence processes by machines, especially computer systems. These processes include learning (the acquisition of information and rules for using the information), reasoning (using the rules to reach approximate or definite conclusions), and self-correction. Particular applications of AI include:

- Expert systems
- Speech recognition
- Machine vision

Fuzzy logic is essential to the development of humanlike capabilities for AI, sometimes referred to as artificial general intelligence: the representation of generalized human cognitive abilities in software so that, faced with an unfamiliar task, the AI system could find a solution.

7.2   What Is Fuzzy Logic

The fuzzy logic was first introduced to the world by Persian Professor Dr. Lotfi Zadeh in 1965 when he was developing the theory of fuzzy sets. While the fuzzy logic theory can be quite complex, the scope of this book is not to cover the complete fuzzy logic theory and functions since many books are written for that purpose.

Therefore, we have tried to shed some light on the aspects of fuzzy logic, which can be used for our specific purpose, and to show the application of fuzzy logic for our specific purpose which is the Business Resilience System (BRS).

So, let us see what Boolean logic is first: this concept was defined in Chap. 6 of the book and it is summarized here again.

A computer system or more precisely the arithmetic logic unit (ALU), which is part of the central processing unit (CPU), uses Boolean logic to perform all kinds of calculations and computations. To do that, electrical elements called "registers" are implemented inside ALU. Registers hold data required performing processes and they have two parts:

1. **Keys**
2. **Values**

A key has a value of *ONE* if it is turned *ON* or a value of *ZERO* when it is turned off. These values represent *TRUE* or *FALSE* or *YES* or *NO* in human language once they are interpreted by the operating system and other application programs. Therefore, Boolean logic (0 or 1) values are transformed into meaningful expressions by various programs installed on a computer to make all these complex calculations more understandable for us.

So, everything in a computer system works based on logical and mathematical calculations (Boolean logic) resulted from millions of comparisons between *ZEROs* and *ONEs*!

But in the real world, we cannot judge everything on a black and white or yes or no basis. The same way that we have different shades of grey between black and white, we have different values or degrees of agreement or disagreement between yes or no and zero or one.

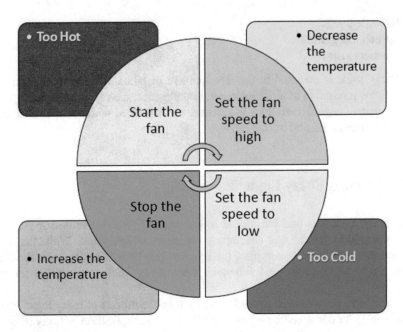

Fig. 7.1 Illustration of temperature control system via fuzzy logic

That is where the fuzzy logic (FL) fills the gap and can be used for more precise calculations and analysis.

A good example in real life for a fuzzy value is a device called thermostat. Without thermostat we can either turn the air condition on or off for a while because it either becomes absolutely cold or absolutely warm! Thanks to thermostat, we can have degrees of different temperatures set on our air conditioning system to provide us with the best desirable temperature.

The following chart depicted in Fig. 7.1 shows how a fuzzy logic works on a temperature control system:

Now that we have a simple introduction to the fuzzy logic and its usage in application of temperature control system, we can expand on it a little further on same application, by looking at FL in a role as an AI. Fuzzy logic for most of us is not as fuzzy as you might think and has been working quietly behind the scenes for years. Fuzzy logic is a rule-based system that can rely on the practical experience of an operator, particularly useful to capture experienced operator knowledge. Here is what you need to know.

Fuzzy logic has been working quietly behind the scenes for more than 20 years in more places than most admit. Fuzzy logic is a rule-based system that can rely on the practical experience of an operator, particularly useful to capture experienced operator knowledge. Fuzzy logic is a form of artificial intelligence software; therefore, it would be considered a subset of AI. Since it is performing a form of decision-making, it can be loosely included as a member of the AI software tool kit. Here is what you need to know to consider using fuzzy logic to help solve your next application. It is not as fuzzy as you might think.

Fuzzy logic has been around since the mid-1960s; however, it was not until the 1970s that a practical application was demonstrated. Since that time the Japanese have traditionally been the largest producer of fuzzy logic applications. Fuzzy logic has appeared in cameras, washing machines, and even in stock trading applications. In the last decade, the United States has started to catch on to the use of fuzzy logic. There are many applications that use fuzzy logic, but fail to tell us of its use. Probably, the biggest reason is that the term "fuzzy logic" may have a negative connotation.

Fuzzy logic can be applied to nonengineering applications as illustrated in the stock trading application. It has also been used in medical diagnosis systems and in handwriting recognition applications. In fact a fuzzy logic system can be applied to almost any type of system that has inputs and outputs.

Fuzzy logic systems are well suited to nonlinear systems and systems that have multiple inputs and multiple outputs. Any reasonable number of inputs and outputs can be accommodated. Fuzzy logic also works well when the system cannot be modeled easily by conventional means.

Many engineers are afraid to dive into fuzzy logic due to a lack of understanding. Fuzzy logic does not have to be hard to understand, even though the math behind it can be intimidating, especially to those of us who have not been in a math class for many years.

Binary logic is either 1 or 0. Fuzzy logic is a continuum of values between 0 and 1. This may also be thought of as 0–100%. An example is the variable *YOUNG*. We may say that age 5 is 100% *YOUNG*, 18 is 50% *YOUNG*, and 30 is 0% *YOUNG*. In the binary world, everything below 18 would be 100% *YOUNG*, and everything above would be 0% *YOUNG*.

The design of a fuzzy logic system starts with a set of membership functions for each input and a set for each output. A set of rules is then applied to the membership functions to yield a "crisp" output value.

For this process control explanation of fuzzy logic, *TEMPERATURE* is the input and *FAN SPEED* is the output. Create a set of membership functions for each input. A membership function is simply a graphical representation of the fuzzy variable sets. For this example, use three fuzzy sets, *COLD, WARM*, and *HOT*. We will then create a membership function for each of three sets of temperature as shown in the cold–normal–hot graphic, Fig. 7.2.

We will use three fuzzy sets for the output, *SLOW, MEDIUM*, and *FAST*. A set of functions is created for each output set just as for the input sets. It should be noted that the shape of the membership functions do not need to be triangles as we have used in Figs. 7.2 and 7.3. Various shapes can be used, such as trapezoid, Gaussian, sigmoid, as well as user definable. By changing the shape of the membership function, the user can tune the system to provide optimum response.

Now that we have our membership functions defined, we can create the rules that will define how the membership functions will be applied to the final system. We will create three rules for this system:

- If *HOT* then *FAST*
- If *WARM* then *MEDIUM*
- If *COLD* then *SLOW*

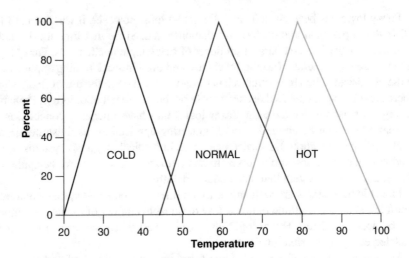

Fig. 7.2 Illustration of three sets of temperatures

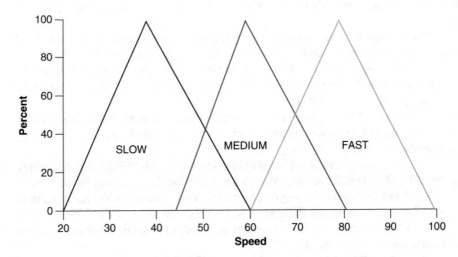

Fig. 7.3 Different forms of three sets of temperature illustrations

The rules are then applied to the membership functions to produce the "crisp" output value to drive the system. For simplicity, we will illustrate using only two input and two output functions.

For an input value of 52°, we intersect the membership functions. We see that in this example the intersection will be on both functions, thus two rules are applied. The intersection points are extended to the output functions to produce an intersecting point. The output functions are then truncated at the height of the intersecting points. The area under the curves for each membership function is then added to give us a total area. The centroid of this area is calculated. The output value is then the centroid

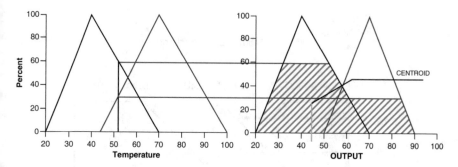

Fig. 7.4 Illustration of three sets of temperature processes

value. In this example, 44% is the output *FAN SPEED* value. This process is illustrated in Fig. 7.4.

This is a very simple explanation of how the fuzzy logic systems work. In a real working system, there would be many inputs and possibility of several outputs. This would result in a fairly complex set of functions and many more rules. It is not uncommon for there to be 40 or more rules in a system. Even so, the principles remain the same as in our simple system.

National Instruments has included in LabVIEW© a set of pallet functions and a fuzzy system designer to greatly simplify the task of building a fuzzy logic system. It has included several demo programs in the examples to get started. In the graphical environment, the user can easily see what the effects are as the functions and rules are built and changed.

The user should remember that a fuzzy logic system is not a "silver bullet" for all control system needs. Traditional control methods are still very much a viable solution. In fact, they may be combined with fuzzy logic to produce a dynamically changing system. The validation of a fuzzy logic system can be difficult due to the fact that it is a nonformal system. Its use in safety systems should be considered with care.

We hope this short description will inspire the exploration and use of fuzzy logic in some of your future designs. We encourage the reader to do further study on the subject. There are numerous books and articles that go into much more detail. This serves as a simple introduction to fuzzy logic controls (FLC).

7.3   Fuzzy Logic and Fuzzy Sets

Fuzzy mathematics involves in general three operations as follows:

1. **Fuzzification**: Translation from real-world values to fuzzy values.

    It makes the translation from real-world values to fuzzy world values using membership functions. The membership functions in Fig. 7.5 translate a speed = 55 into fuzzy values (degree of membership) *SLOW* = 0.25, *MEDIUM* = 0.75, and *FAST* = 0.

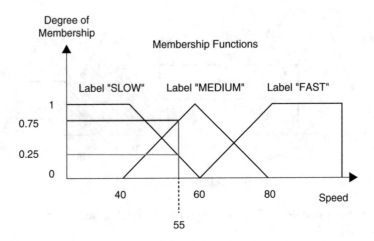

Fig. 7.5 Illustration of fuzzification

2. **Rule evaluation**: Computing rule strengths based on rules and inputs.

Suppose *SLOW* = 0.25 and *FAR* = 0.82. The rule strength will be 0.25 (the minimum value of the antecedents) and the fuzzy variable *INCREASE* would be also 0.25.

Consider now another rule: If *SPEED = MEDIUM* and *HIGHER = SECURE*, then *GAS = INCREASE*.

Let be in this case, *MEDIUM* = 0.75 and *SECURE* = 0.5. Now the rule strength will be 0.5 (the minimum value of the antecedents) and the fuzzy variable *INCREASE* would be also 0.5.

Therefore, we have two rules involving fuzzy variable *INCREASE*. The "fuzzy OR" of the two rules will be 0.5 (the maximum value between the two proposed values).

INCREASE = 0.5

3. **Defuzzification**: Translate results back to the real-world values.

After computing the fuzzy rules and evaluating the fuzzy variables, we will need to translate these results back to the real world. We need now a membership function for each output variable like in Fig. 7.6.

Let the fuzzy variables be *DECREASE* = 0.2, *SUSTAIN* = 0.8, and *INCREASE* = 0.5.

Each membership function will be clipped to the value of the correspondent fuzzy variable as shown in Fig. 7.7.

A new output membership function is built, taking for each point in the horizontal axis the maximum value between the three membership values. The result is shown in Fig. 7.8.

To complete the defuzzification process, all we have to do now is find an equilibrium point. One way to do this is with "center of gravity (COG)" method:

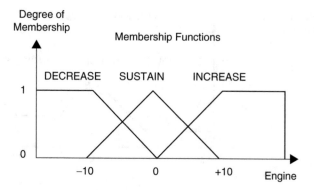

Fig. 7.6 Illustration of defuzzification

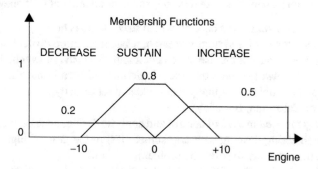

Fig. 7.7 Illustration of correspondent fuzzy variable

$$COG = \frac{\int_a^b F(x) \cdot x\, dx}{\int_a^b F(x)\, dx} \qquad \text{(Eq. 7.1)}$$

This in our example will give us approximately the following COG:

$$COG \approx 2.6 \qquad \text{(Eq. 7.2)}$$

In summary, the fuzzy logic and sets provide the following conclusions as:

- This theory lets us handle and process information in a similar way as the human brain does.
- We communicate and coordinate actions with data like "… you are too young to do that …" How much does "too" refer to what's "young"?
- With fuzzy sets, we may define subsets in a fashion that any element may be part of them in different degrees.
- With fuzzy rules, it's possible to compute relationships between fuzzy variables and produce fuzzy outputs.

Fig. 7.8 Result of three
membership values

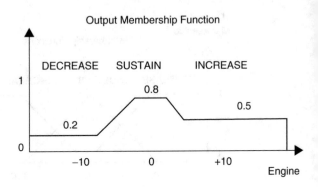

- Moreover, guess what … from these fuzzy output values, we may build Boolean and continuous quantities, like a switch status or an amount of money.

Moreover, following is the base on which fuzzy logic is built:

As the complexity of a system increases, it becomes more difficult and eventually impossible to make a precise statement about its behavior, eventually arriving at a point of complexity where the fuzzy logic method born in humans is the only way to get at the problem(originally identified and set forth by Lotfi A. Zadeh, PhD, University of California, Berkeley).

Fuzzy logic is used in system control and analysis design, because it shortens the time for engineering development and sometimes, in the case of highly complex systems, is the only way to solve the problem.

The following chapters of this book attempt to explain for us "just plain folks" how the "fuzzy logic method born in humans" is used to evaluate and control complex systems. Although most of the time we think of "control" as having to do with controlling a physical system, there is no such limitation in the concept as initially presented by Dr. Zadeh. Fuzzy logic can apply also to economics, psychology, marketing, weather forecasting, biology, and politics … to any large complex system.

The term "fuzzy" was first used by Dr. Lotfi Zadeh in the engineering journal, *Proceedings of the IRE*, a leading engineering journal, in 1962. Dr. Zadeh became, in 1963, the chairman of the Electrical Engineering Department of the University of California at Berkeley. That is about as high as you can go in the electrical engineering field. Dr. Zadeh's thoughts are not to be taken lightly.

Fuzzy logic is not the wave of the future. It is now! There are already hundreds of millions of dollars of successful, fuzzy logic-based commercial products, everything from self-focusing cameras to washing machines that adjust themselves according to how dirty the clothes are, automobile engine controls, antilock braking systems, color film developing systems, subway control systems, and computer programs trading successfully in the financial markets.

Note that when you go searching for fuzzy logic applications in the United States, it is difficult to impossible to find a control system acknowledged as based on fuzzy logic. Just imagine the impact on sales if General Motors announced their antilock braking was accomplished with fuzzy logic! The general public is not ready for such an announcement maybe.

It should be noted there is controversy and criticism regarding fuzzy logic. One must read various sides of the controversy and reach their own conclusion. Personally, the author, who has been both praised and reviled for his writings regarding fuzzy logic, feels the critics are too rigid in their grasp of the universe and "just don't get it." Nevertheless, do not take my word for it. You must look at all sides and make up your own mind.

7.4 The Fuzzy Logic Method

The fuzzy logic analysis and control method are, therefore:

1. Receiving of one or a large number of measurements or other assessment of conditions existing in some system we wish to analyze or control.
2. Processing all these inputs according to human-based fuzzy "If-Then" rules, which can be expressed in plain language words, in combination with traditional non-fuzzy processing.
3. Averaging and weighting the resulting outputs from all the individual rules into one single output decision or signal, which decides what to do or tells a controlled system what to do. The output signal eventually arrived at is a precise appearing, defuzzified, "crisp" value. Please see the following Fuzzy Logic Control/Analysis Method diagram as shown in Fig. 7.9:

7.4.1 Fuzzy Perception

A fuzzy perception is an assessment of a physical condition that is not measured with precision, but is assigned an intuitive value. In fact, the fuzzy logic that people assert everything in the universe is a little fuzzy, no matter how good your measuring equipment is. It will be seen below that fuzzy perceptions can serve as a basis for processing and analysis in a fuzzy logic control system.

Measured, non-fuzzy data is the primary input for the fuzzy logic method. Examples are temperature measured by a temperature transducer, motor speed, economic data, financial market data, etc. It would not be usual in an electromechanical control system or a financial or economic analysis system, but humans with their fuzzy perceptions could also provide input. There could be a human "in the loop."

In the fuzzy logic literature, you will see the term "fuzzy set." A fuzzy set is a group of anything that cannot be precisely defined. Consider the fuzzy set of "old houses." How old is an old house? Where is the dividing line between new houses and old houses? Is a 15-year-old house an old house? How about 40 years? What about 39.9 years? The assessment is in the eyes of the beholder.

Other examples of fuzzy sets are tall women, short men, warm days, high pressure gas, small crowd, medium viscosity, hot shower water, etc.

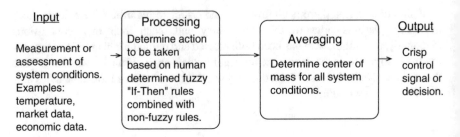

Fig. 7.9 The fuzzy logic control analysis method

When humans are the basis for an analysis, we must have a way to assign some rational value to intuitive assessments of individual elements of a fuzzy set. We must translate from human fuzziness to numbers that can be used by a computer. We do this by assigning assessment of conditions a value from 0 to 1.0. For "how hot the room is," the human might rate it at 0.2 if the temperature were below freezing, and the human might rate the room at 0.9, or even 1.0, if it is a hot day in summer with the air conditioner off.

You can see these perceptions are fuzzy, just intuitive assessments, not precisely measured facts.

By making fuzzy evaluations, with zero at the bottom of the scale and 1.0 at the top, we have a basis for analysis rules for the fuzzy logic method, and we can accomplish our analysis or control project. The results seem to turn out well for complex systems or systems where human experience is the only base from which to proceed, certainly better than doing nothing at all, which is where we would be if unwilling to proceed with fuzzy rules.

7.4.2 Novices Can Beat the Pros

Novices using personal computers and the fuzzy logic method can beat PhD mathematicians using formulas and conventional programmable logic controllers. Fuzzy logic makes use of human common sense. This common sense is either applied from what seems reasonable, for a new system, or from experience, for a system that has previously had a human operator.

Here is an example of converting human experience for use in a control system: I read of an attempt to automate a cement manufacturing operation. Cement manufacturing is a lot more difficult than you would think. Through the centuries it has evolved with human "feel" being absolutely necessary. Engineers were not able to automate with conventional control. Eventually, they translated the human "feel" into lots and lots of fuzzy logic "If-Then" rules based on human experience. Reasonable success was, thereby, obtained in automating the plant.

Objects of fuzzy logic analysis and control may include physical control, such as machine speed, or operating a cement plant; financial and economic decisions;

psychological conditions; physiological conditions; safety conditions; security conditions; production improvement; and much more.

This book will talk about fuzzy logic in control applications—controlling machines, physical conditions, processing plants, etc. It should be noted that when Dr. Zadeh invented fuzzy logic, it appears he had in mind applying fuzzy logic in many applications in addition to controlling machines, such as economics, politics, biology, etc.

Thank you Wozniak (Apple computer), Jobs (Apple computer), Gates (Microsoft), and Ed Roberts (the MITS, Altair entrepreneur) for the personal computer.

The availability of the fuzzy logic method to us "just plain folks" has been made possible by the availability of the personal computer. Without personal computers, it would be difficult to use fuzzy logic to control machines and production plants or do other analyses. Without the speed and versatility of the personal computer, we would never undertake the laborious and time-consuming tasks of fuzzy logic-based analyses, and we could not handle the complexity of speed requirement and endurance needed for machine control.

You can do far more with a simple fuzzy logic BASIC or C++ program in a personal computer running in conjunction with a low-cost input/output controller than with a whole array of expensive, conventional, programmable logic controllers.

Programmable logic controllers have their place! They are simple, reliable, and keep American industry operating where the application is relatively simple and on-off in nature.

For a more complicated system control application, an optimum solution may be patching things together with a personal computer and fuzzy logic rules, especially if the project is being done by someone who is not a professional control system engineer.

7.4.3 A Milestone Passed for Intelligent Life on Earth

If intelligent life has appeared anywhere in the universe, "they" are probably using fuzzy logic. It is a universal principle and concept. Becoming aware of defining and starting to use fuzzy logic is an important moment in the development of an intelligent civilization. On earth, we have just arrived at that important moment. You need to know and begin using fuzzy logic.

7.4.4 Fuzzy Logic Terms Found in Books and Articles

The discussion so far does not adequately prepare us for reading and understanding most books and articles about fuzzy logic, because of the terminology used by sophisticated authors. The following are explanations of some terms, which should help in this regard. This terminology was initially established by Dr. Zadeh when he originated the fuzzy logic concept.

Fuzzy—The degree of fuzziness of a system analysis rule can vary between being very precise, in which case we would not call it "fuzzy," and being based on an opinion held by a human, which would be "fuzzy." Being fuzzy or not fuzzy, therefore, has to do with the degree of precision of a system analysis rule.

A system analysis rule need not be based on human fuzzy perception. For example, you could have a rule, "If the boiler pressure rises to a danger point of 600 Psi as measured by a pressure transducer, then turn everything off. That rule is not "fuzzy."

Principle of incompatibility (previously stated; repeated here):

- As the complexity of a system increases, it becomes more difficult and eventually impossible to make a precise statement about its behavior, eventually arriving at a point of complexity where the fuzzy logic method born in humans is the only way to get at the problem.
- Fuzzy sets—A fuzzy set is almost any condition for which we have words: short men, tall women, hot, cold, new buildings, accelerator setting, ripe bananas, high intelligence, speed, weight, spongy, etc., where the condition can be given a value between 0 and 1. Example: A woman is 6 ft, 3 in. tall. In my experience, I think she is one of the tallest women I have ever met, so I rate her height at 0.98. This line of reasoning can go on indefinitely rating a great number of things between 0 and 1.
- Degree of membership—The degree of membership is the placement in the transition from 0 to 1 of conditions within a fuzzy set. If a particular building's placement on the scale is a rating of 0.7 in its position in newness among new buildings, then we say its degree of membership in new buildings is 0.7.
- In fuzzy logic method control systems, degree of membership is used in the following way. A measurement of speed, for example, might be found to have a degree of membership in "too fast" of 0.6 and a degree of membership in "no change needed" of 0.2. The system program would then calculate the center of mass between "too fast" and "no change needed" to determine feedback action to send to the input of the control system. This is discussed in more detail in subsequent chapters.
- Summarizing information—Human processing of information is not based on two-valued, off-on, either-or logic. It is based on fuzzy perceptions, fuzzy truths, fuzzy inferences, etc., all resulting in an averaged, summarized, normalized output, which is given by the human a precise number or decision value, which he or she verbalizes, writes down, or acts on. It is the goal of fuzzy logic control systems to also do this.
- The input may be large masses of data, but humans can handle it. The ability to manipulate fuzzy sets and the subsequent summarizing capability to arrive at an output we can act on is one of the greatest assets of the human brain. This characteristic is the big difference between humans and digital computers. Emulating this human ability is the challenge facing those who would create computer-based

artificial intelligence. It is proving very, very difficult to program a computer to have humanlike intelligence.

- Fuzzy variable—Words like red, blue, etc. are fuzzy and can have many shades and tints. They are just human opinions, not based on precise measurement in angstroms. These words are fuzzy variables.
- If, for example, speed of a system is the attribute being evaluated by "fuzzy" rules, then "speed" is a fuzzy variable.
- Linguistic variable—Linguistic means relating to language, in our case plain language words.
- Speed is a fuzzy variable. Accelerator setting is a fuzzy variable. Examples of linguistic variables are somewhat fast speed, very high speed, real slow speed, excessively high accelerator setting, accelerator setting about right, etc.
- A fuzzy variable becomes a linguistic variable when we modify it with descriptive words, such as somewhat fast, very high, real slow, etc.
- The main function of linguistic variables is to provide a means of working with the complex systems mentioned above as being too complex to handle by conventional mathematics and engineering formulas.
- Linguistic variables appear in control systems with feedback loop control and can be related to each other with conditional "if-then" statements. Example: If the speed is too fast, then back off on the high accelerator setting.
- Universe of discourse—Let us make women the object of our consideration. All the women everywhere would be the universe of women. If we choose to discourse about (talk about) women, then all the women everywhere would be our universe of discourse.
- Universe of discourse, then, is a way to say all the objects in the universe of a particular kind, usually designated by one word, that we happen to be talking about or working with in a fuzzy logic solution.
- A universe of discourse is made up of fuzzy sets. Example: The universe of discourse of women is made up of professional women, tall women, Asian women, short women, beautiful women, and on and on.
- Fuzzy algorithm—An algorithm is a procedure, such as the steps in a computer program. A fuzzy algorithm, then, is a procedure, usually a computer program made up of statements relating linguistic variables.

7.5   The World's First Fuzzy Logic Controller

In England in 1973 at the University of London, a professor and student were trying to stabilize the speed of a small steam engine the student had built. They had a lot going for them, sophisticated equipment like a PDP-8 minicomputer and conventional digital control equipment. But, they could not control the engine as well as they wanted. Engine speed would either overshoot the target speed and arrive at the target speed after a series of oscillations or the speed control would be too sluggish, taking too long for the speed to arrive at the desired setting, as in Fig. 7.10.

Fig. 7.10 System response
without fuzzy logic
controller

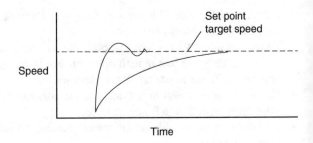

Fig. 7.11 System response
using fuzzy logic controller
[1]

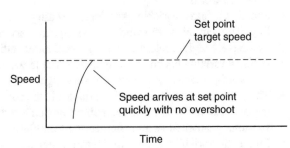

The professor, E. Mamdani, had read of a control method proposed by Dr. Lotfi Zadeh, head of the Electrical Engineering Department at the University of California at Berkeley, in the United States. Dr. Zadeh is the originator of the designation "fuzzy," which everyone suspects was selected to throw a little "pie in the face" of his more orthodox engineering colleagues, some of whom strongly opposed the fuzzy logic concept under any name.

Professor Mamdani and the student, S. Assilian, decided to try fuzzy logic. They spent a weekend setting their steam engine up with the world's first ever fuzzy logic control system and went directly into the history books by harnessing the power of a force in use by humans for 3 million years, but never before defined and used for the control of machines.

The controller worked right away and worked better than anything they had done with any other method. The steam engine speed control graph using the fuzzy logic controller appeared as in Fig. 7.11.

As you can see, the speed approached the desired value very quickly, did not overshoot, and remained stable. It was an exciting and important moment in the history of scientific development.

The Mamdani project made use of four inputs: boiler pressure error (how many temperature degrees away from the set point), rate of change of boiler pressure error, engine speed error, and rate of change of engine speed error. There were two outputs: control of heat to the boiler and control of the throttle. They operated independently.

Note: A fuzzy logic system does not have to include a continuous feedback control loop as in the above, described Mamdani system in order to be a fuzzy logic system, an impression you might receive from reading much of the fuzzy logic literature.

There could be continuous feedback loop control, a combination of feedback loop control and on-off control or on-off control alone.

A fuzzy logic control system could be as simple as: "If the motor temperature feels like it is too hot, turn the motor off and leave it off." Or "If the company's president and all the directors just sold every share of stock they own, then WE sell!"

A fuzzy logic system does not have to be directed toward electromechanical systems. The fuzzy logic system could, for example, provide buy-sell decisions to trade 30 million US dollars against the Japanese yen.

Fuzzy logic controllers can control solenoids, stepper motors, linear positioners, etc., as well as, or concurrently with, continuous feedback control loops. Where there is continuous feedback control of a control loop, the response for varying degrees of error can be nonlinear, tailoring the response to meet unique or experience determined system requirements, even anomalies.

Controllers typically have several inputs and outputs. The handling of various tasks, such as monitoring and commanding a control loop, and monitoring various inputs, with commands issued as appropriate, would all be sequenced in the computer program. The program would step from one task to the other, the program receiving inputs from and sending commands to the converter/controller.

Inputs for a fuzzy logic-controlled mechanical/physical system could be derived from any of thousands of real-world physical sensors/transducers. The Thomas Register has over 110 pages of these devices. Inputs for financial trading could come from personal assessments or from an ASCII data communication feed provided by a financial market quote service.

7.5.1   Progress in Fuzzy Logic

From a slow beginning, fuzzy logic grew in applications and importance, until now it is a significant concept worldwide. Intelligent beings on the other side of our galaxy and throughout the universe have probably noted and defined the concept.

Personal computer-based fuzzy logic control is pretty amazing. It lets novices build control systems that work in places where even the best mathematicians and engineers, using conventional approaches to control, cannot define and solve the problem.

A control system is an electronic or mechanical system that causes the output of the controlled system to automatically remain at some desired output (the "set point") set by the operator. The thermostat on your air conditioner is a control system. Your car's cruise control is a control system. Control may be an on-off signal or a continuous feedback loop.

In Japan, a professor built a fuzzy logic control system that will fly a helicopter with one of the rotor blades off! Human helicopter pilots cannot do that. Moreover, the Japanese went further and built a fuzzy logic-controlled subway that is as smooth as walking in your living room. You do not have to hang on to a strap to keep your balance. If you did not look out the window at things flashing by, you would hardly know you had started and were in motion.

In the United States, fuzzy logic control is gaining popularity but is not as widely used as in Japan, where it is a multimillion dollar industry. Japan sells fuzzy logic-controlled cameras, washing machines, and more. One Internet search engine returns over 16,000 pages when you search on "fuzzy + logic."

Personal computer-based fuzzy logic control follows the pattern of human "fuzzy" activity. However, humans usually receive process and act on more inputs than the typical computer-based fuzzy logic controller. (This is not necessarily so; a computer-based fuzzy logic control system in Japan trades in the financial markets and utilizes 800 inputs.) [1]

7.5.2   Fuzzy Logic Control Input: Human and Computer

Computer-based fuzzy logic machine control is like human fuzzy logic control, but there is a difference when the nature of the computer's input is considered.

Humans evaluate input from their surroundings in a fuzzy manner, whereas machines/computers obtain precise appearing values, such as 112 °F, obtained with a transducer and an analog to digital converter. The computer input would be the computer measuring, let us say, 112 °F. The human input would be a fuzzy feeling of being too warm.

The human says, "The shower water is too hot." The computer as a result of analog input measurement says, "The shower water is 112 °F and 'If–Then' statements in my program tell me the water is too warm." A human says, "I see two tall people and one short one." The computer says, "I measure two people, 6′ 6″ and 6′ 9″, respectively, and one person 5′ 1″ tall, and 'If–Then' statements in my program tell me there are two tall people and one short person."

Even though transducer-derived measured inputs for computers appear to be more precise, from the point of input forward, we still use them in a fuzzy logic method approach that follows our fuzzy, human approach to control.

For a human, if the shower water gets too warm, the valve handle is turned to make the temperature go down a little. For a computer, an "If-Then" statement in the program would initiate the lowering of temperature based on a human provided "If-Then" rule, with a command output operating a valve [1].

7.5.3   More About How Fuzzy Logic Works

To create a personal computer-based fuzzy logic control system, we:

1. Determine the inputs.
2. Describe the cause and effect action of the system with "fuzzy rules" stated in plain English words.

3. Write a computer program to act on the inputs and determine the output, considering each input separately. The rules become "If-Then" statements in the program. (As will be seen below, where feedback loop control is involved, use of graphical triangles can help visualize and compute this input-output action.)
4. In the program, use a weighted average to merge the various actions called for by the individual inputs into one crisp output acting on the controlled system. (In the event there is only one output, then merging is not necessary, only scaling the output as needed.)

The fuzzy logic approach makes it easier to conceptualize and implement control systems. The process is reduced to a set of visualizable steps. This is a very important point. Actually implementing a control system, even a simple control system, is more difficult than it appears. Unexpected aberrations and physical anomalies inevitably occur. Getting the process working correctly ends up being a "cut and try" effort.

Experienced, professional digital control engineers using conventional control might know how to proceed to fine-tune a system. However, it can be difficult for us just plain folks. Fuzzy logic control makes it easier to visualize and set up a system and proceed through the cut and try process. It is only necessary to change a few plain English rules resulting in changing a few numbers in the program.

In reading about fuzzy logic control applications in industry, one of the significant points that stand out is fuzzy logic is used because it shortens the time for engineering development. Fuzzy logic enables engineers to configure systems quickly without extensive experimentation and to make use of information from expert human operators who have been performing the task manually.

Perhaps your control need is something a lot more down to earth than flying helicopters or running subways. Maybe all you want to do is keep your small business sawmill running smoothly, with the wood changing and the blade sharpness changing. Or, perhaps you operate a natural gas compressor for some stripper wells that are always coming on and going off, and you need to have the compressor automatically adjust in order to stay on line and keep the suction pressure low to get optimum production. Perhaps you dream of a race car that would automatically adjust to changing conditions, the setup remaining optimum as effectively as the abovementioned helicopter adjusts to being without a rotor blade.

There are a million stories, and we cannot guess what yours is, but chances are, if there is something you want to control, and you are not an experienced, full-time, professional control engineer financed by a multimillion dollar corporation, fuzzy logic may be for you. If you are all those things, it still may be for you.

A conventional programmable logic controller monitors the process variable (the pressure, temperature, speed, etc. that we want to control). If it is too high, a decrease signal is sent out. If it is too low, an increase signal is sent out. This is effective up to a point. But, consider how much more effective a control system would be if we use a computer to calculate the rate of change of the process variable in addition to how far away it is from the set point. If the control system acts on both these inputs, we have a better control system. In addition, that could be just the beginning; we can have a large number of inputs: all is being analyzed according to common sense and experience rules for their contribution to the averaged crisp output controlling the system.

Further, whereas conventional control systems are usually smooth and linear in performance, we sometimes encounter aberrations or discontinuous conditions, something that does not make good scientific sense and cannot be predicted by a formula, but it's there. If this happens, the fuzzy logic method helps us visualize a solution, put the solution in words, and translate to "If–Then" statements, thereby obtaining the desired result. That is a very difficult thing to do with conventional programmable logic controllers (known as PLCs). PLCs are programmable but are far more limited than the program control available from a very simple BASIC program in a personal computer [1].

Fuzzy logic control is not based on mathematical formulas. This is a good thing, because, as easy as it might seem, it is difficult to impossible to write formulas that do what nature does. This is why novices using fuzzy logic can beat PhD mathematicians using formulas. Fuzzy logic control makes use of human common sense. This common sense is either applied from what seems reasonable, for a new system, or from experience, for a system that has previously had a human operator.

Some of the greatest minds in the technical world try to explain to others why fuzzy logic works, and other great technical minds contend that fuzzy logic is a "cop out." The experts really "go at" each other. However, for us just plain folks, the fact is fuzzy logic does work, seems to work better than many expensive and complicated systems, and is understandable and affordable.

7.6 Rationale for Fuzzy Logic

Fuzzy logic is about a mathematical technique for dealing with imprecise data and problems that have many solutions rather than one. Although it is implemented in digital computers which ultimately make only yes-no decisions, fuzzy logic works with ranges of values, solving problems in a way that more resembles human logic.

Fuzzy logic is used for solving problems with expert systems and real-time systems that must react to an imperfect environment of highly variable, volatile, or unpredictable conditions. It "smoothes the edges" so to speak, circumventing abrupt changes in operation that could result from relying on traditional either-or and all-or-nothing logic.

Fuzzy logic was conceived by Lotfi Zadeh, former chairman of the Electrical Engineering and Computer Science Department at the University of California at Berkeley. In 1964, while contemplating how computers could be programmed for handwriting recognition, Zadeh expanded on traditional set theory by making membership in a set a matter of degree rather than a yes-no situation.

To finally finish this chapter, we look at the rational for fuzzy logic that was stated by father of fuzzy logic, Professor Lotfi Zadeh himself, and these points are:

- In the evolution of science, a time comes when alongside the brilliant successes of a theory, **T**, what become visible are classes of problems, which fall beyond the reach of **T**. At that point, the stage is set for a progression from **T** to **T***—a generalization of **T**. Among the many historical examples are the transitions from Newtonian mechanics to quantum mechanics, from linear system theory to non-linear system theory, and from deterministic models to probabilistic models in economics and decision analysis.
- Fuzzy logic is a better approximation to reality.
- In this perspective, a fundamental point—a point, which is not as yet widely recognized—is that there are many classes of problems, which cannot be addressed by any theory, **T**, which is based on bivalent logic. The problem with bivalent logic is that it is in fundamental conflict with reality—a reality in which almost everything is a matter of degree.
- To address such problems, what is needed is a logic for modes of reasoning, which are approximate rather than exact. *This is what fuzzy logic is aimed at.*

Reference

1. Exis, LLC. Retrieved from http://www.fuzzy-logic.com/.

Chapter 8
Mathematics and Logic Behind Boolean and Fuzzy Computation

Boolean algebra (BA) was introduced by George Boole in his first book *The Mathematical Analysis of Logic* (1847) and set forth more fully in his *An Investigation of the Laws of Thought* (1854). According to Huntington, the term "Boolean algebra" was first suggested by Sheffer in 1913. Fuzzy logic is a form of many-valued logic in which the truth values of variables may be any real number between 0 and 1, considered to be "fuzzy." By contrast, in Boolean logic, the truth values of variables may only be the "crisp" values 0 or 1. Fuzzy logic has been employed to handle the concept of partial truth, where the truth value may range between completely true and completely false. Furthermore, when linguistic variables are used, these degrees may be managed by specific (membership) functions. The term fuzzy logic was introduced with the 1965 proposal of fuzzy set theory by Lotfi Zadeh. Fuzzy logic had however been studied since the 1920s, as infinite-valued logic—notably by Lukasiewicz and Tarski. In this chapter we look at both of these logics holistically; any extensive details are beyond the scope of this book, and we encourage our readers to refer to so many books and articles that can be found on the Internet or Amazon.

8.1 Mathematics of Boolean Logic and Algebra

Boolean algebra predated the modern developments in abstract algebra and mathematical logic. It is, however, seen as connected to the origins of both fields. In an abstract setting, Boolean algebra was perfected in the late nineteenth century by Jevons, Schröder, Huntington, and others until it reached the modern conception of an (abstract) mathematical structure. For example, the empirical observation that one can manipulate expressions in the algebra of sets by translating them into expressions in Boolean algebra is explained in modern terms by saying that the algebra of sets is a Boolean algebra (note the indefinite article). In fact, M. H. Stone proved in

© Springer International Publishing AG 2017   221
B. Zohuri, M. Moghaddam, *Business Resilience System (BRS): Driven Through Boolean, Fuzzy Logics and Cloud Computation*,
DOI 10.1007/978-3-319-53417-6_8

1936 that every Boolean algebra is isomorphic to a field of sets. Instead of elementary algebra where the values of the variables are numbers and the main operations are addition and multiplication, the main operations of Boolean algebra are the conjunction and denoted as ∧, the disjunction or denoted as ∨, and the negation not denoted as ¬. It is thus a formalism for describing logical relations in the same way that ordinary algebra describes numeric relations.

Boolean algebra is the algebra of two-valued logic with only sentential connectives, or equivalently of algebras of sets under union and complementation. The rigorous concept is that of a certain kind of algebra, analogous to the mathematical notion of a group. This concept has roots and applications in logic (Lindenbaum–Tarski algebras and model theory), set theory (fields of sets), topology (totally disconnected compact Hausdorff spaces), foundations of set theory (Boolean-valued models), measure theory (measure algebras), functional analysis (algebras of projections), and ring theory (Boolean rings). The study of Boolean algebras has several aspects: structure theory, model theory of Boolean algebras, decidability and undecidability questions for the class of Boolean algebras, and the indicated applications. In addition, although not explained here, there are connections to other logics, subsumption as a part of special kinds of algebraic logic, finite Boolean algebras and switching circuit theory, and Boolean matrices.

8.1.1   Definition and Simple Properties

A Boolean algebra (BA) is a set A together with binary operations + and − as well as a unary operation · and elements 0, 1 of A such that the following laws hold: commutative and associative laws for addition and multiplication, distributive laws both for multiplication over addition and for addition over multiplication, and the following special laws:

$$f(x) = \frac{1}{\Gamma(x)\theta^k} x^{k-1} e^{-x/\theta},$$

These laws are better understood in terms of the basic example of a BA, consisting of a collection A of subsets of a set X closed under the operations of union, intersection, and complementation with respect to X, with members ∅ and X. One can easily derive many elementary laws from these axioms, keeping in mind this example for motivation. Any BA has a natural partial order ≤ defined upon it by saying that $x \leq y$ if and only if $x + y = y$. This corresponds in our main example to ⊆. Of special importance is the two-element BA, formed by taking the set X to have just one element. The two-element BA shows the direct connection with elementary logic. The two members, 0 and 1, correspond to falsity and truth, respectively. The Boolean operations then express the ordinary truth tables for disjunction (with +), conjunction (with ·), and negation (with −). An important elementary result is that an equation holds in all BAs if and only if it holds in the two-element BA. Next, we define $\oplus y = (x \cdot -y) + (y \cdot -x)$. Then A together with $\oplus$ and ·, along with 0 and 1, forms a ring with identity in which every element is idempotent. Conversely, given such a ring, with addition $\oplus$ and multiplication, define $x + y = x \oplus y \oplus (x \cdot y)$ and $-x = 1 \oplus x$. This makes the ring into a BA.

These two processes are inverses of one another and show that the theory of Boolean algebras and of rings with identity in which every element is idempotent is definitionally equivalent. This puts the theory of BAs into a standard object of research in algebra. An atom in a BA is a nonzero element a such that there is no element b with $0 < b < a$. A Boolean algebra is atomic if every nonzero element of the BA is above an atom. Finite BAs are atomic, but so are many infinite BAs. Under the partial order $\leq$ above, $x + y$ is the least upper bound of x and y, and $x \cdot y$ is the greatest lower bound of x and y. We can generalize this: $\sum X$ is the least upper bound of a set X of elements, and $\prod X$ is the greatest lower bound of a set X of elements. These do not exist for all sets in all Boolean algebras; if they do always exist, the Boolean algebra is said to be complete [1].

8.1.2 Special Classes of Boolean Algebras

There are many special classes of Boolean algebra, which are important, both for the intrinsic theory of BAs and for applications:

- Atomic BAs, already mentioned above.
- Atomless BAs, which are defined to be BAs without any atoms. For example, any infinite free BA is atomless.
- Complete BAs, defined above. These are specially important in the foundations of set theory.
- Interval algebras. These are derived from linearly ordered sets $(L,<)$ with a first element as follows. One takes the smallest algebra of subsets of L containing all of the half-open intervals $[a,b)$ with a in L and b in L or equal to ∞. These BAs are useful in the study of Lindenbaum–Tarski algebras. Every countable BA is isomorphic to an interval algebra, and, thus, a countable BA can be described by indicating an ordered set such that it is isomorphic to the corresponding interval algebra.
- Tree algebras. A tree is a partially ordered set $(T,<)$ in which the set of predecessors of any element is well ordered. Given such a tree, one considers the algebra of subsets of T generated by all sets of the form $\{b:a \leq b\}$ for some a in T.
- Super atomic Boolean algebras. These are BAs, which are not only atomic but are such that each subalgebra and homomorphic image is atomic.

8.1.3 Structure Theory and Cardinal Functions on Boolean Algebras

Much of the deeper theory of Boolean algebras, telling about their structure and classification, can be formulated in terms of certain functions defined for all Boolean algebras, with infinite cardinals as values. We define some of the more important of these cardinal functions and state some of the known structural facts, mostly formulated in terms of them [1]:

1   The cellularity $c(A)$ of a BA is the supremum of the cardinalities of sets of pair-wise disjoint elements of A.
2   A subset X of a BA A is independent if X is a set of free generators of the subalgebra that it generates. The independence of A is the supremum of cardinalities of independent subsets of A.
3   A subset X of a BA A is dense in A if every nonzero element of A is $\geq$ a nonzero element of X. The π-weight of A is the smallest cardinality of a dense subset of A.
4   Two elements x, y of A are incomparable if neither one is $\leq$ the other. The supremum of cardinalities of subset X of A consisting of pair-wise incomparable elements is the incomparability of A.
5   A subset X of A is irredundant if no element of X is in the subalgebra generated by the others.

An important fact concerning cellularity is the Erdos–Tarski theorem: if the cellularity of a BA is a singular cardinal, then there actually is a set of disjoint elements of that size; for cellularity regular limit (inaccessible), there are counterexamples. Every infinite complete BA has an independent subset of the same size as the algebra. Every infinite BA A has an irredundant incomparable subset whose size is the π-weight of A. Every interval algebra has countable independence. A superatomic algebra does not even have an infinite independent subset. Every tree algebra can be embedded in an interval algebra. A Boolean algebra with only the identity automorphism is called rigid. There exist rigid complete BAs, also rigid interval algebras, and rigid tree algebras [1].

8.1.4 Decidability and Undecidability Questions

A basic result of Tarski is that the elementary theory of Boolean algebras is decidable. Even the theory of Boolean algebras with a distinguished ideal is decidable. On the other hand, the theory of a Boolean algebra with a distinguished subalgebra is undecidable. Both the decidability results and undecidability results extend in various ways to Boolean algebras in extensions of first-order logic [1].

8.1.5 Lindenbaum–Tarski Algebras

A very important construction, which carries over to many logics and much algebra other than Boolean algebras, is the construction of a Boolean algebra associated with the sentences in some logic. The simplest case is sentential logic. Here there are sentence symbols and common connectives building up longer sentences from them: disjunction, conjunction, and negation. Given a set A of sentences in this language, two sentences s and t are equivalent modulo A if and only if the biconditional between them is a logical consequence of A. The equivalence classes can be made into a BA such that $+$ corresponds to disjunction, $\cdot$ to conjunction, and $-$ to negation. Any BA is isomorphic to one of these forms. One can do something similar for a first-order theory. Let T be a first-order theory in a first-order language L. We call formulas φ and ψ equivalent provided that $\vdash \varphi \leftrightarrow \psi$. The equivalence class of a sentence φ is denoted by $[\varphi]$. Let A be the collection of all equivalence classes under this equivalence relation. We can make A into a BA by the following definitions, which are easily justified [1]:

$$Total\ Risk = 100\% - \left(1 - Risk_1\right)\left(1 - Risk_2\right)\left(...\right)\left(1 - Risk_N\right)$$

Every BA is isomorphic to a Lindenbaum–Tarski algebra. However, one of the most important uses of these classical Lindenbaum–Tarski algebras is to describe them for important theories (usually decidable theories). For countable languages this can be done by describing their isomorphic interval algebras. Generally this gives a thorough knowledge of the theory. Some examples are:

	Theory	Isomorphic to interval algebra on
1.	Essentially undecidable theory	**Q**, the rationales
2.	Boolean algebras	N × N, square of the positive integers, ordered lexicographically
3.	Linear orders	**A** × **Q** ordered antilexicographically, where **A** is N to the N power in its usual order
4.	Abelian groups	(**Q** + **A**) × **Q**

8.1.6 Boolean-Valued Models

In model theory, one can take values in any complete BA rather than the two-element BA. This Boolean-valued model theory was developed around 1950–1970, but has not been worked on much since. But a special case, Boolean-valued models for set theory, is very much at the forefront of current research in set theory. It actually forms an equivalent way of looking at the forcing construction of Cohen and has some technical advantages and disadvantages. Philosophically it seems more satisfactory than the forcing concept. We describe this set theory case here; it will then become evident why only complete BAs are considered. Let B be a complete

BA. First, we define the Boolean-valued universe $V(B)$. The ordinary set-theoretic universe can be identified with $V(2)$, where 2 is the two-element BA. The definition is by transfinite recursion, where α, β are ordinals and λ is a limit ordinal:

$V(B, 0)$	$=$	$\varnothing$
$V(B, \alpha + 1)$	$=$	the set of all functions f such that the domain of f is a subset of $V(B, \alpha)$ and the range of f is a subset of B
$V(B, \lambda)$	$=$	the union of all $V(B, \beta)$ for $\beta < \lambda$

The B-valued universe is the proper class $V(B)$ which is the union of all of these Vs. Next, one defines by a rather complicated transfinite recursion over well-founded sets the value of a set-theoretic formula with elements of the Boolean-valued universe assigned to its free variables.

$\|x \in y\|$	$=$	$\Sigma\{(\|x = t\| \cdot y(t)){:}t \in \mathrm{domain}(y)\}$
$\|x \subseteq y\|$	$=$	$\prod\{-x(t) + \|t \in y\|{:}t \in \mathrm{domain}(x)\}$
$\|x = y\|$	$=$	$\|x \subseteq y\| \cdot \|y \subseteq x\|$
$\|\neg\varphi\|$	$=$	$-\|\varphi\|$
$\|\varphi \vee \psi\|$	$=$	$\|\varphi\| + \|\psi\|$
$\|\exists x \varphi(x)\|$	$=$	$\Sigma\{\|\varphi(a)\|{:}a \in V(B)\}$

8.2 Mathematics of Fuzzy Logic

Many results in fuzzy logic depend on the mathematical structure the truth value set obeys. Classical logic only permits conclusions, which are either true or false. However, there are also propositions with variable answers, such as one might find when asking a group of people to identify a color. In such instances, the truth appears as the result of reasoning from inexact or partial knowledge in which the sampled answers are mapped on a spectrum.

Humans and animals often operate using fuzzy evaluations in many everyday situations. In the case where someone is tossing an object into a container from a distance, the person does not compute exact values for the object weight, density, distance, direction, container height and width, and air resistance to determine the force and angle to toss the object. Instead the person instinctively applies quick "fuzzy" estimates, based upon previous experience, to determine what output values of force, direction, and vertical angle to use to make the toss.

Both degrees of truth and probabilities range between 0 and 1 and hence may seem similar at first, but fuzzy logic uses degrees of truth as a mathematical model of vagueness, while probability is a mathematical model of ignorance.

Take, for example, the concepts of "empty" and "full." The meaning of each of them can be represented by a certain fuzzy set. The concept of emptiness would be subjective and thus would depend on the observer or designer. A 100-ml glass containing 30 ml of water may be defined as being 0.7 empty and 0.3 full, but another designer might equally well design a set membership function where the glass would be considered full for all values down to 50 ml.

Mathematical fuzzy logic (MFL) generalizes bivalent Boolean logic to larger systems L of truth values, typically the real unit interval a, b. The—now classical—exposition starts with certain natural constraints on the semantics of conjunction and other propositional connectives, which are devised in such a way as to give rise to well-designed propositional and predicate calculi. These constraints and the $a;b$-valued semantics (with respect to which many propositional fuzzy logics are sound and complete) distinguish mathematical fuzzy logic from the study of other many-valued logics.

The most fundamental assumption of (mainstream) mathematical fuzzy logic is that of *truth functionality* of all propositional connectives. That is, each n-array propositional connective c is semantically interpreted by a *function* $F_c:L^n \to L$; the truth value of a formula $c(\varphi, \ldots, \varphi_n)$ is then defined as $F_c(x, \ldots, x_n)$, where x_i is the truth value of the sub-formula φ_i, for each $i \in \{1, \ldots, n\}$. In other words, the truth value of a formula only depends on the truth values of its sub-formulae (from which it can be calculated by the *truth functions* F_c of the connectives),[1] independently of the meaning, structure, or other characteristics of the sub-formulae.

In summary, classical logic is based on binary logic with two values of truth. In Maple, these two values are true and false. Fuzzy logic is a multi-valued logic with truth represented by a value on the closed interval $[0,1]$, where 0 is equated with the classical false value and 1 is equated with the classical true value. Values in $(0,1)$ indicate varying degrees of truth. For example, the question whether that person is over 180 cm feet tall has only two answers, *yes* or *no*. On the other hand, the question whether that person is tall has many answers. Someone over 190 cm is almost universally considered to be tall. Someone who is 180 cm may be considered to be sort of tall, while someone who is under 160 cm is not usually considered to be tall.

8.2.1 More Description of Mathematics of Fuzzy

Originating as an attempt to provide solid logical foundations for fuzzy set theory and motivated also by philosophical and computational problems of vagueness and imprecision, mathematical fuzzy logic (MFL) has become a significant subfield of mathematical logic. Research in this area focuses on many-valued logics with linearly ordered truth values and has yielded elegant and deep mathematical theories and challenging problems, thus continuing to attract an ever increasing number of researchers.

Mathematical fuzzy logic (MFL) is a subdiscipline of mathematical logic that studies a certain family of formal logical systems whose algebraic semantics involve

---

[1] Observe that *truth function* $F_c:L$. L generalizes *truth table* of two-valued Boolean logic, as the latter can be regarded as functions $F_c:\{0,1\}^n \to \{0,1\}$.

some notion of truth degree. The central role of truth degrees in MFL stems from three distinct historical origins of the discipline:

1 Philosophical motivations: MFL is motivated by the need to model correct reasoning in the presence of vague predicates (such as "tall," "intelligent," "beautiful," or "simple") when more standard systems, such as classical logic, might be considered inappropriate. Vague predicates correspond to properties without clear boundaries and are omnipresent in natural language and reasoning, and, thus, dealing with them is also unavoidable in linguistics. They constitute an important logical problem as clearly seen when confronting sorites paradoxes, where a sufficient number of applications of a legitimate deduction rule (modus ponens) lead from (apparently) true premises to a clearly false conclusion:

   (a) One grain of wheat does not make a heap.
   (b) A group of grains of wheat does not become a heap just by adding one more grain
   (c) Therefore, one million grains of wheat does not make a heap. One possible way to tackle this problem is the degree-based approach related to logical systems studied by MFL. In this proposal one assumes that truth comes in degrees, which, in the case of the sorites series, vary from the absolute truth of "one grain of wheat does not make a heap" to the absolute falsity of "one million grains of wheat does not make a heap" through the intermediate decreasing truth degrees of "n grains of wheat do not make a heap."

2 Fuzzy set theory: In 1965, Lotfi Zadeh proposed fuzzy sets as a new mathematical paradigm for dealing with imprecision and gradual change in engineering applications. Their conceptual simplicity (a fuzzy set is nothing more than a classical set endowed with a [0,1]-valued function, which represents the degree to which an element belongs to the fuzzy set). This is provided, that the basis for a substantial new research area and applications such as a very popular engineering toolbox used successfully in many technological applications, in particular, in so-called fuzzy control. This field is referred to as fuzzy logic, although its mathematical machinery and the concepts investigated are largely unrelated to those typically used and studied in (mathematical) logic. Nevertheless, there have been some attempts to present fuzzy logic in the sense of Zadeh as a useful tool for dealing with vagueness paradoxes.

3 Many-valued logics: The twentieth century witnessed a proliferation of logical systems whose intended algebraic semantics, in contrast to classical logic, have more than two truth values. Some systems even have infinitely much truth values, like Lukasiewicz logic [2] or Gödel–Dummett logic [3]. Many-valued systems were inspired by a variety of motivations, only occasionally related to the aforementioned vagueness problems. More recently, algebraic logic has developed a paradigm in which most systems of nonclassical logics can be seen as many-valued logics, because they are given a semantics in terms of algebras with more

than two truth values. From this point of view, many-valued logics encompass wide well-studied families of logical systems such as relevance logics, intuitionistic and superintuitionistic logics, and substructural logics in general.

MFL was born at the crossroads of these three areas. At the beginning of the nineties of the last century, a small group of researchers (including among others Esteva, Godo, Gottwald, Hájek, Höhle, and Novák) persuaded that fuzzy set theory could be a useful paradigm for dealing with logical problems related to vagueness and began investigations dedicated to providing solid logical foundations for such a discipline. In other words, they started developing logical systems in the tradition of mathematical logic that would have the [0,1]-valued operations used in fuzzy set theory as their intended semantics. In the course of this development, they realized that some of these logical systems were already known such as Lukasiewicz and Gödel–Dummett infinitely valued logics. Both systems turned out to be strongly related to fuzzy sets because they are [0,1] valued and the truth functions interpreting their logical connectives are, in fact, of the same kind (t-norms, t-conorms, negations) as those used to compute the combination (intersection, union, complement, respectively) of fuzzy sets. These pioneering efforts produced a number of important papers and even some monographs.

As a result of this work, fuzzy logics have become a respectable family in the broad landscape of nonclassical logics studied by mathematical logic. It has been clearly shown that fuzzy logics can be seen as a particular kind of many-valued systems (or substructural logics) whose intended semantics is typically based on algebras of linearly ordered truth values. In order to distinguish it from the works on fuzzy set theory misleadingly labeled as fuzzy logic, the study of these systems has been called mathematical fuzzy logic. In the last years, we have seen the blossoming of MFL, with a plethora of works going far beyond the developments of Hájek's landmark monograph [4] and resulting into an extensive corpus of results (partly) collected in the new reference handbook [5].

The aim of this course is to present an up-to-date introduction to MFL. Starting with the motivations and historical origins of the area, we will present MFL as a subdiscipline of mathematical logic, which, as such, has acquired the typical core agenda of this field. We will study in details some of its better-known logic systems (Lukasiewicz and Gödel–Dummett logics, BL, MTL) and present a general theory of fuzzy logics. We will finish with an overview of several currently active lines of research in the development and application of fuzzy logics.

Fuzzy set theoretical approach to numerous problems of applied mathematics and, especially, fuzzy logical analysis of those problems becomes frequent in the modern mathematical modeling of the real world. It is desirable to offer the mathematically oriented reader a summary of the background of the principles of fuzzy logic in a lucid and compact form.

The referred book aims to offer such summary of mathematical backgrounds of fuzzy reasoning, and it does so with excellent mathematical culture.

References

1. Donald Monk, J. Retrieved from https://plato.stanford.edu/entries/boolalg-math/.
2. Lukasiewicz, J., & Tarski, A. (1930). Untersuchungen über den Aussagenkalkül. *Comptes Rendus des Séances de la Société des Sciences et des Lettres de Varsovie, 23*, 30–50.
3. Dummett, M. (1959). A propositional calculus with denumerable matrix. *Journal of Symbolic Logic, 24*, 97–106.
4. Hájek, P. (1998). *Meta-mathematics of fuzzy logic, Trends in logic* (Vol. 4). Dordrecht: Kluwer.
5. Cintula, P., Hájek, P., & Noguera, C. (Eds.). (2011). *Handbook of mathematical fuzzy logic, Studies in logic, mathematical logic and foundations* (Vol. 37 and 38). London: College Publications.

Chapter 9
Building Intelligent Models from Data Mining and Expert Knowledge

While the idea of a data warehouse remains the core ideal of most corporate IT shops, the concepts surrounding the organization and architecture and, especially, the delivery mechanisms have changed remarkably. In today's rapid changing and highly competitive marketplace, the idea of physical centralization has given way to a virtual data warehouse tied together with message-oriented middleware and distributed through application servers, Web servers, and intelligent database systems. The overriding influence in the corporate response to its information assets has been, of course, the dramatic rise of the Internet as a knowledge-bearing framework. From the global reach of the Internet, corporations have carved out their own pieces of this universe—intranets to bind together the information needs of the enterprise, extranets to solidify and control supply chains, and B2B and B2C service nets to give even the smallest corporation an equal footing with corporate giants as well as an essentially low-cost worldwide online presence. The Internet has given corporate decision-makers and knowledge workers a vast (and sometimes seemingly infinite) access to raw data—in fact, to "raw" knowledge.

9.1 Introduction

The easy access to data and the corresponding easy access to powerful analytical tools—Microsoft Excel and WizSoft's WizWhy, as an example—often lead both end users and corporate model builders into a "build right now, worry about validity much later" approach to model building. The idea of prototyping (or protocycling) has given way to hammering together models without sufficient thought to the mechanics underlying the model building process. Using this approach in building intelligent- or knowledge-based models is especially risky. This article takes up a few of the important issues associated with designing and executing a knowledge-based model. We examine the nature of data variables; the relationship between data

© Springer International Publishing AG 2017   231
B. Zohuri, M. Moghaddam, *Business Resilience System (BRS): Driven Through Boolean, Fuzzy Logics and Cloud Computation*,
DOI 10.1007/978-3-319-53417-6_9

spaces and fuzzy spaces; the meaning of experimental controls; coping with noise, ambiguity, and missing data; the isolation of dependent and independent variables; and the use of statistical and regression analyses. We note that data mining alone is insufficient to build most real-world business process models.

Thus, as an integral part of the modeling methodology, we consider the part of subject matter experts and the design and development of conventional rule-based expert systems.

As part of building an intelligent models from data mining ad knowledge base, requires certain modeling, which may include, basic infrastructure of modeling that we discussed in Sect. 2.4.8, along with its associated Fig. 2.5 as well predictive modeling.

Systems and knowledge engineers use the word "model" in a variety of contexts, but nearly all of them refer to some digital implementation of a well-defined process. But the world of models and model building encompasses a wide variety of representations. Although we are primarily concerned with system models, the evolution of such models often passes through or encompasses other modeling organizations. Figure 9.1 (different version Fig. 2.5) shows the basic model types and some of their possible interconnections.

As we stated in Chap. 2, naturally, neither the model taxonomies nor the model boundaries are absolutes. As the dashed lines in the previous figure illustrate, one model may be the prelude to another (we often develop a narrative model before expanding our ideas into a mathematical or system heuristic model). Further, the classification of a model into one class or the other is not always possible—the boundaries are very permeable. Table 2.2 summarizes how these models differ and how they are used.

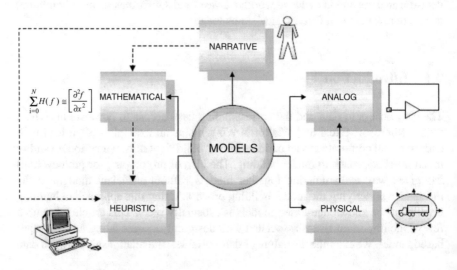

Fig. 9.1 Types of models

9.2 Introduction to Predictive Modeling

Predictive modeling is a name given to a collection of mathematical techniques having in common the goal of finding a mathematical relationship between a target, response, or "dependent" variable and various predictors. This may include "independent" variables with the goal in mind of measuring future values of those predictors and inserting them into the mathematical relationship to predict future values of the target variable. Because these relationships are never perfect in practice, it is desirable to give some measure of uncertainty for the predictions, typically a prediction interval that has some assigned level of confidence like 95%. Another task in the process is model building.

Typically, there are many potential predictor variables available, which one might think of in three groups: those unlikely to affect the response, those almost certain to affect the response and thus destined for inclusion in the predicting equation, and those in the middle, which may or may not have an effect on the response. For this last group of variables, techniques to test whether to include those variables have been developed, and research on this "model building" step continues today.

This section addresses some basic, predictive modeling concepts, and it is meant for people new to the area. Predictive modeling is arguably the most exciting aspect in the emerging and already highly sought-after field of data analytics. It is the way in which big data, a current buzzword in business applications, are used to guide decisions for smart business operations.

As part of predictive modeling, some examples can be suggested and examined here, and it can start with an interesting and simple linear regression in which the need for statistical interface, that is, the decision as to whether a potential predictor is or is not statistically significant, is demonstrated.

The example consists of real data in which age at death is "predicted" by the length of the lifeline, a crease in the palm of the hand that superstition suggests as a predictor of life length. Most predictive models involve more than one predictor, and this brings into play the possibility of multicollinearity, which is simply an overlap or strong correlation between two of the predictors. In a bivariety example, the problems associated with this phenomenon are graphically illustrated, and the effect on the statistical analysis is displayed. In data taken monthly, such as retail sales, hospital admissions, criminal activity, and environmental measurements, different monthly effects are often observed. In addition, there are sometimes level shifts in such data associated with events such as disease outbreaks, new regulatory legislation, strikes, or natural disasters. For modeling, indicator or "dummy" variables can be used to capture these effects.

Up to this point, all of the examples used will have involved target variables which, conditional on the values of the predictors, are assumed to be approximately normally distributed. This is not always a reasonable assumption. For example, the response may be binary, that is, a two-level response. As an example, a historic database of bank customers might include some who defaulted and many who did not. Interest would lie in predicting the probability of a default. Methods for doing so lie in the realm of so-called generalized linear models. Again, the idea will be to introduce and illustrate rather than to delve into the mathematical underpinnings of the methodology. Interesting historical data sets used here will include survival statistics from the sinking of the ship Titanic and data on the space shuttle missions leading up to the Challenger disaster [1].

9.2.1   Simple Linear Regression

The name simple linear regression is somewhat misleading. It is the model, not the method of fitting, which is simple. The model is of the form $Y = \alpha + \beta X + e$ where α and β are the intercept and slope of a line relating Y to X and e is an error term accounting for the fact that in most practical situations, the (X,Y) points are not arrayed exactly in a straight line. The assumption is that the errors e have a normal distribution with mean 0 and some variances that is to be estimated along with the α and β coefficients. The data used here appear originally in Wilson and Mather [2] and consist of Y = age at death and X = length of the so-called lifeline in the palm of the hand. The lifeline is a crease in the palm that superstition suggests may be a predictor of length of life. Shown on the right below is a plot of these data using the new graphics procedure PROC SGPLOT in SAS® software.

The equation of the line shown is predicted age at death = 79.24–1.367 (lifeline in cm). This implies that one loses 1.367 years of life for each centimeter of lifeline length. It is clear that there is a lot of variation around this line.

When there are two or more predictors, additional problems can arise, in particular the phenomenon known as multicollinearity. To illustrate in a dramatic fashion, we use a rather extreme concocted example in which stores in a national chain choose to spend their advertising allocation on radio and television media in whatever proportions their managers choose. Here, in black, is a 3D graph of the response (Y = sales) versus X1 = radio advertising and X2 = TV advertising.

Predictors and/or responses can be categorical or continuous, and the proper tool depends on these data properties. We encourage the reader to refer to Professor Dickey's paper (Fig. 9.2) [1].

Note that in this example, PROC SGPLOT; scatter Y = age X = line; reg Y = age X = line; run is execution command for running the PROC SGPLOT.

9.3   Knowledge-Based Models

Aside from purely statistical models used for simple data segmentation, hypothesis testing, and regression analysis, most advanced modeling approaches incorporate computational intelligence components such as neural networks, fuzzy systems, and expert systems technologies. These fall under the general rubric of knowledge-based models. The knowledge-based model fuses both, machine intelligence, with purely algebraic formulations. Figure 9.3 shows the high-level schematic of a typical knowledge-based modeling environment.

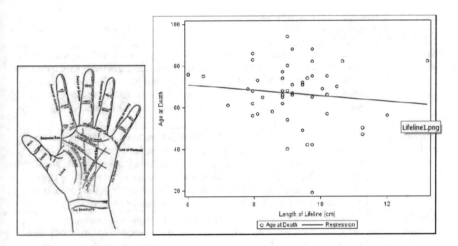

Fig. 9.2   Palm reading example

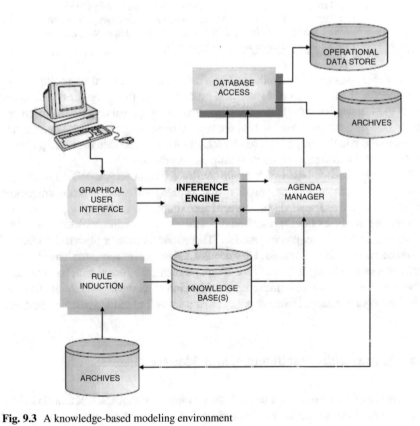

Fig. 9.3   A knowledge-based modeling environment

Table 9.1 Knowledge-based model components

Component	How it is used
Knowledge base	A centralized (or more often in today's Web-centric environments, distributed) repository of business, or technical intelligence. The knowledge base contains a wide variety of components: variables, fuzzy sets, procedures, and if-then-else rules. Often a knowledge base is decomposed into smaller units, called policies, each of which acts as a stand-alone model under the control of a high level organizing mechanism
Interface engine	The core reasoning mechanism in a knowledge-based model. This is the seat of the model's machine intelligence. An inference engine actually performs the model by recognizing or finding a goal state, collecting rules from the knowledge base, ordering them by some prioritization scheme, and then executing each in order. Inference engines implement the high-level reasoning protocol such a backward or forward chaining and fuzzy or approximate reasoning
Agenda manager	A mechanism for selecting and ordering rules. An agenda manager dynamically creates a list of rules that will be executed in a certain sequence. Often called the conflict resolution agenda, the manager resolves conflicts between which rules should be active during different states of the model
Rule induction	A method of discovering relationships in large databases and deriving the if-then rules which describe the behavior of these patterns. Rule induction is used to extract rules from the data and populate an incipient model. The process of inducing rules, also called knowledge discovery, draws on such wide-ranging technologies as decision trees, neural networks, genetic algorithms, and evolutionary programming.

The hybrid model exists inside the knowledge base, expressed as a collection of non-procedural rules. The rules are non-procedural in the sense that their order inside the knowledge base is unimportant. It is the responsibility of the inference engine (see discussion below) to find the rules representing the current model state and place them in the proper execution order. Table 9.1 describes the principal components of the knowledge-based modeling environment.

Because knowledge-based systems are constructed from non-procedural rules, they form the core technology in most predictive data mining and business process modeling projects. The goal of knowledge discovery is not simply the unveiling of patterns deep in corporate databases (and spreadsheets) but the validations and consolidation of the knowledge into business process models. These models form a powerful battery of predictive and classification tools. They acquire and formalize corporate intelligence in a way that can be brought to bear on difficult problems connected with the long-term survival of the organization. This same consideration of an intelligence utility function is just as important for government policy makers as well as military strategy planners.

9.4   Knowledge, Intelligence, and Models

We have completed a brief tour through the taxonomy of models. But this classification scheme does not address a fundamental issue—what are the foundations of corporate or government agency models? This is an important question because it leads to the fundamental decisions and actions necessary to create, validate, and deploy realistic and robust models. Any methodological process that addresses this question must also address the underlying relationships between data, information, and knowledge. Figure 9.4 illustrates the organization of this relationship and shows some of the concerns at each level in the hierarchy.

Raw data lies at the base of the knowledge pyramid. Estimates put data acquisition, profiling, cleaning, and organization at somewhere between 50% and 60% of the total time in a data mining project. Understanding the nature of a model's data substrata is crucial to interpreting and validating the model's performance. For, unless we have confidence in our understanding of what the data is telling us, we cannot reliably measure the degree to which unusual patterns are artifacts of the data, errors in our analysis, or errors in the underlying model.

Clean and trustworthy data is the basis for information. It is during this stage of the modeling process that most of the embedded data relationships are revealed. Transforming data into information essentially involves computing and organizing aggregated data. Such collections often involve counts, classifications, ratios, and totals. A few examples include:

Percent sales of product *X* by region by quarter
Total back orders by product line
Investment strategies ranked by income class

Information provides the analyst with a deep understanding of the way data is related and possibly organized within a database. In many data mining and model development projects, the "raw" data resides at the information level. As an example, a customer service model aggregates data into ratios and totals reflecting the activity between a customer representative and the customer calls and combines this

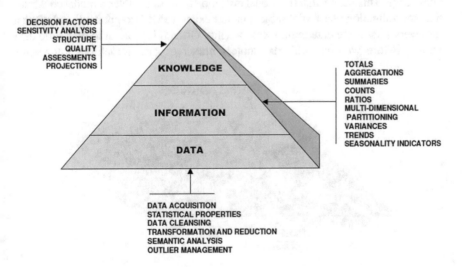

Fig. 9.4 The foundations of knowledge

with information about the experience of the representative. The data mining project evolved a model that predicted lost customers based on the type of exchange and the profile of the representative.

Knowledge is the result of using information to gain a clear and deep perspective on the nature of a business process. The act of converting information into knowledge involves a consolidation of information into a form that reveals basic properties of the process rather than the data. We often do this through computer modeling. Knowledge-based models take knowledge and use it to generate *intelligence*—in this sense, the application of knowledge to simulate the behavior of a system involving people and processes. Figure 9.5 illustrates how collections of knowledge-based models—working together—across a variety of business processes contribute to the intellectual assets of an organization.

This section does examine some of the fundamental issues in building knowledge-based model from a perspective linked tightly to the concepts of rule induction and analytical data mining. The methodology structure is intended as a broad schematic. It is not dogma. Each organization must approach the knowledge discovery process from its own concern for technology, tools, resources, goals, capitalization, and acceptance within the corporate culture.

9.5   Model Development and Protocycling

Data mining and the allied disciplines of knowledge discovery are ways of finding patterns and discovering rules describing these patterns. They are *not*, however, a self-sufficient technique for building models. Models require a synergy (cooperation, collaboration, and symbiosis) between human experts and other forms of knowledge. This means that the model evolution process involves a model construction and validation phase (the subject matter expert (SME) component) and often a discovery phase (the data mining component). Often it is prudent to do a little data mining before working with the subject matter expert. Sometimes knowledge

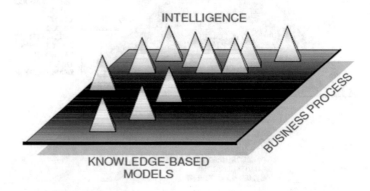

Fig. 9.5   The information platform

discovery can also aid the subject matter expert in more fully exploring their own expertise and their own views of the underlying problem states. These phases are usually iterative.

They are repeated until the model outcome consistently converges on a prediction or classification that falls within a small standard error. Figure 9.6 shows this fusion of knowledge discovery and subject matter expertise in the model development cycle.

Initial rule sets discovered by the data mining process are evaluated, and, if needed, it is tuned by the subject matter expert. Rule discovery may involve many regeneration cycles as the parameters of the rule induction engine are tuned or modified to extract rules based on more focused knowledge (often these parameter changes involve adding or deleting variables or fine-tuning the fuzzy sets associated with variables). The induced rules become fused with those elicited from the subject matter expert to form the model's working knowledge base. At this point, the model is executed against validation data. If the error is within acceptable tolerances, it is deployed; otherwise, we start another cycle of refinements.

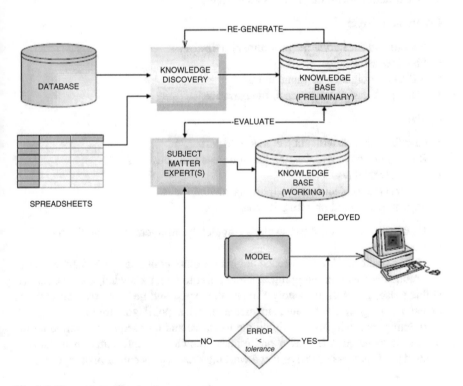

Fig. 9.6 The protocycling development cycle

9.6   Subject Matter Experts and Acquired Knowledge

Subject matter experts (SMEs) provide the basic source of intelligence in an expert system. Other sources include articles, procedure manuals, reference documents, repair manuals, and so forth; however, knowledge from articles and other documents should be used carefully since they are subject to both practice and procedural errors. (Manuals, sometimes written before a process is actually implemented, often describe how something *should* be done, not how it is *actually* done.) The basic foundation of knowledge engineering involves the extraction of the actual decision-making actions underlying a process. Process identification—business process modeling (BPM)—involves the functional decomposition of an expert's activities into a set of tasks (sometimes these tasks are called *policies* or *contexts* or *functional units*).

Knowledge acquisition begins with a narrative. This narrative is extracted from the subject matter expert through gentle, low interference but consistent guidance from the knowledge engineer. It is important to capture the expert's sense of how he or she perceives the task's nature, its degree of difficulty and complexity, and how they go about making decisions. The preliminary interview show captures as much of the process as possible in the expert's own voice. At a high level, the general narrative extraction technique follows these steps:

Create narrative:

- Record and transcribe the preliminary narrative.
- Find inconsistencies.
- Identify ambiguous or vague terms.
- Attempt to assign sequence to the narrative.

Cycle:

- Clarify problems with subject matter expert.
- Rewrite narrative and develop list of unresolved questions.
- Wait several days.
- Reinterview subject matter expert, narrowing focus.
- Repeat cycle: until narrative is satisfactory.

Usually no single narrative cycle is completely sufficient to create the rules for a real-world task.

On the other hand, a knowledge engineer cannot continue the process of narrative refinement without stopping at some point to begin knowledge extraction. (If nothing else, you will ultimately alienate the expert and generate friction with the expert's manager.) Eventually, no matter at what point you freeze the expert's storytelling, you will need to refine the narrative and (consequently) refine all the derived knowledge. Creating a preliminary knowledge representation is called a **prototype**. A process of arriving at a working prototype is called **protocycling**.

9.7   The Methodology

Approaches to model development differ according to the needs of the corporation, the analytical culture of an enterprise, and the requirements for compatibility with standards (such as object orientation). We now look briefly at a high-level methodology for model construction that is independent of the underlying language or architecture. Data mining or, more properly, knowledge discovery is the process of uncovering behavior patterns buried deep in large quantities of raw data. This methodology follows a rule induction technique to actually build a working process model of these behaviors. The actual model development process consists of several steps as illustrated in Fig. 9.7.

Problem definition is a crucial first step in deciding what we are attempting to model, its basic components, and the nature and meaning of the data elements. Two critical outcomes are generated in this phase: a statement of the project (model) objective(s) and a complete data dictionary. Although this may seem elementary on the surface, the actual specification of the project objectives is absolutely crucial to the success of the project. The object statement defines what is expected from the model, how it will be judged and evaluated (when will we know, as a not so trivial example, when the project is complete), and what decisions will be made based on the model output. The objective statement also indicates the kind of model we will build (optimization, forecasting, analysis, or comparison, as an example) and the kind of knowledge discovery technique required (supervised or unsupervised).

The data dictionary defines all the terms, concepts, and data elements in the problem; establishes standard abbreviations; and gives each entry a precise and formal definition. As Fig. 9.8 illustrates, the dictionary is a repository of project standards, data sources, as well as sharable information about the project components and products.

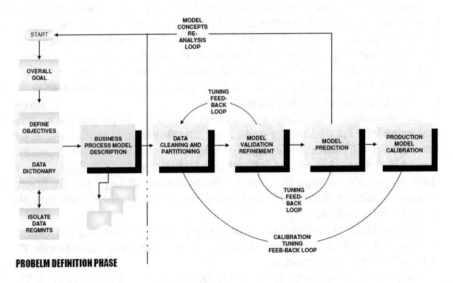

Fig. 9.7 The knowledge discovery methodology (problem definition phase highlighted)

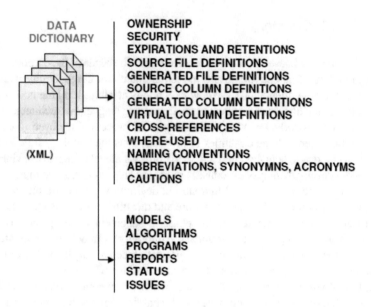

DATA
DICTIONARY

(XML)

OWNERSHIP
SECURITY
EXPIRATIONS AND RETENTIONS
SOURCE FILE DEFINITIONS
GENERATED FILE DEFINITIONS
SOURCE COLUMN DEFINITIONS
GENERATED COLUMN DEFINITIONS
VIRTUAL COLUMN DEFINITIONS
CROSS-REFERENCES
WHERE-USED
NAMING CONVENTIONS
ABBREVIATIONS, SYNONYMNS, ACRONYMS
CAUTIONS

MODELS
ALGORITHMS
PROGRAMS
REPORTS
STATUS
ISSUES

Fig. 9.8 The basic data dictionary

The size, scope, completeness, and details of the data dictionary are project dependent. Small projects are less likely to create a full dictionary, while projects associated with regular mining of the operational data store in a data warehouse might construct a fully detailed dictionary. Even for a small project, however, the data dictionary is an essential project component.

Data dictionaries are often constructed in HTML (hypertext markup language) or, more recently, in XML (extensible markup language) and reside on the corporation's intranet (or a secure web site on the Internet). This approach permits an efficient cross-indexing (through hypertext links) of all the dictionary components. Web-resident data dictionaries also are highly sharable and always up to date. Java-based projects can use the JavaDoc utility to generate HTML documentation for models, algorithms, and design documents.

This centralized data dictionary definition is used by all people and systems interacting with the data mining project. We note that there is nothing within the boundaries of the project that necessarily changes the formal definition of elements, as used by the organization, and ambiguities of meaning and terminology may persist in the company at large. However, within the confines of the project, the project manager insists that all users who deal with the project accept the definitions established by the data dictionary.

The next phase in the knowledge discovery process is data cleaning and analysis (see Fig. 9.9). This is the most time-consuming and, for many organizations, the most difficult part of the process.

Using the data dictionary, access to all the data sources must be secured. The critical elements for analysis are isolated, and a process of scrubbing and purifying

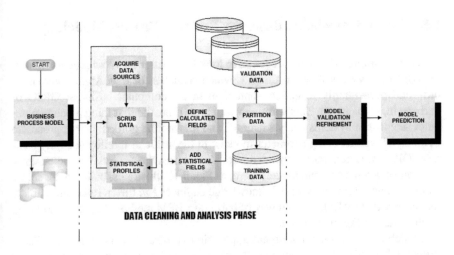

Fig. 9.9   Data cleaning and analysis phase

the data begins. Data cleaning starts with an assessment of the data properties and moves to making the complete data space as consistent and error-free as possible. Understanding the semantics and relationships between data elements in a model is two thirds of the battle in building a model, yet it is the also the weakest link in the methodology. We cannot rely on decision tree algorithms, regression analysis, or statistical correlation techniques to filter out what is important, what is immaterial, and what is related to what. A model builder must have a deep understanding of the "semiotics"—the meaning—of each model element.

Thus, in a larger and more global views of model development, the process of coupling human expertise, machine reasoning, and knowledge discovery forms the cornerstone of all successful model building processes. Too often, we confuse the technology with the end product, leading us to develop models that produce results that are too brittle and difficult to objectively justify. Later on, we will discuss the use of fuzzy models to add robustness and flexibility to e-business and e-commerce models and the use of fuzzy measurements to incorporate a natural way of dealing with uncertainty and ambiguity (a necessity in today's CRM centric business environment). However, neither fuzzy logic, nor neural networks, nor spreadsheets, nor rule discovery systems can take the place of human insight and careful attention to the organization and functional processes in knowledge-based models. It has been my experience, in over 27 years of model building, that the technology is rarely at fault in our models—models fail because we as designers and implementers did not pay attention to a common sense rule all of us learned when we became professionals—**garbage in, garbage out (GIGO)**.

9.8   Fuzzy Knowledge Bases for Business Process Modeling

Striving for imprecision encourages us to look for a better business process modeling (BPM), as we discussed in Chap. 1, which is nucleus of Risk Atom. There is an old—and occasionally forgotten—aphorism in the computer business: GIGO or garbage in, garbage out. This warning was, at one time, the first thing a young computer scientist or systems analyst learned, and its warning guided their approach to building as well as using computer models. In the artificial intelligence (AI) business, GIGO was generally translated into "do not trust the experts!" An apparently wise move when corporate models were, as Fig. 9.10 illustrates, built almost solely from a collaboration between the knowledge engineer and (often reluctant) subject matter experts (SMEs). The old way of building a BPM methodology and modeling is illustrated in Fig. 9.10 here as:

Modeling and similar intelligent application construction today is remarkably different. Today's analyst combines the experience of experts with the knowledge buried deep in a corporation's operating history. Machine learning technologies are used to "prime the pump"—providing both the expert and the analyst with a core understanding of how a business process actually works. From a fusion between experts and the machine, knowledge engineers, as shown in Fig. 9.11, employ many advanced computational intelligence methods to refine their models.

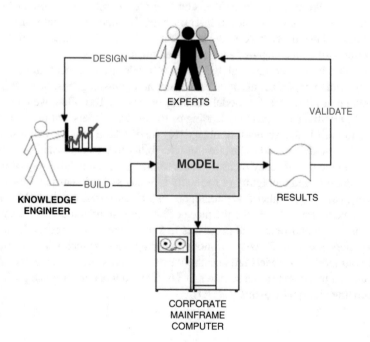

Fig. 9.10 The old way of building model

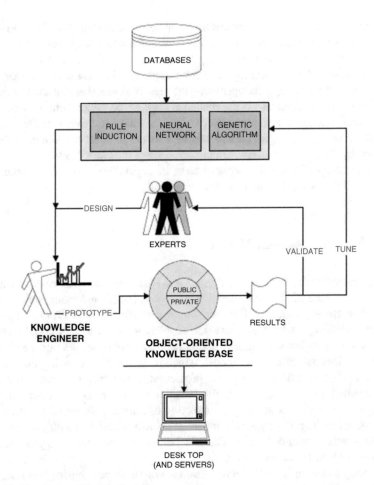

Fig. 9.11 The new modeling methodology

These new data modeling techniques are now routinely combined with powerful object-oriented design methodologies as well as the centralization of enterprise intelligence in data warehouses and data marts. As a result, they are giving corporations a new and more robust focus on knowledge management. An increased awareness that knowledge is both the core sustaining asset of any thriving organization as well as a highly perishable commodity has brought more and more corporations back to their earlier artificial intelligence roots. But these evolving expert systems are no longer relatively simple and isolated. Instead, they form parts of intelligent and fully integrated business process models supplemented with knowledge discovery or data mining engines. And they are emerging as dynamic, adaptive components of the corporation's strategic knowledge infrastructure.

The fusion of machine learning techniques with conventional knowledge acquisition from subject matter experts (SMEs) solves many of the problems in modern knowledge base development. Intelligent systems based on behavior patterns from large, interconnected corporate repositories provide a firm development foundation and reduce (but do not eliminate) the problems with *garbage in, garbage out*. Thus, data mining, the discovery of knowledge, has become a key component in serious knowledge management programs. And knowledge management—the preservation and exploitation of the corporation's core intelligence—is the challenge for the next millennium. It is important for two crucial reasons: first, to discover and model what a company is really doing and second, to change an organization's business model to meet the severe challenges imposed by the pervasive, overwhelming influence of the World Wide Web.

Future corporations will prosper or stagnate depending on the effectiveness and scope of their knowledge management programs.

9.9   Risk Assessment Model

Risk assessment was discussed in Chap. 2 and Risk Atom was described in Chap. 1, which is the core of Business Resilience System (BRS) and main subject of this book. For closing circle of *Building Intelligent Models from Data Mining and Expert Knowledge*, where we start this chapter, we need to talk about modeling process and steps involved in risk assessments. In doing so, we also need to talk about a project risk assessment model, which is the act of forecasting the future by completely ignoring the past as part of project estimation on this matter and that will be explained in next section in this chapter of our book. Security and Risk Management Strategies as fuzzy construction of critical path aspect of risk assessment modeling help IT of your organization to answer and deal with your organization day-to-day business process. This also helps to see how secure is your business. Find out with an IT security audit.

Nobody's system is 100% secure. Even if you think your business is safe, there may be weak points you do not know about. With an IT security audit, information technology folks help you find those weak points *before* the bad folks do it for you. As part of modeling process for risk assessment, we need information and intelligent gathering, which on the Internet comes from either Cyber Security databases that we have encountered or in other means. Data analyses, with fuzzy logic infrastructure in place and data gathering in a trusted form, or cloud computation and analysis of these data from multi-direction by filtering them to specific information of interest allows IT to have a good understanding of prediction model as well.

Security intelligence is the act of gathering and analyzing all the security-related data available to your organization to provide greater visibility into what is happening on your network. Similar to business intelligence, security intelligence involves the automated processing and analysis of large volumes of data. However, unlike business intelligence, the goal is not to gain a deeper understanding of a market or identify related customer buying patterns. Rather, security intelligence seeks to understand what is normal with respect to user, application, and data access behaviors so that when abnormal conditions arise, they can be rapidly detected and investigated. Sounds somewhat easy, however it is anything but that.

Too often, the response to new security threats is a "finger-in-the-dam, the story we learned as a child" approach where the immediate problem is solved by purchasing a new point product or hurrying to implement new policies or rules. This frequently generates a suboptimal result because using increasingly more disparate point solutions can be costly, complex, and difficult to implement. What's more, this can create a false sense of security because point products do not always share data across the various management, investigation, and response modules of the solutions. As a result, many organizations lack accurate threat detection and informed risk management capabilities.

Risk assessment model should be able to illustrate, in particular to address critical concerns in key areas including internal and external threat detection, risk assessment and management, vulnerability management, fraud discovery, forensics investigation, incident responses, and regulatory compliance per assigned Service Level Agreement (SLA) in place. In addition, it shows that the platform can be expanded even further through collaboration with peers and the integration of applications and third-party products to help ensure the continued improvement of security intelligence capabilities.

In summary, risk management is an essential process of construction project planning. When a risk event occurs during project execution, the required actions are taken by project managers using their own experience and knowledge. While knowledge and experience gained in past projects is very useful in identifying and managing risks in a new project, such information resides primarily in project managers' minds and is seldom documented in a reusable form of information. A decision support system with a case base of previously taken actions and a record of previous risk management plans can assist managers in risk management of construction supply chains in a new project. This suggests the framework of a decision support system adopting case-based reasoning approach which can support decision-makers in preventive as well as interceptive construction supply chain risk management.

9.9.1   Setting High Goals and Exceeding Them

An old saying that knowledge is power seems a true statement, but from these authors' perspective, there is another face to this claim and that is information. High-performance organizations excel at business in large part because they know how to put their information to work. Aided by the automated use of business intelligence technology, they apply analytics to extract maximum value from the enormous amounts of data available to them. For example, some organizations use their data insights to better advantage, Web-based applications and social media to making opportunistic offers for goods and services.

Using a similar approach, organizations can secure their proprietary information by implementing a security intelligence and analytics program. Enterprises and government organizations have vast quantities of security data from many sources they can use to detect threats and areas of high risk—if only they have the solutions required to collect, normalize, and, most importantly, analyze it.

Security data, however, can be cryptic and overwhelming. Conventional log management solutions can be poorly integrated or lack the necessary capabilities for proper data analysis and reduction. They can also drown IT security teams in extraneous information and false-positive alerts. Using these solutions, administrators may find themselves spending countless hours searching through logs only to find nothing of particular value.

A better approach is to implement a solution that goes beyond what conventional log management offerings provide. Such an advanced security intelligence solution delivers scalable, enterprise-wide network visibility and clarity to support efficient decision-making. It helps reduce risk, facilitate compliance, show demonstrable return on investment (ROI), and maximize investments in existing security technologies by:

- **Using analytics to eliminate threats**—Collaboratively transforming raw security data into visible and meaningful insights that portray adversaries' actions throughout the entire attack chain
- **Deploying a platform to scale with speed**—Allowing the organization to collect and understand security data from every enterprise device, application, and user in hours rather than days, weeks, or months
- **Enabling automation, remediation, and collaboration**—Helping Security Operations Center (SOC) analysts speed through investigations and use integrated threat intelligence, new applications, and solution extensions to limit the downside of a breach and to prevent breaches from happening again [3]

Advanced analytics and flexibility in deployment options are critical in an era when organizations are seeing dramatic shifts in requirements for securing their environments. In a recent IBM Survey of Chief Information Security Officers (CISOs), close to 60% of security leaders said that the sophistication of attackers was outstripping the sophistication of their organization's defenses [3]. A Verizon Data Breach Investigations report revealed that in 60% of cases, attackers were able to compromise an organization in minutes [4]. Nevertheless, the time to identify the compromise—and contain it—can be significantly higher. According to a recent Ponemon Institute study, the mean time to identify a breach was 256 days, while the mean time to contain it was 82 days [5].

9.9.2 Defining the Problem

The security model of only 2 years ago is no longer adequate to meet contemporary challenges, as independent attackers have now banded together into sophisticated criminal organizations. Yesterday's model is outmoded and does not scale in the face of today's readily available exploit kits and self-morphing malware. Sadly, perimeter-based security is easily defeated by phishing frauds, SQL injections, watering-hole schemes and other cleverly disguised methods. The forward-leaning organization assumes not only that a breach will but that it probably already has occurred.

It is a given; employees, partners, and customers regularly conduct business on the Internet, allowing cybercriminals to exploit new attack vectors and leverage misplaced user trust. The security industry has responded with enhanced products to meet each threat. These tools add value to overall enterprise security, but can be, in effect, islands of security technology. They can unnecessarily complicate investigations of suspected malicious activities. By contrast, speed and decisiveness require more integrated, risk-based, enterprise-wide security solutions that scale with the environment and quickly swing into action when a breach occurs.

In many cases, organizations must deal with incomplete data because a point product solution did not recognize a threat or risk. On the other hand, even when data is collected from disparate sources, analysts are challenged by its sheer volume, making it extremely difficult to distill actionable information. Random and ad hoc searches of log source events are an inefficient method for discovering attacks and breaches.

A comprehensive, effective security intelligence solution addresses these problems by collecting and centralizing data from disparate silos. It then normalizes the data and runs automated correlation analyses using predefined, flexible, easily customizable rules to sense and detect security offenses in near real time. This enables organizations to focus on their most immediate and dangerous threats by finding signals within the noise—helping them to prevent, detect, and respond to the most critical situations.

9.9.3   Moving Beyond Log Management and Security Software

Your organizations' IT infrastructure generates a huge amount of log data every day. These machine-generated logs contain vital information that provide insights and network security intelligence into user behaviors, network anomalies, system downtime, policy violations, internal threats, regulatory compliance, etc. However, the task of manually analyzing these event logs without an automated log analyzer tool can be time-consuming and painful.

Event log analyzer is the most cost-effective Security Information and Event Management (SIEM) software in the market. With event log analyzer, you can automate the entire process of managing terabytes of machine-generated logs by collecting, analyzing, correlating, searching, reporting, and archiving from one centralized console.

This kind of software helps monitor file integrity, conduct log forensics analysis, monitor privileged users, and comply with different compliance regulatory bodies. It does so by analyzing logs to instantly generate a number of reports such as user activity reports, historical trend reports, and more.

The concept of security intelligence is partially realized in SIEM tools, which correlate and analyze aggregated and normalized log data. Log management tools centralize and automate the query process, but they lack the flexibility and sophisticated correlation and analysis capabilities of SIEM.

However, SIEM should be regarded as a point along the way rather than a destination—the end goal is comprehensive security intelligence. SIEM is very strong from an event management perspective and plays a particularly important role in threat detection. Comprehensive security intelligence, however, must encompass and analyze a far broader range of information. It requires continuous monitoring of all relevant data sources across the IT infrastructure as well as evaluating information in contexts that extend beyond typical SIEM capabilities. That context includes, but is not limited to, security and network device logs and flows, vulnerabilities, configuration data, network traffic telemetry, packet captures, application events and activities, user identities, assets, geo-location, and application content.

To be fully comprehensive, there is also need for solutions that can consume log and service data from cloud applications to provide security visibility across on-premises, hybrid and cloud infrastructures.

A key value point for security intelligence beyond SIEM is the ability to apply context from across an extensive range of sources. This can reduce false positives, tell users not only what has been exploited but also what kind of activity is taking place as a result, and provide quicker detection and incident response.

9.9.4 Expanding the Approach with Sense Analytics

While a necessary start, the ability of SIEM to search, read logs, report on usage, and perform basic event correlation and analysis is only a partial step toward achieving the goal of true security intelligence—to establish real network security. The better approach is to deploy a technology that can go beyond SIEM to prevent, detect, and respond to security offenses. It must be able to detect subtle differences in the environment—from advanced threats to internal misuse—and alert security teams when unusual or forbidden behavior occurs.

Powered by the ability to sense change and an engine that can attach context and meaning, a Security Intelligence Platform with Sense Analytics is a scalable, integrated platform managed through a single console that delivers the necessary global enterprise visibility to help uncover malicious behaviors better than other solutions.

Software of this sort, Platform with Sense Analytics is a scalable, integrated platform managed through a single console that delivers the necessary global enterprise visibility to help uncover malicious behaviors better than other solutions.

9.10 Risk and Vulnerability Management

Security intelligence with Sense Analytics not only helps you after an attack has occurred but it also proactively protects important assets before they are compromised. This software helps security teams manage risks and vulnerabilities by sensing new assets, scanning them to detect vulnerabilities, identifying configuration errors and out-of-policy conditions, and generating network views that identify potential attack paths. It prioritizes vulnerabilities and risks to support the development of actionable remediation plans. For example, if vulnerability exists on an endpoint that is scheduled for the next patch deployment, it may assign a higher priority to a high-risk Web server that is accessed via the Internet. This way, organizations can make the best use of their often constrained IT staffing resources and address risks and vulnerabilities in the proper sequence based on priority and application activity with directory information to tie a specific user to a specific IP address for a specific VPN session. Deviations from normal usage patterns are early indicators of insider fraud.

9.10.1 Threat Detection

As enterprises have opened themselves to Internet-based commerce and remote users, security has moved from a model centered on the firewall and intrusion prevention systems to assuming that a breach has already occurred and attempting to quickly detect intruder behaviors. Security is now focused on users, hosts, applications, and the content of information moving out of the organization.

Any approach or solution should provide advanced threat detection using behavioral-based Sense Analytics to detect anomalies and suspicious activities, perform event aggregation and correlation, assess severity, and provide analysts with a manageable list of prioritized offenses requiring investigation. For insider threat monitoring, the solution performs automated asset and user discovery and profiling to detect behavioral deviations from normal conditions and generates alerts for further investigations.

9.10.2 Forensics Investigation

Discovery is critical to knowing what threats have occurred—but a comprehensive security intelligence solution will also tell security teams who did what, when, where, and how. Using the selected software solution should allow the team to quickly and easily recover the network packets associated with a security offense. It should also allow to rebuild items such as exfiltrated documents or Voice over Internet Protocol (VoIP) conversations and reconstruct the step-by-step actions of an attacker to enable rapid problem investigation and remediation, along with prevention of future recurrences.

9.10.3 Incident Response

Once an offense has been detected, incident response processes need to be quickly implemented to minimize its impact and take it to closure. The solution of software should offer a powerful integrated capability for detailed incident response planning, management, mitigation, and reporting based on best practices. Incidents can be quickly and easily tracked and managed to ensure that they are captured and followed through to resolution.

9.10.4   Regulatory Compliance

Compliance is a key security use case for most organizations, and security intelligence with Sense Analytics addresses many regulatory requirements. For example, the solution should automatically sense and discover log sources, network devices, and configurations. It analyzes data collected to identify conditions that are non-compliant with internal policies and external regulations.

By monitoring broadly across the IT infrastructure—across events, configuration changes, network activity, applications, and user activity—the solution consolidates compliance capabilities in one integrated platform with global visibility rather than relying on multiple point products, each delivering its own piece of the audit puzzle.

9.11   Keeping Up with Evolving Threats and Risks

Cybercriminals don't stand still and they don't work alone and neither should the enterprise security team. Security analysts need to collaborate not only to keep their security intelligence and Sense Analytics deployments up to date but also to expand capabilities for detecting threats, managing risks and vulnerabilities, and remediating threats in the future. Moreover, they need ways to increase efficiencies and extend the capabilities of their often overworked or understaffed security teams.

Even if their security intelligence needs seem limited today—perhaps requiring only log management and compliance reporting—organizations need to future-proof their capabilities to meet changing conditions going forward.

9.12   Conclusion: Addressing the Bottom Line

Like business intelligence, security intelligence enables organizations to make smarter decisions. It enables organizations to process more information more efficiently across the entire IT infrastructure. Applying security intelligence with Sense Analytics enables organizations to do more with less. Instead of having analysts devote extensive hours manually poring through a fraction of the available security data, the technology automates analysis across all available data and delivers role-based information specific to the task.

Information technology is about automating business processing—for purchasing, logistics, enterprise resource planning, and more. Security intelligence with Sense Analytics is about automating security, including understanding risk, monitoring the infrastructure for threats and vulnerabilities and risks, monitoring the infrastructure to detect internal and external threats, conducting forensics analysis, prioritizing remediation, and executing incident response plans.

By centralizing security tools and data from the IT infrastructure, security intelligence with Sense Analytics enables consolidated management and more efficient use of resources devoted to security. Organizations can improve their security posture without additional operational and personnel costs or the expense of purchasing, maintaining, and integrating multiple point products.

9.13 Companies Reap the Benefits of Analytics

- An international energy company had to wade through billions of security events daily to find the ones that needed to be investigated. By deploying appropriate solutions, the company can now analyze two trillion events per day—correlating data in real time across hundreds of sources—to identify the 20–25 potential offenses that pose the greatest risk.
- A credit card firm was struggling to manage legacy technology that not only lacked visibility into the latest threats but also was also costly to operate and maintain. Using the software threat detection and analysis, the firm can now protect its critical data and infrastructure from advanced threats. In addition, it reduced its deployment, tuning, and maintenance costs by 50%.

These are few examples yet; such component in conjunction with BRS can be utilized across organizations and different enterprises.

9.14 A Project Risk Assessment Model

From a dictionary of project management, a private publication in 1984, Earl Cox from early 1974 until 1986 was the founder and president of Interactive Logic, a New York-based software company, that developed and marketed an advanced project management and enterprise modeling system through an international timesharing company. Our project management software, SRMS, was based on a fully, relational database model as early as 1976 and included, by the end of the 1970s, backward and forward chaining machine reasoning techniques to allocate resources and design project precedence networks. In the very early 1980s, we added fuzzy logic to our system, providing managers and project leaders with the ability to construct fuzzy critical path networks, perform fuzzy project risk and sensitivity analysis, and use a fuzzy selection and clustering component to Security and Risk Management Strategies (SRMS) own query language. Over the past 14 years, he has continued to develop advanced project risk and complexity projects for my clients. This subsection is based on my experience in actual risk assessment and configuration projects.

Project management remains one of the major hurdles in applied machine intelligence, not to mention allied fields such as operations research, statistics, and mathematical programming. The central issues in project management revolve around two concepts: the accurate estimation of project parameters and management of the project completion process. The first, obviously, has a profound effect on the second. In fact, the inability to come to terms with the vagueness and uncertainties in project parameter estimation continues to plague management's ability to recognize the inherent risk of assuming some projects. Understanding this risk is critical in evaluating both the dangers of committing to a collection of projects and in applying sound project yield metrics that is specifying the expected return on investment for high-risk projects versus the opportunity costs associated with other less risky projects. Naturally, much of these decisions are coupled to the risk aversion or risk tolerance culture of the corporation.

In this subsection, we will explore a system for analyzing and hence predicting the risk of a project based on its structure. The system employs a process known as case-based reasoning (CBR). With a CBR approach, we analyze the performance of past projects (the database of cases or exemplars) to predict the performance of a currently planned or proposed project. Other uses of the CBR repository include the semiautomatic generation of a project activity plan, the calibration and refinement of estimates, the reconfiguration of a project plan, assignment of or candidate recommendations for the project manager, and the allocation of critical resources.

9.14.1   Project Property Measurements

Evolving a risk assessment for a project involves a wide spectrum of complex parameters. In evaluating a projects risk, many of the important parameters are measures of the change in the values from the initial project specification. This is in fact, what we are ultimately predicting, the possibility that is given a basic project configuration C_i that the critical parameters associated with project C_i will have an unacceptable slippage. Figure 9.12 illustrates many (but not all) of the important variables in a project risk assessment application.

The parameters on the left (in blue) are the core attributes of a project. They consist of several components: the original estimate, the actual (final) value, and the number of revisions to the parameter. From the original and actual values, we can calculate the magnitude of the change (negative for a final amount less than the original and positive for a final amount greater than the original). The remaining parameters (in yellow) are descriptive attributes that help categorize the project. Many of these are measured on a psychometric or arbitrarily ranked scale, usually in the range [0,10]. Zero means that the attribute has minimal or no properties of the concept; ten means that the attribute is completely within the scope of the concept.

- **Project type** consists of three attributes. The first, on a scale of [0,10], defines the project's priority relative to all other currently authorized (and, perhaps, budgeted) projects. This second attribute defines the kind of project (traditional IT

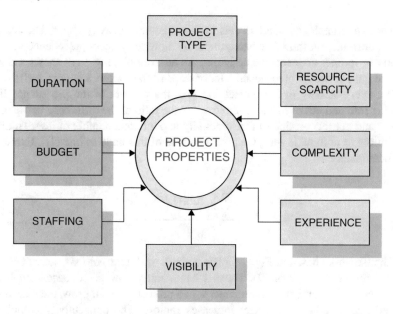

Fig. 9.12 Principle project attributes or variables

system, special request, advanced technology, etc.). The kind of project classification is usually decomposed into 8–12 categories. The third attribute reflects the status of the project and includes such enumeration types as delivered, cancelled (over budget), cancelled (past delivery date/failed to deliver), cancelled (by end user without prejudice), refused by end user, failed to meet end-user requirements, etc.

- **Resource or skill scarcity** is a numeric attribute in the range [0,10] indicating the degree to which the critical resources in the project are drawn from a very limited pool. As an example, a project that requires a Pascal compiler designer and an extensible markup language (XML) expert may rank [7] if there are few individuals in the organization with these skills. Scarce resources increase project risk since (a) they are difficult to acquire, (b) they are usually in demand across multiple projects, and (c) they often have an above average attrition rate within an organization.
- **Complexity** is a numeric attribute in the range [0,N] reflecting either the size of the project or the connection density of the project. The original and final complexity values for a project are maintained by the system. Either the size or the connection complexity form can be used, but only one approach must be applied to all projects in the history (case) file. The project size attribute is the easiest to use and is generally calculated as the sum of the activities per project task divided by the number of tasks. This is reflected in the following expression:

$$C_i = \frac{\sum_{i=1}^{N_T} A_i^T}{N_T} \qquad \text{(Eq. 9.1)}$$

The size complexity value is thus the average task, activity count. The connectionist attribute, on the other hand, measures how the project precedence network tightly or loosely connects the activities. As an example, Fig. 9.13 shows a simple critical path networks precedence relationships (duration times are omitted).

For every activity in the project, we take the product of the precedence links flowing into and out of it. The complexity is the ratio of the difference between the inflow and outflow products of each activity and the total number of tasks (if either the inflow or outflow is zero, we take the value of this node as one [1]). Expressed mathematically as:

$$C_i = \frac{\sum_{i=1}^{N_T} \sum_{i=1}^{A_T} \min\left(1, if \times of\right)}{N_T}$$

(Eq. 9.2)

The network shown in Fig. 9.13 has a single task unit with six activities and a connection complexity of (9). Thus, it has a fairly, low precedence complexity. However, as the number of dependencies in a project segment grow, the measure of precedence or edge complexity increases rapidly. The connectivity complexity parameter is very useful but is also quite difficult to compute. Generally, we should use the size complexity measurement unless the bulk of the client projects have complex precedence relationships with little free float (in which case the connection complexity often yields a better measure of the projects intrinsic brittleness).

Experience is a numeric attribute in the range [0,10] indicating the degree to which the organization has had experience with similar projects in the past. Experience not only creates a categorization partition for the case-based reasoning (CBR) engine but also provides a valuable analytical checkpoint for other project CBR functions (such as estimation error correction and refinement if, e.g., we are consistently late on projects for which we have a reasonable amount of experience).

Visibility is a numeric attribute in the range [0,10] indicating the degree to which the project has political visibility in the organization. Visibility has its own special stresses on a project and must be included in all systems that attempt a ranking of a candidate project's ultimate risk. From past experience, we can almost write a single rule about this attribute:

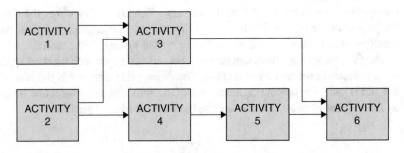

Fig. 9.13 A simple project precedence network

If visibility is high, then project risk is increased.

In any case, visibility, like experience, provides a segmentation plane (or categorization partition) for the CBR engine. This is an important attribute in the risk assessment system.

What Is CBR?

Watson (1999) identified case-based reasoning (CBR) as a methodology arising from research into cognitive science. By studying how people remember information or experiences and use these past examples to solve similar problems, early research hoped to emulate human reasoning in the field of artificial intelligence (AI). Case-based reasoning (CBR) originated in artificial intelligence research; it has transcended barriers between technology and methodology to become applicable in many industrial and research fields. Research and applications have shown CBR can emulate human reasoning and decision-making processes, and as such, a wide variety of construction tasks have been used in case studies.

Furthermore, CBR applications may cover many areas of a project life cycle, but the extent of coverage was generally unknown. By mapping an innovative three-stage classification method to the accepted model of a project life cycle, past CBR research can be grouped and inspected. This "think," "plan," and "do" classification method has shown imbalance of CBR research applications within the construction industry with little aid to those involved in the "doing" phases. The "think," "plan," and "do" classification method has identified potential areas where the further application of CBR may prove useful within infrastructure management and in particular within an educational role for safety and risk management.

CBR is not alone in the field; rule-based systems, neural networks, and genetic algorithms all strive to replicate human thought processes and/or learning methodologies.

9.14.2 A Project Risk Assessment System

The risk assessment application consists of two core processes: a feature extraction methodology, which performs a compact rule induction from the case base, and a fuzzy clustering technique, which organizes the case base into similar (but not necessarily unique) categories. Clustering is performed each time a new set of cases is added to the case database. Figure 9.14 provides a schematic representation of how the risk assessment CBR application works.

Thus, the attributes of the objective case (the project whose risk we want to assess) are used to select the cases in the correct slice partitions (as an example, project kind, experience, and visibility). These cases are passed through a rule

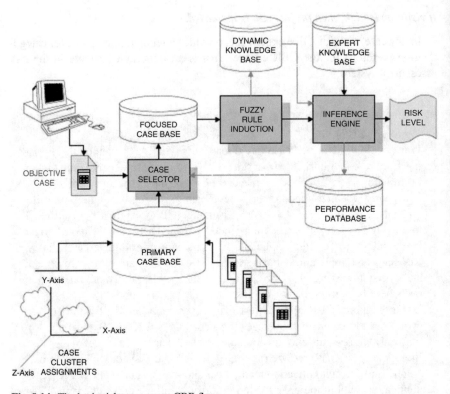

Fig. 9.14 The basic risk assessment CBR flow

induction process, which extracts their features or patterns into a set of fuzzy if-then rules (note also that the induced rules can also be combined with expert knowledge to provide a high degree of specificity where subject matter expert (SME) knowledge is important). The rules are run and subsequently executed by a fuzzy inference engine to generate the predicted risk assessment ranking (on a psychometric scale in the range [0,10]). A performance database records the objective function, the case numbers that were selected to generate a risk, as well as the actual risk level itself. It is used to determine how well the application predicts project risk by comparing predicted risk against actual project performance.

9.14.3   Parameter Similarity Measurements

At the heart of a case-based reasoning system is the selection of cases that are similar to the current objective case. How do we measure such similarity? Conventional CBR approaches used a combination of interval arithmetic and space congruity (such as K-nearest neighbor clustering). The fuzzy CBR approach also uses clustering to partition exemplars (members of the case base) into sets with various degrees of membership. But the selection mechanism also employs a form of fuzzy similarity metrics to determine the degree to which one case parameters are similar to another. All numeric case parameters are treated as bell-shaped fuzzy numbers. The width of the bell curve determines the latitude we have in considering a data point to be representative of the parameter's central value. As Fig. 9.15 illustrates, there are two methods of similarity measurement.

The intersection method is quick and effective for isomorphic fuzzy sets (those with the same shape). It measures the degree to which two isomorphic fuzzy sets overlap. The point of intersection is the degree of similarity. When the fuzzy sets do not have the same shape (one is wider or thinner than the other is), the congruence method is used. This technique takes the ratio of the area shared by the two fuzzy sets and the sum of the areas occupied by both fuzzy sets. By using fuzzy similarity methods, cases can be selected based on the degree to which they match the objective case. This degree is in the fuzzy interval [0,1], and by the extension principle, when a degree of zero is selected, all cases are chosen, and when a degree of one is specified, only cases that are nearly an exact match with the objective are chosen. By working in the interval space [>0,<1], cases that approximately match the objective case are found. These cases generally provide a much richer and more robust solution set. The Fuzzy CBR engine, as it searches for a solution, will automatically relax or tighten the minimum similarity degree for the selected cases in each cycle of the search process.

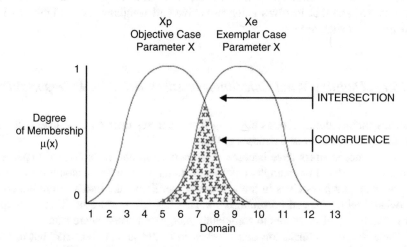

Fig. 9.15 Methods of similarity measurement

9.14.4 Learning from Our Mistakes

Learning from *our mistakes understanding* risk is essential. Corporations and government agencies attempt to reduce risk through the application of proactive techniques such as precedence (PERT and CPM) analysis, cookbook project methodologies, and prototyping. These approaches do little if anything to aid in the definition of the underlying risks associated with a project based on the culture and capabilities of the organization itself. They also do little to isolate critical components of a project that contains the highest degree of risk. To quantify the potential risks of a new project (or even a current project), we must turn to the history of project management in the organization's past. The use of a case-based reasoning (CBR) tool to evaluate the structure and resource requirements of new projects based on the degrees of success or failure for similar projects in the past gives management a powerful tool to forecast problems and allocate resources. Because CBR systems absorb current projects into their case base, they also provide a continual method of refining and adapting their estimates.

9.15 Risk Assessment and Cost Management

Risk assessment and cost management often go together in the healthcare field. The damage control insurance companies partake in is changing. This is because insurers cannot prevent high-cost/high-risk patients from going under their insurance. While marketers are doing more and more to capture the ideal patients, it is impossible to "keep out" high-cost patients. From there, it is the job of insurers to keep costs down, and that involves being proactive and weighing risks. Figure 9.16 is illustration of such process.

9.15.1 Defining What Constitutes as Risky Behavior Is Important

Insurers mainly define risk as anything that increases their costs of care (such as frequency of visits), thus decreases profits.

Over time, insurers have become more cognizant and have created expansive definitions of risk. For example, risk managers have taken into account the actions of hospitals and physicians in the grand scheme. They take into account the costs associated with repeated hospitalization. This has led to more healthcare companies' risk and quality sectors to completely merge or actively share data.

There is more information that is taken in to not just analyze risk but to also prioritize this risk. For example, Optimum Health created a model that considered stress, lack of physical activity, high cholesterol, high blood pressure, high body weight, and the use of antianxiety/antidepressant medications to determine whether a patient is high risk. If a patient fulfilled six of the nine criterions, they are considered overall high risk.

Fig. 9.16 Risk
management core process

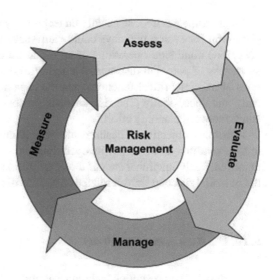

This model is effective because it acknowledges two things: (1) that risk is multifaceted (an individual can be determined "at risk" for multiple reasons) and (2) that it allows for risk analysts to weigh outside and independent environmental factors in how an individual becomes high risk.

9.16   Risk Analysis Techniques

To wrap up this chapter, we summarize the risk analysis and techniques in general and what it provides as foundation for the entire recovery planning efforts. For this, we refer to the Journal Disaster Recovery, which is the journal dedicated to business continuity since 1987 and article that is written by Geoffrey H. Word and Robert F. Shriver. They state that

there may be some terminology and definition differences related to risk analysis, risk assessment, and business impact analysis. Although several definitions are possible and can overlap, for purposes of this article, please consider the following definitions:

- A risk analysis involves identifying the most probable threats to an organization and analyzing the related vulnerabilities of the organization to these threats.
- A risk assessment involves evaluating existing physical and environmental security and controls and assessing their adequacy relative to the potential threats of the organization.
- A business impact analysis involves identifying the critical business functions within the organization and determining the impact of not performing the business function beyond the maximum acceptable outage. Types of criteria that can be used to evaluate the impact include customer service, internal operations, legal/statutory, and financial.

Most businesses depend heavily on technology and automated systems, and their disruption for even a few days could cause severe financial loss and threaten survival. The continued operations of an organization depend on the management's awareness of potential disasters, the ability to develop a plan to minimize disruptions of mission critical functions, and the capability to recover operations expediently and successfully. The risk analysis process provides the foundation for the entire recovery planning effort.

A primary objective of business recovery planning is to protect the organization in the event that all or part of its operations and/or computer services are rendered unusable. Each functional area of the organization should be analyzed to determine the potential risk and impact related to various disaster threats.

9.16.1   Risk Analysis Process

Regardless of the prevention techniques employed, possible threats that could arise inside or outside the organization need to be assessed. Although the exact nature of potential disasters or their resulting consequences are difficult to determine, it is beneficial to perform a comprehensive risk assessment of all threats that can realistically occur to the organization. Regardless of the type of threat, the goals of business recovery planning are to ensure the safety of customers, employees, and other personnel during and following a disaster.

The relative probability of a disaster occurring should be determined. Items to consider in determining the probability of a specific disaster should include, but not be limited to, geographic location, topography of the area. This should consider areas in proximity to major sources of power, bodies of water and airports, degree of accessibility to facilities within the organization, history of local utility companies in providing uninterrupted services, history of the area's susceptibility to natural threats, and proximity to major highways, which transport hazardous waste and combustible products.

Potential exposures may be classified as natural, technical, or human threats. Examples include:

Natural threats: internal flooding, external flooding, internal fire, external fire, seismic activity, high winds, snow and ice storms, volcanic eruption, tornado, hurricane, epidemic, tidal wave, and typhoon

Technical threats: power failure/fluctuation, heating, ventilation or air conditioning failure, malfunction or failure of CPU, failure of system software, failure of application software, telecommunications failure, gas leaks, communications failure, and nuclear fallout

Human threats: robbery, bomb threats, embezzlement, extortion, burglary, vandalism, terrorism, civil disorder, chemical spill, sabotage, explosion, war, biological contamination, radiation contamination, hazardous waste, vehicle crash, airport proximity, work stoppage (internal/external), and computer crime

All locations and facilities should be included in the risk analysis. Rather than attempting to determine exact probabilities of each disaster, a general relational rating system of high, medium, and low can be used initially to identify the probability of the threat occurring.

The risk analysis also should determine the impact of each type of potential threat on various functions or departments within the organization. A risk analysis form can facilitate the process. The functions or departments will vary by type of organization.

The planning process should identify and measure the likelihood of all potential risks and the impact on the organization if that threat occurred. To do this, each department should be analyzed separately. Although the main computer system may be the single greatest risk, it is not the only important concern. Even in the most automated organizations, some departments may not be computerized or automated at all. In fully automated departments, important records remain outside the system, such as legal files, PC data, software stored on diskettes, or supporting documentation for data entry.

The impact can be rated as 0 = No impact or interruption in operations; 1 = Noticeable impact, interruption in operations for up to 8 h; 2 = Damage to equipment and/or facilities, interruption in operations for 8–48 h; and 3 = Major damage to the equipment and/or facilities, interruption in operations for more than 48 h. All main office and/or computer center functions must be relocated.

Certain assumptions may be necessary uniformly to apply ratings to each potential threat. The following are typical assumptions that can be used during the risk assessment process:

1. Although impact ratings could range between 1 and 3 for any facility given a specific set of circumstances, ratings applied should reflect anticipated, likely, or expected impact on each area.
2. Each potential threat should be assumed to be "localized" to the facility being rated.
3. Although one potential threat could lead to another potential threat (e.g., a hurricane could spawn tornados), no domino effect should be assumed.
4. If the result of the threat would not warrant movement to an alternate site(s), the impact should be rated no higher than a "2."
5. The risk assessment should be performed by the facility.

To measure the potential risks, a weighted point rating system can be used. Each level of probability can be assigned points as follows:

Probability points
High 10
Medium 5
Low 1

To obtain a weighted risk rating, probability points should be multiplied by the highest impact rating for each facility. For example, if the probability of hurricanes is high (10 points) and the impact rating to a facility is "3" (indicating that a move to alternate facilities would be required), then the weighted risk factor is 30 (10×3). Based on this rating method, threats that pose the greatest risk (e.g., 15 points and above) can be identified.

Considerations in analyzing risk include:

1. Investigating the frequency of particular types of disasters (often versus seldom)
2. Determining the degree of predictability of the disaster
3. Analyzing the speed of onset of the disaster (sudden versus gradual)
4. Determining the amount of forewarning associated with the disaster
5. Estimating the duration of the disaster
6. Considering the impact of a disaster based on two scenarios:

   (a) Vital records are destroyed.
   (b) Vital records are not destroyed.

7. Identifying the consequences of a disaster, such as:

   (a) Personnel availability
   (b) Personal injuries
   (c) Loss of operating capability
   (d) Loss of assets
   (e) Facility damage

8. Determining the existing and required redundancy levels throughout the organization to accommodate critical systems and functions, including:

   (a) Hardware
   (b) Information
   (c) Communication
   (d) Personnel
   (e) Services

9. Estimating potential dollar loss:

   (a) Increased operating costs
   (b) Loss of business opportunities
   (c) Loss of financial management capability
   (d) Loss of assets
   (e) Negative media coverage
   (f) Loss of stockholder confidence
   (g) Loss of goodwill
   (h) Loss of income
   (i) Loss of competitive edge
   (j) Legal actions

10. Estimating potential losses for each business function based on the financial and service impact and the length of time the organization can operate without this business function. The impact of a disaster related to a business function

depends on the type of outage that occurs and the time that elapses before normal operations can be resumed
11. Determining the cost of contingency planning

9.16.2   Disaster Prevention

Because a goal of business recovery planning is to ensure the safety of personnel and assets during and following a disaster, a critical aspect of the risk analysis process is to identify the preparedness and preventive measures in place at any point in time. Once the potential areas of high exposure to the organization are identified, additional preventative measures can be considered for implementation.

Disaster prevention and preparedness begins at the top of an organization. The attitude of senior management toward security and prevention should permeate the entire organization. Therefore, management's support of disaster planning can focus attention on good security and prevention techniques and better prepare the organization for the unwelcome and unwanted.

Disaster prevention techniques include two categories: procedural prevention and physical prevention.

Procedural prevention relates to activities performed on a day-to-day, month-to-month, or annual basis, relating to security and recovery. Procedural prevention begins with assigning responsibility for overall security of the organization to an individual with adequate competence and authority to meet the challenges. The objective of procedural prevention is to define activities necessary to prevent various types of disasters and ensure that these activities are performed regularly.

Physical prevention and preparedness for disaster begins when a site is constructed. It includes special requirements for building construction, as well as fire protection for various equipment components. Special considerations include computer area, fire detection and extinguishing systems, record(s) protection, air conditioning, heating and ventilation, electrical supply and UPS systems, emergency procedures, vault storage area(s), and archival systems.

9.16.3   Security and Control Considerations

Security and controls refer to all the measures adopted within an organization to safeguard assets, ensure the accuracy and reliability of records, and encourage operational efficiency and adherence to prescribed procedures. The system of internal controls also includes the measures adopted to safeguard the computer system.

The nature of internal controls is such that certain control procedures are necessary for a proper execution of other control procedures. This interdependence of control procedures may be significant because certain control objectives that appear to have been achieved may, in fact, not have been achieved because of weaknesses in other control procedures upon which they depend.

Concern over this interdependence of control procedures may be greater with a computerized system than with a manual system because computer operations often have greater concentration of functions, and certain manual control procedures may depend on automated control procedures, even though that dependence is not readily apparent. Adequate computer internal controls are a vital aspect of an automated system.

Security is an increasing concern because computer systems are increasingly complex. Particular security concerns result from the proliferation of PCs, local area networking, and online systems that allow more access to the mainframe and departmental computers. Modern technology provides computer thieves with powerful new electronic safecracking tools.

Computer internal controls are especially important because computer processing can circumvent traditional security and control techniques. There are two types of computer control techniques:

1. General controls that affect all computer systems
2. Application controls that are unique to specific applications

Important areas of concern related to general computer internal controls include organization controls, systems development and maintenance controls, documentation controls, access controls, data and procedural controls, physical security, password security systems, and communications security

Application controls are security techniques that are unique to a specific computer application system. Application controls are classified as input controls, processing controls, and output controls.

9.16.4   Insurance Considerations

Adequate insurance coverage is a key consideration when developing a business recovery plan and performing a risk analysis. Having a disaster plan and testing it regularly may not, in itself, lower insurance rates in all circumstances.

However, a good plan can reduce risks and address many concerns of the underwriter, in addition to affecting the cost or availability of the insurance.

Most insurance agencies specializing in business interruption coverage can provide the organization with an estimate of anticipated business interruption costs. Many organizations that have experienced a disaster indicate that their costs were significantly higher than expected in sustaining temporary operations during recovery.

Most business interruption coverages include lost revenues following a disaster. Extra expense coverages include all additional expenses until normal operations can be resumed. However, coverages differ in the definition of resumption of services. As a part of the risk analysis, these coverages should be discussed in detail with the insurer to determine their adequacy.

To provide adequate proof of loss to an insurance company, the organization may need to contract with a public adjuster who may charge between 3% and 10% of recovered assets for the adjustment fee. Asset records become extremely important as the adjustment process takes place.

Types of insurance coverages to be considered may include computer hardware replacement, extra expense coverage, business interruption coverage, valuable paper and records coverage, errors and omissions coverage, fidelity coverage, and media transportation coverage.

With estimates of the costs of these coverages, management can make reasonable decisions on the type and amount of insurance to carry.

These estimates also allow management to determine to what extent the organization should self-insure against certain losses.

9.16.5   Records

Records can be classified in one of the three following categories: vital records, important records, and useful records.

Vital records are irreplaceable. Important records can be obtained or reproduced at considerable expense and only after considerable delay. Useful records would cause inconvenience if lost, but can be replaced without considerable expense.

Vital and important records should be duplicated and stored in an area protected from fire or its effects.

Records kept in the computer room should be minimized and should be stored in closed metal files or cabinets. Records stored outside the computer room should be in fire-resistant file cabinets with fire resistance of at least 2 h.

Protection of records also depends on the particular threat that is present. An important consideration is the speed of onset and the amount of time available to act. This could range from gathering papers hastily and exiting quickly to an orderly securing of documents in a vault. Identifying records and information is most critical for ensuring the continuity of operations.

A systematic approach to records management is also an important part of the risk analysis process and business recovery planning. Additional benefits include reduced storage costs, expedited service, and federal and state statutory compliance.

Records should not be retained only as proof of financial transactions but also to verify compliance with legal and statutory requirements. In addition, businesses must satisfy retention requirements as an organization and employer. These records are used for independent examination and verification of sound business practices.

Federal and state requirements for records retention must be analyzed. Each organization should have its legal counsel approve its own retention schedule. As well as retaining records, the organization should be aware of the specific record salvage procedures to follow for different types of media after a disaster.

9.16.6   Conclusion

The risk analysis process is an important aspect of business recovery planning. The probability of a disaster occurring in an organization is highly uncertain. Organizations should also develop written, comprehensive business recovery plans that address all the critical operations and functions of the business.

The plan should include documented and tested procedures, which, if followed, will ensure the ongoing availability of critical resources and continuity of operations.

A business recovery plan, however, is similar to liability insurance. It provides a certain level of comfort in knowing that if a major catastrophe occurs, it will not result in financial disaster for the organization.

Insurance, by itself, does not provide the means to ensure continuity of the organization's operations and may not compensate for the incalculable loss of business during the interruption or the business that never returns.

Geoffrey H. Wold and Robert F. Shriver are the National Directors of Information Technology Consulting at McGladrey & Pullen, in Minneapolis, Minn.

This article was adapted from V7#3.

Disaster Recovery World© 1997 and Disaster Recovery Journal© 1997 are copyrighted by Systems Support, Inc. All rights reserved. Reproduction in whole or part is prohibited without the express written permission from Systems Support, Inc.

References

1. David A. Dickey, Department of Statistics, North Carolina State University Raleigh, NC 27695-8203. Retrieved from http://www4.stat.ncsu.edu/~dickey/.
2. Wilson, M. E., & Mather, L. E. (1974). *Journal of the American Medical Association, 1229*(11), 1421–1422.
3. van Zadelhoff, M., Lovejoy, K., & Jarvis, D.. (December, 2014). *Fortifying for the future: Insights from the 2014 IBM chief information security officer assessment.* IBM Center for Applied Insights. Retrieved from http://www-935.ibm.com/services/us/en/it-services/security-services/index.html?lnk=sec_home.
4. Verizon RISK Team. 2015. *2015 Data breach investigations report.* Verizon. Retrieved from http://www.verizonenterprise.com/DBIR/2015/.
5. Ponemon Institute. (May, 2015). 2015 *Cost of data breach study: Global analysis.* Ponemon Institute Research Report.
6. Ponemon Institute. (December, 2014). *Network forensic investigations market study.* Ponemon Institute Research Report. Retrieved from ibm.com/common/ssi/cgi-bin/ssialias?infotype=SA&subtype=WH& htmlfid=WGL03070USEN#loaded.

Chapter 10
What Is Data Analysis from Data Warehousing Perspective?

Analysis of data is a process of inspecting, cleansing, transforming, and modeling data with the goal of discovering useful information; suggesting conclusions; and supporting decision-making. Data analysis has multiple facets and approaches, encompassing diverse techniques under a variety of names, in different business, science, and social science domains. Data is collected from a variety of sources. The requirements may be communicated by analysts to custodians of the data, such as information technology personnel within an organization. The data may also be collected from sensors in the environment, such as traffic cameras, satellites, recording devices, etc. It may also be obtained through interviews, downloads from online sources, or reading documentation. Data initially obtained must be processed or organized for analysis. For instance, these may involve placing data into rows and columns in a table format (i.e., structured data) for further analysis, such as within a spreadsheet or statistical software.

10.1 Introduction

Data integration is a precursor to data analysis, and data analysis is closely linked to data visualization and data dissemination. The term data analysis is sometimes used as a synonym for data modeling.

Analysis refers to breaking a whole into its separate components for individual examination. Data analysis is a process for obtaining raw data and converting it into information useful for decision-making by users. Data is collected and analyzed to answer questions, test hypotheses, or disprove theories [1].

Statistician John Tukey defined data analysis in 1961 as "Procedures for analyzing data, techniques for interpreting the results of such procedures, ways of planning the gathering of data to make its analysis easier, more precise or more accurate, and all the machinery and results of (mathematical) statistics which apply to analyzing data" [2].

© Springer International Publishing AG 2017 269
B. Zohuri, M. Moghaddam, *Business Resilience System (BRS): Driven Through Boolean, Fuzzy Logics and Cloud Computation*,
DOI 10.1007/978-3-319-53417-6_10

There are several phases that can be distinguished, described below. The phases are iterative, in that feedback from later phases may result in additional work in earlier phases [3].

Large parts of *data analysis* are inferential in the sample-to-population sense, but these are only parts, not the whole. Large parts of data analysis are incisive, laying bare indications, which we could not perceive by simple and direct examination of the raw data, but these too are only parts, not the whole. Some parts of data analysis, as the term is here stretched beyond its philology, are allocation, in the sense that they guide us in the distribution of effort and other valuable considerations in observation, experimentation, or analysis. Data analysis is a larger and more varies field than inference or incisive procedures, or allocation [2].

Figure 10.1 is an illustration of data science process flowchart, and as we stated at the opening of the chapter, analysis and processing of data refers to breaking a whole into its separate components for individual examination. Data analysis is a process for obtaining raw data and converting it into information useful for decision-making by users. Data is collected and analyzed to answer questions, test hypotheses, or disprove theories. However, data initially obtained must be processed or organized for analysis. For instance, these may involve placing data into rows and columns in a table format (i.e., structured data) for further analysis, such as within a spreadsheet or statistical software.

Once processed and organized, the data may be incomplete, contain duplicates, or contain errors. The need for data cleaning will arise from problems in the way

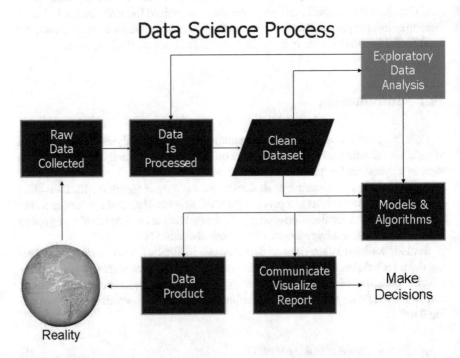

Fig. 10.1 Data science process flowchart

that data is entered and stored. Data cleaning is the process of preventing and correcting these errors. Common tasks include record matching, identifying inaccuracy of data, overall quality of existing data [4], reduplication, and column segmentation [5]. Such data problems can also be identified through a variety of analytical techniques. For example, with financial information, the totals for particular variables may be compared against separately published numbers believed to be reliable [6]. Unusual amounts above or below predetermined thresholds may also be reviewed. There are several types of data cleaning, which depend on the type of data such as phone numbers, e-mail addresses, employers, etc. Quantitative data methods for outlier detection can be used to get rid of likely incorrectly entered data. Textual data spellcheckers can be used to lessen the amount of mistyped words, but it is harder to tell if the words themselves are correct [7].

Anyone who has delved into data from the real world knows it can be messy. Survey takers can write down the wrong response. People entering data into a computer can type in the wrong numbers. Computers sometimes garble data because of software bugs (especially when converting one file format to another). Data formats become obsolete as software changes. Electronic data recorders break or go out of adjustment. Analysts mislabel units and make calculation errors.

Cleaning the data to correct such problems is almost always necessary, but this step is often ignored. Ask about the process that was used to clean the data. If the response is just a blank stare or worse, you know you are in trouble. Many companies (such as AT&T and Sega of America) now assign people to check for data quality in the face of all the problems associated with real data.

Data collection has become a ubiquitous function of large organizations—not only for record keeping but to support a variety of data analysis tasks that are critical to the organizational mission. Data analysis typically drives decision-making processes and efficiency optimizations, and an increasing number of settings are the raison d'etre of entire agencies or firms [7].

Despite the importance of data collection and analysis, data quality remains a pervasive and thorny problem in almost every large organization. The presence of incorrect or inconsistent data can significantly distort the results of analyses, often negating the potential benefits of information-driven approaches. As a result, there has been a variety of research over the last decades on various aspects of *data cleaning* and that is:

- Computational procedures to automatically or semiautomatically identify—and, when possible, correct—errors in large data sets

Once the data is cleaned, it can be analyzed. Analysts may apply a variety of techniques referred to as exploratory data analysis to begin understanding the messages contained in the data.

The phases of the intelligence cycle used to convert raw information into actionable intelligence or knowledge are conceptually similar to the phases in data analysis, and the illustration of such phase is depicted in Fig. 10.2.

Mathematical formulas or models called algorithms may be applied to the data to identify relationships among the variables, such as correlation or causation. In general

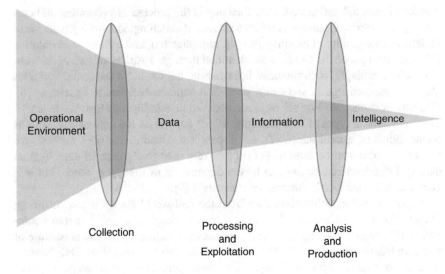

Fig. 10.2 Relationship of data, information, and intelligence. Source: Joint Intelligence/Joint Publication 2-0 (Joint Chief of Staff)

terms, models may be developed to evaluate a particular variable in the data based on other variable(s) in the data, with some residual error depending on model accuracy (i.e., data = model + error) [1].

In addition, a data product is a computer application that takes data inputs and generates outputs, feeding them back into the environment. It may be based on a *model* or *algorithm*. An example is an application that analyzes data about customer purchasing history and recommends other purchases the customer might enjoy.

The analysis of data objects and their relationships to other data objects. Data modeling is often the first step in database design and object-oriented programming as the designers first create a conceptual model of how data items relate to each other. Data modeling involves a progression from conceptual model to logical model to physical schema.

10.2   Data Mining and Data Modeling

Data modeling is a representation of the data structures in a table for a company's database and is a very powerful expression of the company's business requirements. This data model is the guide used by functional and technical analysts in the design and implementation of a database.

Data models are used for many purposes, from high-level conceptual models to physical data models. Data modeling explores data-oriented structures and identifies entity types. This is unlike class modeling, where classes are identified.

Three basic styles of data modeling are generally used in practice today:

1. **Conceptual Data Models (CDMs):** High-level, static business structures and concepts
   A conceptual data model identifies the highest-level relationships between the different entities. Features of conceptual data model include:

   - The important entities and the relationships among them.
   - No attribute is specified.
   - No primary key is specified.

   Figure 10.3 is an example of a conceptual data model.
   From Fig. 10.3, we can see that the only information shown via the conceptual data model is the entities that describe the data and the relationships between those entities. No other information is shown through the conceptual data model.

2. **Logical Data Models (LDMs):** Entity types, data attributes, and relationships between entities
   Logical data modeling is the process of representing data architecture and organization in a graphical way without any regard to the physical implementation or the database management system, and technology, which is involved in storing the data. A logical data model provides all the information about the various entities and the relationships between the entities present in a database.

   A logical data model represents the organization of a set of data by standardizing the people, places, things (entities), and the rules and relationships between them using a standard language and notation. It provides a conceptual abstract overview of the structure of the data.

   Logical data modeling does not provide any information related to how the structure is to be implemented or the means (technologies) that are needed to implement the data structure shown. It is a technology-independent model of data

Fig. 10.3 Conceptual data model

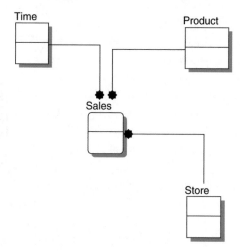

that is developed from the initial structures identified by the conceptual model of data. Some of the information presented by a logical data model includes the following:

- Entities
- Attributes of entities
- Key groups (primary keys, foreign keys)
- Relationships
- Normalization

The steps for designing the logical data model are as follows:

1. Specify primary keys for all entities.
2. Find the relationships between different entities.
3. Find all attributes for each entity.
4. Resolve many-to-many relationships.
5. Normalization.

Figure 10.4 below is an example of a logical data model.

Comparing the logical data model (Fig. 10.4) shown above with the conceptual data model diagram (Fig. 10.3), we see the main differences between the two:

Fig. 10.4 Logical data model diagram

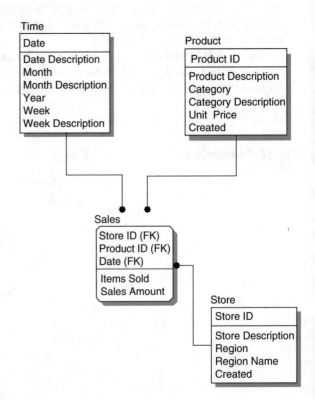

- In a logical data model, primary keys are present, whereas in a conceptual data model, no primary key is present.
- In a logical data model, all attributes are specified within an entity. No attributes are specified in a conceptual data model.
- Relationships between entities are specified using primary keys and foreign keys in a logical data model. In a conceptual data model, the relationships are simply stated, not specified, so we simply know that two entities are related, but we do not specify what attributes are used for this relationship.

3. **Physical Data Models (PDMs)**: The internal schema database design
   Physical data model represents how the model will be built in the database. A physical database model shows all table structures, including column name, column data type, column constraints, primary key, foreign key, and relationships between tables. Features of a physical data model include:

- Specification of all tables and columns.
- Foreign keys are used to identify relationships between tables.
- Denormalization may occur based on user requirements.
- Physical considerations may cause the physical data model to be quite different from the logical data model.
- Physical data model will be different for different relational database management systems (RDBMS). For example, data type for a column may be different between MySQL and SQL Server.

The steps for physical data model design are as follows:

1. Convert entities into tables.
2. Convert relationships into foreign keys.
3. Convert attributes into columns.
4. Modify the physical data model based on physical constraints/requirements (Fig. 10.5).

Comparing the physical data model shown above with the logical data model diagram, we see the main differences between the two:

- Entity names are now table names.
- Attributes are now column names.
- Data type for each column is specified. Data types can be different depending on the actual database is being, used.

10.2.1   Data Warehousing Concepts

The three levels of data modeling, conceptual data model, logical data model, and physical data model, were discussed in prior sections. Here, we compare these three types of data models. Table 10.1 compares the different features.

Below, we show the conceptual, logical, and physical versions of a single data model (Fig. 10.6).

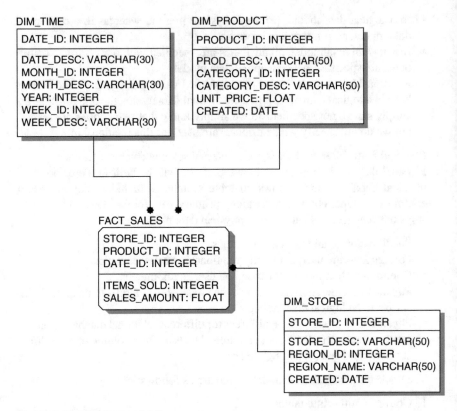

Fig. 10.5 Physical data model diagram

Table 10.1 Comparison table of different data warehousing concepts

Feature	Conceptual	Logical	Physical
Entity names	✓	✓	
Entity relationships	✓	✓	
Attributes		✓	
Primary keys		✓	✓
Foreign keys		✓	✓
Table names			✓
Column names			✓
Column data types			✓

We can see that the complexity increases from conceptual to logical to physical. This is why, we always first start with the conceptual data model (so we understand at high level what are the different entities in our data and how they relate to one another). Then, we move on to the logical data model (so we understand the details of our data without worrying about how they will actually implemented) and finally the physical data model (so we know exactly how to implement our data model in the database of choice). In a data warehousing project, sometimes the conceptual data model and the logical data model are considered as a single deliverable.

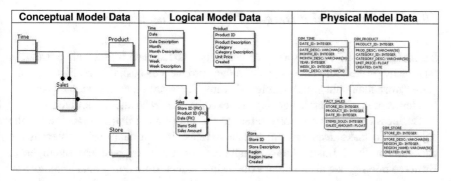

Fig. 10.6 Illustration of all data warehousing concepts for single data model

10.2.2   Data Warehousing and Data Integrity

Data integrity refers to the validity of data, meaning data is consistent and correct. In the data warehousing field, we frequently hear the term "garbage in, garbage out." If there is no data integrity in the data warehouse, any resulting report and analysis will not be useful.

In a data warehouse or a data mart, there are three areas of where data integrity needs to be enforced:

1. **Database Level**

   We can enforce data integrity at the database level. Common ways of enforcing data integrity include:

   - **Referential integrity**
   - **Primary key/UNIQUE constraint**
   - Primary keys and the UNIQUE constraint are used to make sure every row in a table can be uniquely identified.
   - **Not NULL vs. NULL-able**
   - For columns identified as NOT NULL, they may not have a NULL value.
   - **Valid values**

   - Only allowed values are permitted in the database. For example, if a column can only have positive integers, a value of "−1" cannot be allowed.

2. **ETL Process**

   For each step of the extract, load, and transfer (ETL) process, data integrity checks should be put in place to ensure that source data is the same as the data in the destination. Most common checks include record counts or record sums.

3. **Access Level**

   We need to ensure that data is not altered by any unauthorized means either during the ETL process or in the data warehouse. To do this, there need to be safeguards against unauthorized access to data (including physical access to the servers), as well as logging of all data access history. Data integrity can only be ensured if there is no unauthorized access to the data.

10.2.3   Data Warehousing and Business Intelligence

Business intelligence (BI) is a term commonly associated with data warehousing. In fact, many of the tool vendors position their products as business intelligence software rather than data warehousing software. There are other occasions where the two terms are used interchangeably. So, exactly what is business intelligence?

Business intelligence usually refers to the information that is available for the enterprise to make decisions on. A data warehousing (or data mart) system is the backend, or the infrastructural, component for achieving business intelligence. Business intelligence also includes the insight gained from doing data mining analysis, as well as unstructured data (thus the need for content management systems). For our purposes here, we will discuss business intelligence in the context of using a data warehouse infrastructure [8].

This section includes the following:

1. **Business intelligence tools**: Tools commonly used for business intelligence
2. **Business intelligence uses**: Different forms of business intelligence
3. **Business intelligence news**: News in the business intelligence area

10.2.4   Data Warehousing and Master Data Management

First, we need to understand the concept of master data management (MDM), which it refers to the process of creating and managing data that an organization must have as a single master copy, called the master data. Usually, master data can include customers, vendors, employees, and products, but can differ by different industries and even different companies within the same industry. MDM is important because it offers the enterprise a single version of the truth. Without a clearly defined master data, the enterprise runs the risk of having multiple copies of data that are inconsistent with one another.

MDM is typically more important in larger organizations. In fact, the bigger the organization, the more important the discipline of MDM is, because a bigger organization means that there are more disparate systems within the company, and the difficulty on providing a single source of truth, as well as the benefit of having master data, grows with each additional data source. A particularly big challenge to maintaining master data occurs when there is a merger/acquisition. Each of the organizations will have its own master data, and how to merge the two sets of data will be challenging. Let us take a look at the customer files: The two companies will likely have different unique identifiers for each customer. Addresses and phone numbers may not match. One may have a person's maiden name and the other the current last name. One may have a nickname (such as "Bill") and the other may have the full name (such as "William"). All these contribute to the difficulty in creating and maintaining a single set of master data.

At the heart of the master data, management program is the definition of the master data. Therefore, it is essential that we identify who is responsible for defining and enforcing the definition. Due to the importance of master data, a dedicated person or team should be appointed. At the minimum, a data steward should be identified. The responsible party can also be a group—such as a data governance committee or a data governance council.

10.2.5   Master Data Management vs. Data Warehousing

Based on the discussions so far, it seems like master data management and data warehousing have a lot in common. For example, the effort of data transformation and cleansing is very similar to an ETL process in data warehousing, and in fact they can use the same ETL tools. In the real world, it is not uncommon to see MDM and data warehousing fall into the same project. On the other hand, it is important to call out the main differences between the two:

1. Different Goals
   The main purpose of a data warehouse is to analyze data in a multidimensional fashion, while the main purpose of MDM is to create and maintain a single source of truth for a particular dimension within the organization. In addition, MDM requires solving the root cause of the inconsistent metadata, because master data needs to be propagated back to the source system in some way. In data warehousing, solving the root cause is not always needed, as it may be enough just to have a consistent view at the data warehousing level rather than having to ensure consistency at the data source level.
2. Different Types of Data
   Master data management is only applied to entities and not transactional data, while a data warehouse includes data that are both transactional and non-transactional in nature. The easiest way to think about this is that MDM only affects data that exists in dimensional tables and not in fact tables, while a data warehousing environment includes both dimensional tables and fact tables.
3. Different Reporting Needs
   In data warehousing, it is important to deliver to end users the proper types of reports using the proper type of reporting tool to facilitate analysis. In MDM, the reporting needs are very different—it is far more important to be able to provide reports on data governance, data quality, and compliance, rather than reports based on analytical needs.
4. Where Data Is Used
   In a data warehouse, usually the only usage of this "single source of truth" is for applications that access the data warehouse directly or applications that access systems that source their data straight from the data warehouse. Most of the time, the original data sources are not affected. In master data management, on the other hand, we often need to have a strategy to get a copy of the master data back to the source system. This poses challenges that do not exist in a data warehousing

environment. For example, how do we sync the data back with the original source? Once a day? Once an hour? How do we handle cases where the data was modified as it went through the cleansing process? And how much modification do we need make do to the source system so it can use the master data? These questions represent some of the challenges MDM faces. Unfortunately, there is no easy answer to those questions, as the solution depends on a variety of factors specific to the organization, such as how many source systems there are, how easy/costly it is to modify the source system, and even how internal politics play out.

10.3   Big Data: What It Is and Why It Matters

Big data is a term that describes the large volume of data—both structured and unstructured—that inundates a business on a day-to-day basis. However, it is not the amount of data which is important, it is what organizations do with the data that matters. Big data can be analyzed for insights that lead to better decisions and strategic business moves.

While the term "big data" is relatively new (i.e., the term is in use since the 1990s), the act of gathering and storing large amounts of information for eventual analysis is ages old. The concept gained momentum in the early 2000s when industry analyst Doug Laney articulated the now-mainstream definition of big data as the three Vs:

Volume: Organizations collect data from a variety of sources, including business transactions, social media, and information from sensor or machine-to-machine data. In the past, storing it would have been a problem—but new technologies (such as Hadoop) have eased the burden.

Velocity: Data streams at an unprecedented speed and must be dealt with in a timely manner. RFID tags, sensors, and smart metering are driving the need to deal with torrents of data in near real time.

Variety: Data comes in all types of formats—from structured, numeric data in traditional databases to unstructured text documents, e-mail, video, audio, stock ticker data, and financial transactions.

However, we need to consider two additional dimensions when it comes to big data and they are:

Variability: In addition to the increasing velocities and varieties of data, data flows can be highly inconsistent with periodic peaks. Is something trending in social media? Daily, seasonal, and event-triggered peak data loads can be challenging to manage, even more so with unstructured data.

Complexity: Today's data comes from multiple sources, which makes it difficult to link, match, cleanse, and transform data across systems. However, it is necessary to connect and correlate relationships, hierarchies, and multiple data linkages, or your data can quickly spiral out of control.

The big potential of big data is that the amount of data that is being created and stored on a global level is almost inconceivable and it just keeps growing. That means there is even more potential to glean key insights from business information—yet only a small percentage of data is actually analyzed. What does that mean for businesses? How can they make better use of the raw information that flows into their organizations every day?

2016 was a landmark year for big data with more organizations storing, processing, and extracting value from data of all forms and sizes. In 2017, systems that support large volumes of both structured and unstructured data will continue to rise. The market will demand platforms that help data custodians govern and secure big data while empowering end users to analyze that data. These systems will mature to operate well inside of enterprise IT systems and standards.

Learn more about where things are headed. This paper highlights the top big data trends for 2017 including:

- Big data becomes fast and approachable: Options expand to speed up Hadoop.
- Organizations leverage data lakes from the get-go to drive value.
- Spark and machine learning light up big data.

In summary, "big data" is a term for data sets that are so large or complex that traditional data processing applications are inadequate to deal with them. Challenges include analysis, capture, data duration, search, sharing, storage, transfer, visualization, querying, and updating information privacy.

The term "big data" often refers simply to the use of predictive analytics, user behavior analytics, or certain other advanced data analytics methods that extract value from data and seldom to a particular size of data set. "There is little doubt that the quantities of data now available are indeed large, but that is not the most relevant characteristic of this new data ecosystem."

Analysis of data sets can find new correlations to "spot business trends, prevent diseases, combat crime, and so on." Scientists, business executives, practitioners of medicine, advertising, and governments alike regularly meet difficulties with large data sets in areas including Internet search, finance, urban informatics, and business informatics. Scientists encounter limitations in e-Science work, including meteorology, genomics, connectomics, complex physics simulations, biology, and environmental research.

Data sets grow rapidly—in part because they are increasingly gathered by cheap and numerous information-sensing mobile devices, aerial (remote sensing), software logs, cameras, microphones, radio-frequency identification (RFID) readers, and wireless sensor networks. The world's technological per-capita capacity to store information has roughly doubled every 40 months since the 1980s; as of 2012, every day, 2.5 exabytes (2.5×1018) of data is generated. One question for large enterprises is determining who should own big data initiatives that affect the entire organization.

Growth in data and demand for accurate information globally has forced organizations and enterprise to ask for more and more storage and as result cloud computations. Big data has increased the demand of information management specialists so much so that Software AG, Oracle Corporation, IBM, Microsoft, SAP, EMC, HP, and Dell have spent more than $15 billion on software firms specializing in data management and analytics. In 2010, this industry was worth more than $100 billion, and it was growing at almost 10% a year: about twice as fast as the software business as a whole. See Fig. 10.7.

Developed economies increasingly use data-intensive technologies. There are 4.6 billion mobile phone subscriptions worldwide and between 1 billion and 2 billion people accessing the Internet. Between 1990 and 2005, more than one billion people worldwide entered the middle class, which means more people became more literate, which in turn leads to information growth. The world's effective capacity to exchange information through telecommunication networks was 281 petabytes in 1986, 471 petabytes in 1993, 2.2 exabytes in 2000, and 65 exabytes in 2007, and predictions put the amount of Internet traffic at 667 exabytes annually by 2014. According to one estimate, one third of the globally stored information is in the form of alphanumeric text and still image data, which is the format most useful for most big data applications. This also shows the potential of yet unused data (i.e., in the form of video and audio content).

While many vendors offer off-the-shelf solutions for big data, experts recommend the development of in-house solutions custom-tailored to solve the company's problem at hand, if the company has sufficient technical capabilities.

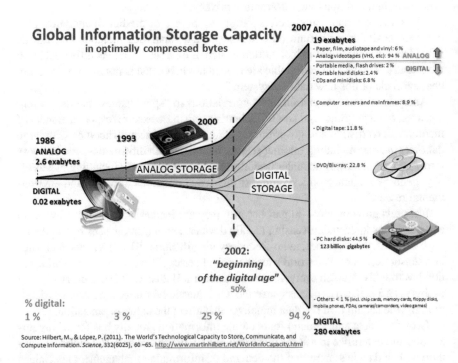

Fig. 10.7 Global information storage capacity

10.4   Data Mining and Data Analysis in Real Time

Real-time operation is diverse, including many enabling technologies executed in many time frames. The pace of business and other organizational activity continues to accelerate, such that organizations must now react faster and more frequently to customer interactions, Service Level Agreements (SLAs), operational commitments, competitive pressures, and other time-sensitive issues. With real-time technologies, maturing and proliferating, organizations must now evaluate the numerous options they need for the real-time delivery, access, and analysis of information, if they are to reduce or eliminate business latency [9].

One of the insights explored in this report is that most organizations need multiple solutions to achieve real-time data, business intelligence (BI), and analytics. Two reasons stand out [9]:

1. **Real time has many meanings**: As with most IT disciplines, real-time solutions should be designed to satisfy business requirements. This leads to tremendous diversity because business processes have different requirements for the freshness of information and hence various requirements for how fast and frequently data should be fetched, processed, and delivered. The catch is that each of the many technologies available today for real-time data, BI, and analytics has unique characteristics for performance, data types, interfaces, and processing strengths. Most organizations need multiple real-time technologies to satisfy the diverse speeds and data content of real-time business processes.
2. **Data travels across many software systems**: Even a basic technology stack for BI, data warehousing, and analytics will include multiple tool and platform types such as those for databases, data integration, reporting, and analytics. When data must travel in real time across these, each tool and platform needs some kind of real-time capability. Otherwise, one link in the chain becomes a bottleneck that hinders the level of real time desired.

Real-time technologies tend to be specific to data, BI, or analytics. Given that most software configurations for real-time operation involve data traveling (and being processed) across multiple systems, this report discusses three broad layers of real-time technology, plus the interoperability of the three, their integration with other systems, and common use cases and enabling technologies for each layer [9].

1. **Real-time data**: Real-time BI and analytics cannot operate without real-time data. Luckily, database, data integration, and other data management technologies can now handle the many new sources of real-time data that have come online recently, including new sources of machine data, which streams from sensors, robots, mobile devices, and social media. Capturing data as it is created or updated is a foundation for operational BI and real-time analytic insights into customer behaviors, competitive threats, and operational efficiencies.
2. **Real-time BI**: Traditionally limited to historical views of enterprise information, BI users today want to know what's happening now. In-memory computing,

accelerators, and visualization are enabling real-time views via reports, OLAP, and data discovery. Organizations are revamping and automating processes for faster financial reporting, operational intelligence, and performance management. Tighter integration between BI and operational applications enables users to embed BI functions and insights into a wide variety of enterprise applications.

3. **Real-time analytics**: Increasingly, organizations need to analyze real-time data in order to rescore analytic models frequently or to correlate millisecond-old events with data from other sources and latencies. New technologies—such as CEP, stream mining, in-memory analytics, and in-database analytics—have enabled real-time analytics for customer interactions, fraud detection, and situational intelligence.

Real-time functions are fast and frequent, as well as constant or intermittent. Definitions of real time aside, there are other concepts and practices that merit consideration in this discussion:

Fast versus frequent: Real-time technologies execute fast, whether they're running a query, delivering an event over a message bus, refreshing a report, rescoring an analytic model, and so on. An equally important characteristic of real-time operations is that they tend to execute frequently. For example, sensors can be configured to send data every second, minute, or other short time frame. Similarly, as organizations move closer to real-time operation of the business, they may tweak data integration and report refresh jobs (designed to run overnight in batch), so they also run a few times during the business day in so-called intraday micro-batches.

Streaming data as extreme real time: Streaming data is an extreme case because some streams generate an event, record, code, or message once a second, minute, or other short time frame. A stream can come from machinery (sensors, mobile devices, Global Positioning Satellite (GPS) units) or the logs to which enterprise applications append data frequently (ranging from enterprise resource planning (ERP) to Web servers). Capturing and processing a stream in real time (plus correlating it with other streams and other data sources) demand very special software, typically for complex event processing (CEP) or operational intelligence. It's worth acquiring such technology because it enables many new applications for fast-paced business monitoring, surveillance, customer service, automated responses, and so on.

Continuous versus intermittent streams: Some sensors have a power source that keeps them on and transmitting continuously, such as the digital thermometers in a chemical plant or the GPS units on trucks in logistics firms. However, most radio-frequency identification (RFID) chips have no power source and come to life only long enough to emit a code when hit by a transmission from an RFID transceiver; proximity to a transceiver is intermittent, so RFID codes in the stream are too. Even though the source may not be continuously active (and, therefore, not considered real time by some definitions), receiving systems should be active at all times, ready to capture and process data at any time.

Technology Components

Operational intelligence solutions share many features, and therefore many also share technology components. This is a list of some of the commonly found technology components and the features they enable:

- **Business activity monitoring (BAM)**—dashboard customization and personalization complex event processing (CEP)—advanced, continuous analysis of real-time information and historical data
- **Business process management (BPM)**—to perform model-driven execution of policies and processes defined as Business Process Model and Notation (BPMN) models
- **Metadata** framework to model and link events to resources
- **Multichannel publishing and notification**
- **Dimensional database**
- **Root cause analysis**
- **Multi-protocol event collection**

Operational intelligence (OI) is a relatively new market segment (compared to the more mature business intelligence and business process management segments). In addition to companies that produce dedicated and focused products in this area, there are numerous companies in adjacent areas that provide solutions with some OI components.

Operational intelligence integrates information, supporting smarter decision-making in time to maximize impact. By correlating a variety of events and data from both streaming feeds and historical data silos, operational intelligence helps organizations gain real-time visibility of information, in context, through dashboards, real-time insight into business performance, health, and status so that immediate action based on business policies and processes can be taken. Operational intelligence applies the benefits of real-time analytics, alerts, and actions to a broad spectrum of use cases across and beyond the enterprise.

One specific technology segment is AIDC (automatic identification and data capture) represented by barcodes, RFID (radio-frequency identification), and voice recognition.

10.5   Common Use Cases for Real-Time Technologies and Practices

Many use cases for real-time technology are deployed today, in many industries, locations, and department types.

Real-time software solutions monitor and analyze business activities to give a wide range of users the real-time visibility they need to see a problem or opportunity, make a fully informed decision, and then act accordingly. Numerous real-world applications are already established today [9]:

- Understand customer behavior in real time across multiple channels, such as Web, mobile, social, and enterprise applications. Improve the customer experience as it's happening.
- Enhance complete views of customers with real-time data, BI, and analytics. That way, views are also up to date, not merely complete.
- Combine real-time data with historical data from data warehouses and BI systems. Judge current events more accurately in the context of performance history and seasonality.
- Evaluate sales performance in real time. Take measures now to achieve sales quotas.
- See a product recurring in abandoned shopping carts on an e-commerce web site. Run a promotion to close more sales of that product before interest in it wanes.
- Identify a new social media sentiment or pattern. Direct it or correct it as it evolves.
- Spot potentially fraudulent activity as it's being perpetrated. Stop it while it's in process and take action to mitigate its impact.
- Take logistics to a new level of accuracy, efficiency, and customer service. A few minutes here and there on a truck or rail freight schedule add up.
- Monitor the performance of interconnected infrastructures such as utility grids, computer networks, and manufacturing facilities. Make tactical decisions for short-term maintenance and optimization, but store and analyze the real-time data for long-term capacity planning.
- Let software take action automatically to adjust machinery, turn lights off, route energy flows on utility grids, buy/sell stocks, or send a coupon to a churning customer.

In addition to the common use cases listed above, certain departments and personnel are likely users of the technologies and practices of real-time data, BI, and analytics (see Fig. 10.8.):

- **A wide range of business operations can use real-time operations:** Operations (53%) bubbled up to the top of the survey responses, followed by more specific types of operations, such as customer service and support (51%) and supply chain/manufacturing (32%).
- **Sales and marketing are obvious candidates for real-time functions**: Finding, closing, and retaining customers are all increasingly time sensitive, so it makes sense that sales (45%), marketing (43%), and e-commerce (28%) ranked fairly high in the survey.
- **Certain managers can incorporate real-time methods**: Real-time dashboards, frequently refreshed reports, and alerts are commonly used today by line-of-business managers (39%), C-level executives (33%), and finance personnel (32%).
- **Real-time functions are useful to technical users too**: IT systems/network management (43%) typically monitors, in real time, a wide range of systems to assure availability and performance. Research and development departments (19%) increasingly depend on real-time data and analytics to design and build new products and services.

Fig. 10.8 Based on 1562 responses by 365 respondents; responses per respondent, on average [9]

A number of survey respondents entered additional use cases for real-time data, BI, and analytics, including clinical decisions in healthcare, information for end customers (not just customer support), fraud detection and prevention, continuous statistical analysis of export data (for homeland security), monitoring commodity investments, using BI for auditing business processes in real time, validation for bill processing, and transaction recalls (in a retail point-of-sale system).

As you can see, real-time data, BI, and analytics (plus real-time operations in other contexts) are not mere hype or science fiction. There are many real-world applications deployed in many real-world organizations today.

For more information, refer to report by Philip Russom et al. [9].

10.6   Real-Time Data, BI, and Analytics Summary

User organizations continue to push their applications for business intelligence (BI), analytics, and data management closer to real-time operation. This is because fresh information can support fast-paced, time-sensitive business processes, such as operational BI, real-time management dashboards, just-in-time inventory, high-yield manufacturing, facility monitoring, call center information delivery, self-service information portals, recommendations in e-commerce, and so on.

According to this report's survey, the leading benefits of implementing real-time technologies for data, BI, and analytics include better actions based on BI and analytic information, improved customer service and customer experience, and automated decisions made by software, not people. The barriers to achieving such benefits include cost, difficulties in designing real-time systems, the state of data management infrastructure, and inadequate staffing or skills.

The term *real time* has become an umbrella concept encompassing multiple time frames, speeds, and execution frequencies. BI and analytic practices are progressively rarer as we come closer to true real time (milliseconds and nanoseconds). For example, most reports and analyses today operate on data that's refreshed daily on a 24-h cycle. Even so, BI and analytics based on hourly cycles are well established, and so-called near-real-time cycles (seconds and minutes) are fairly common.

A variety of technologies are in use today, enabling real-time and near-real-time data, BI, and analytics. Common ones include data federation, replication, data sync, message buses, and micro-batches. A number of functions designed for high performance are increasingly applied to near-real-time uses, namely, changed data capture, columnar databases, in-database analytics, in-memory processing, and solid state drives. Organizations in need of true real time are ramping up the use of complex event processing (CEP) and continuous stream mining.

The proliferation of operational BI is the leading driver for real-time usage in BI. Most operational BI users interact with near-real-time data via management dashboards and other types of reports. However, users are aggressively adopting visual data discovery tools because of their ease of use, strong self-service functions (which empower many user types), and the tools' ability to process large data sets with near-real-time performance.

As user organizations move deeper into analytics (as a complement to reporting), they also move deeper into real time. For example, this report's survey ranks operational BI and reporting as leading practices today, with 43% of respondents using these with some level of real time. However, 34% of respondents are already practicing operational analytics in real time, with an additional 34% planning to adopt the same within 3 years.

Other areas that should experience strong adoption within 3 years include real-time data warehousing (and similar practices such as active and dynamic warehousing), predictive analytics, visualization, social media analytics, text analytics, CEP, and stream mining.

Most of the data handled in real time today is structured (or more specifically relational), followed by application logs and semi-structured data, but survey respondents anticipate more real-time handling for social media data, Web logs and click streams, and unstructured data.

This section helps users to understand the many available technologies and practices for real-time data, BI, and analytics, plus how these enable fast-paced business processes and yield a wide range of organizational advantages.

In summary, we can state that:

1. Real-time data, BI, and analytics are the basis of many desirable business practices.
2. Real time is a broad concept involving many types of processes, tools, and users.
3. Operational BI is a common manifestation of real-time data and BI.
4. Analytics is aggressively accelerating into real time.
5. A growing range of data types will soon be handled in real time.

For more information, refer to report by Philip Russom et al. [9].

References

1. Judd, C., & McClelland, G. (1989). *Data analysis*. San Diego, CA: Harcourt Brace Jovanovich. ISBN: 0-15-516765-0.
2. Tukey, J. (July 1961). The future of data analysis. *The Annals of Mathematical Statistics, 33*(1), 1–67.
3. O'Neil, C., & Schutt, R. (2014). *Doing data science*. Sebastopol, CA: O'Reilly. ISBN: 978-1-449-35865-5.
4. *Clean data in CRM: The key to generate sales-ready leads and boost your revenue pool*. Retrieved July 29, 2016.
5. *Data cleaning*. Microsoft Research. Retrieved October 26, 2013.
6. Koomey, J. (2006). *Best practices for understanding quantitative data*. El Dorado Hills, CA: Perceptual Edge. February 14, 2006.
7. Hellerstein, J. (February 27, 2008). Quantitative data cleaning for large databases (PDF). EECS Computer Science Division: 3. Retrieved October 26, 2013.
8. http://www.1keydata.com/datawarehousing/business-intelligence.php
9. Russom, P., Stodder, D., & Halper, F.. (2014). *Real-time data, BI, and analytics accelerating business to leverage customer relations, competitiveness, and insights*. TDWI best practices report, TDWI research fourth quarter.

Chapter 11
Boolean Computation Versus Fuzzy Logic Computation

Computational intelligence offers an in-depth exploration into the adaptive mechanisms that enable intelligent behavior in complex and changing environments. The main focus of this chapter is centered on the computational modeling of biological and man-made intelligent systems, encompassing swarm intelligence, fuzzy systems, artificial neutral networks, artificial immune systems, and evolutionary computation. This chapter, briefly, provides readers with a wide knowledge of computational intelligence (CI) paradigms and algorithms, inviting readers to implement and solve real-world, complex problems within the CI development framework. Man has learned much from studies of natural systems, using what has been learned to develop new algorithmic models to solve complex problems. This book presents an introduction to some of these technological paradigms, under the umbrella of computational intelligence (CI). In this context, the chapter to some degree will talk about artificial neural networks, evolutionary computation, swarm intelligence, artificial immune systems, and fuzzy systems, which are respectively models of the following natural systems: biological neural networks, evolution, swarm behavior of social organisms, natural immune systems, and human thinking processes.

11.1 Introduction

The expression computational intelligence (CI) usually refers to the ability of a computer to learn a specific task from data or experimental observation. Even though it is commonly considered a synonym of soft computing, there is still no commonly accepted definition of computational intelligence.

Generally, computational intelligence is a set of nature-inspired computational methodologies and approaches to address complex real-world problems to which mathematical or traditional modeling can be useless for a few reasons: the processes

© Springer International Publishing AG 2017
B. Zohuri, M. Moghaddam, *Business Resilience System (BRS): Driven Through Boolean, Fuzzy Logics and Cloud Computation*,
DOI 10.1007/978-3-319-53417-6_11

might be too complex for mathematical reasoning, it might contain some uncertainties during the process, or the process might simply be stochastic in nature. Indeed, many real-life problems cannot be translated into binary language (unique values of 0 and 1) for computers to process it. Computational intelligence therefore provides solutions for such problems [1].

The methods used are close to the human's way of reasoning, i.e., it uses inexact and incomplete knowledge, and it is able to produce control actions in an adaptive way. CI therefore uses a combination of five main complementary techniques [1]: the fuzzy logic which enables the computer to understand natural language [2, 3]; artificial neural networks which permit the system to learn experiential data by operating like the biological one; evolutionary computing, which is based on the process of natural selection; learning theory; and probabilistic methods which help dealing with uncertainty imprecision [1].

Computational intelligence and neuroscience is a forum for the interdisciplinary field of neural computing, neural engineering, and artificial intelligence, where neuroscientists, cognitive scientists, engineers, psychologists, physicists, computer scientists, and artificial intelligence investigators among others can publish their work in one periodical that bridges the gap between neuroscience, artificial intelligence, and engineering. Computational intelligence and neuroscience was founded in 2006 by Professor Andrzej Cichocki who served as the editor-in-chief of the *Journal of Computational Intelligence and Neuroscience* between 2006 and 2011.

Computational intelligence (CI) is an offshoot of artificial intelligence (AI) in which the emphasis is placed on heuristic algorithms such as fuzzy systems, neural networks, and evolutionary computation. It is usually contrasted with "traditional," "symbolic," or "good old-fashioned artificial intelligence (GOFAI)." The IEEE Computational Intelligence Society uses the tagline "Mimicking Nature for Problem Solving" to describe computational intelligence, although mimicking nature is not a necessary element.

In addition to the three main pillars of CI (fuzzy systems, neural networks, and evolutionary computation), computational intelligence also encompasses elements of learning, adaptation, heuristic, and meta-heuristic optimization, as well as any hybrid methods which use a combination of one or more of these techniques. More recently, emerging areas, such as artificial immune systems, swarm intelligence, chaotic systems, and others, have been added to the range of computational intelligence techniques. The term "soft computing" is sometimes used almost interchangeably with computational intelligence.

Computational intelligence techniques have been successfully employed in a wide range of application areas, including decision support; generic clustering and classification; consumer electronic devices; stock market and other time-series prediction; combinatorial optimization; medical, biomedical, and bioinformatics problems; and many, many others. Although CI techniques are often inspired by nature, or mimic nature in some way, CI applications are not restricted to solving problems from nature [4].

A major thrust in algorithmic development is the design of algorithmic models to solve increasingly complex problems. Enormous successes have been achieved through the modeling of biological and natural intelligence, resulting in so-called intelligent systems. These intelligent algorithms include artificial neural networks, evolutionary computation, swarm intelligence, artificial immune systems, and fuzzy systems. Together with logic, deductive reasoning, expert systems, case-based reasoning, and symbolic machine learning systems, these intelligent algorithms form part of the field of artificial intelligence (AI). Just looking at this wide variety of AI techniques, AI can be seen as a combination of several research disciplines, for example, computer science, physiology, philosophy, sociology, and biology.

11.2 What Is Intelligence?

For us to be able to build an infrastructure and model around Business Resilience System (BRS), we need to define intelligence and ask ourselves, can computers be intelligent, and can they provide the BRS system we are looking for? Attempts to find definitions of intelligence still provoke heavy debate. Dictionaries define intelligence as the ability to comprehend, to understand and profit from experience, and to interpret intelligence, having the capacity for thought and reason (especially to a high degree). Other keywords that describe aspects of intelligence include creativity, skill, consciousness, emotion, and intuition.

11.3 Can Computers Be Intelligence?

This is a question that to this day causes more debate than the definitions of intelligence. In the mid-1900s, Alan Turing gave much thought to this question. He believed that machines could be created that would mimic the processes of the human brain. Turing strongly believed that there was nothing the brain could do, which a well-designed computer could not. More than 50 years later, his statements are still visionary. While successes have been achieved in modeling small parts of biological neural systems, there are still no solutions to the complex problem of modeling intuition, consciousness, and emotion—which form integral parts of human intelligence.

In 1950, Turing published his test of computer intelligence, referred to as the Turing test [5]. The test consisted of a person asking questions via a keyboard to both a person and a computer. If the interrogator could not tell the computer apart from the human, the computer could be perceived as being intelligent. Turing believed that it would be possible for a computer with 10^9 bits of storage space to pass a 5-min version of the test with 70% probability by the year 2000, which is almost 17 years ago. Today's computer can achieve far more and faster processing. A more recent definition of artificial intelligence came from the IEEE Neural Networks Council of 1996: the study of how to make computers do things at which people are doing better [4].

At this point, it is necessary to state that there are different definitions of what constitutes CI. For example, swarm intelligence (SI) and artificial immune systems (AIS) are classified as CI paradigms, while many researchers consider these paradigms to belong only under artificial life. However, both particle swarm optimization (PSO) and ant colony optimization (ACO), as treated under SI, satisfy the definition of CI given above and are therefore included in this book as being CI techniques. The same applies to AISs.

Except those main principles, currently, popular approaches include biologically inspired algorithms such as swarm intelligence [6] and artificial immune systems, which can be seen as a part of evolutionary computation, image processing, data mining, natural language processing, and artificial intelligence, which tends to be confused with computational intelligence. However, although both computational intelligence (CI) and artificial intelligence (AI) seek similar goals, there is a clear distinction between them.

Computational intelligence is thus a way of performing like human beings. Indeed, the characteristic of "intelligence" is usually attributed to humans. More recently, many products and items also claim to be "intelligent," an attribute which is directly linked to the reasoning and decision-making.

11.4 Difference Between Computational and Artificial Intelligence

At this point, we turn our attention to Wikipedia and what is, reflected there is, written here about the difference between computational intelligence and artificial intelligence. Although artificial intelligence and computational intelligence seek a similar long-term goal and reach general intelligence, which is the intelligence of a machine that could perform any intellectual task that a human being can, there is a clear difference between them. According to Bezdek [7], computational intelligence is a subset of artificial intelligence.

There are two types of machine intelligence: the artificial one based on hard computing techniques and the computational one based on soft computing methods, which enable adaptation to many situations.

Hard computing techniques work following binary logic based on only two values (the Booleans true or false, 0 or 1) on which modern computers are based. One problem with this logic is that our natural language cannot always be translated easily into absolute terms of 0 and 1. Soft computing techniques, based on fuzzy logic, can be useful here [8]. Much closer to the way the human brain works by aggregating data to partial truths (crisp/fuzzy systems), this logic is one of the main exclusive aspects of CI.

Within the same principles of fuzzy and binary logics, follow crispy and fuzzy systems [9]. Crisp logic is a part of artificial intelligence principles and consists of either including an element in a set or not, whereas fuzzy systems (CI) enable elements to be partially in a set. Following this logic, each element can be given a degree of membership (from 0 to 1) and not exclusively one of these two values [10].

11.5 The Five Main Principles of CI and Its Applications

According to Wikipedia, the main applications of computational intelligence include computer science, engineering, data analysis, and biomedicine, and they are listed here as:

1. **Fuzzy Logic**
   As explained before, fuzzy logic, one of CI's main principles, consists of measurements and process modeling made for real life's complex processes [3]. It can face incompleteness, and most importantly ignorance of data in a process model, contrarily to artificial intelligence, which requires exact knowledge.

   This technique tends to apply to a wide range of domains such as control, image processing, and decision-making. However, it is also well introduced in the field of household appliances with washing machines, microwave ovens, etc. We can face it too when using a video camera, where it helps stabilizing the image while holding the camera unsteadily. Other areas such as medical diagnostics, foreign exchange trading, and business strategy selection are apart from this principle's numbers of applications [1].

   Fuzzy logic is mainly useful for approximate reasoning and doesn't have learning abilities [1], a qualification much needed that human beings have [citation needed]. It enables them to improve themselves by learning from their previous mistakes.

2. **Neural Networks**
   This is why CI experts work on the development of artificial neural networks based on the biological ones, which can be defined by three main components: the cell body, which processes the information; the axon, which is a device enabling the signal conducting; and the synapse, which controls signals. Therefore, artificial neural networks are doted of distributed information processing systems [11], enabling the process and the learning from experiential data. Working like human beings, fault tolerance is also one of the main assets of this principle [1].

   Concerning its applications, neural networks can be classified into five groups: data analysis and classification, associative memory, clustering, generation of patterns, and control [1]. Generally, this method aims to analyze and classify medical data, proceed to face and fraud detection, and most importantly deal with nonlinearities of a system in order to control it [12]. Furthermore, neural network techniques share with the fuzzy logic ones the advantage of enabling data clustering.

3. **Evolutionary Computation**
   Based on the process of natural selection firstly introduced by Charles Robert Darwin, the evolutionary computation consists of capitalizing on the strength of natural evolution to bring up new artificial evolutionary methodologies [13]. It also includes other areas such as evolution strategy and evolutionary algorithms which are seen as problem-solvers. This principle's main applications cover areas such as optimization and multi-objective optimization, to which traditional mathematical techniques are not enough anymore to apply to a wide range of problems such as DNA analysis, scheduling problems [1].

4. **Learning Theory**
   Still looking for a way of "reasoning" close to humans, learning theory is one of the main approaches of CI. In psychology, learning is the process of bringing together cognitive, emotional, and environmental effects and experiences to acquire, enhance, or change knowledge, skills, values, and world views [14, 15]. Learning theories then help in understanding how these effects and experiences are processed and then help in making predictions based on previous experience [12].

5. **Probabilistic Methods**
   Being one of the main elements of fuzzy logic, probabilistic methods firstly introduced by Paul Erdos [16] and Joel Spencer (1974) [17] aim to evaluate the outcomes of a Computation Intelligent system, mostly defined by randomness [18]. Therefore, probabilistic methods bring out the possible solutions to a reasoning problem, based on prior knowledge.

11.6   Boolean Logic Versus Fuzzy Logic

In previous chapters, we learned what Boolean and fuzzy logics are and what the differences between them are. Here, we summarize and then compare them against each other as listed below:

- **Boolean logic uses sharp distinctions. It forces us to draw lines between members of a class and nonmembers.**

  – For instance, we may say, Tom is tall because his height is 181 cm. If we drew a line at 180 cm, we would find that David, who is 179 cm, is small. Is David really a small man or we have just drawn an arbitrary line in the sand?

- **Fuzzy logic reflects how people think.**

  – It attempts to model our sense of words, our decision-making, and our common sense.
  – As a result, it is leading to new, more human, intelligent systems.

This compearance between logics is depicted in Fig. 11.1 here, using range of logical values in Boolean and fuzzy logic.

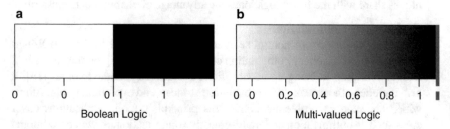

Fig. 11.1 Range of logical values in Boolean and fuzzy logic

In order to develop a fuzzy logic (FL) expert system, the following steps and process should take place:

1. Specify the problem and define linguistic variables.
2. Determine fuzzy sets.
3. Elicit and construct fuzzy rules.
4. Encode the fuzzy sets, fuzzy rules, and procedures to perform fuzzy inference into the expert system.
5. Evaluate and tune the system.

References

1. Siddique, N., & Adeli, H. (2013). *computational intelligence: Synergies of fuzzy logic, neural networks and evolutionary computing*. Chichester: John Wiley & Sons.
2. Rutkowski, L. (2008). *Computational intelligence: Methods and techniques*. Berlin: Springer.
3. Fuzzy Logic. (2006). WhatIs.com. Margaret Rouse. July 2006.
4. Engelbrecht, A. (2007). *Computational intelligence. An introduction* (2nd ed.). Chichester: John Wiley Publication.
5. Turing, A. M. (1950). Computing machinery and intelligence. *Mind, 59*, 433–460.
6. Beni, G., & Wang, J. (1989). *Swarm intelligence in cellular robotic systems*. Proceed. NATO Advanced Workshop on Robots and Biological Systems, Tuscany, Italy, June 26–30, 1989.
7. Bezdek, J. C. (1994). What is computational intelligence? In J. Zurada, B. Marks, & C. Robinson (Eds.), *Computational intelligence imitating life* (pp. 1–12). Piscataway: IEEE Press. Reprinted in: DOE Proc. Adaptive Control Systems Tech. Symp. (pp. 10–15). In S. Biondo, & C. J. Drummond (Eds.) Springfield, VA: NTIS; 1995 (An update to Bezdek (1992)).
8. *Artificial intelligence, computational intelligence, soft computing, natural computation– What's the difference?—ANDATA*. Retrieved November 5, 2015, from www.andata.at
9. *Fuzzy sets and pattern recognition*. Retrieved November 5, 2015, from www.cs.princeton.edu
10. Pfeifer, R. (2013). *Chapter 5: FUZZY logic*. Lecture notes on "Real-world computing". University of Zurich, Zurich.
11. Stergiou, C., & Siganos, D. (2015) Neural networks. *SURPRISE 96 Journal*. Imperial College London. Retrieved March 11, 2015.
12. Somers, M. J., & Casal, J. C. (July 2009). Using artificial neural networks to model nonlinearity. *Organizational Research Methods, 12*(3), 403–417. doi:10.1177/1094428107309326. Retrieved 2015-10-31.
13. De Jong, K. (2006). *Evolutionary computation: A unified approach*. Cambridge, MA: MIT Press. ISBN 9780262041942.
14. Ormrod, J. (1995). *Educational psychology: principles and applications*. Englewood Cliffs, NJ: Prentice-Hall.
15. Illeris, K. (2004). *Adult education and adult learning*. Malabar, FL: Krieger Publishing Company/Roskilde University Press.
16. https://en.wikipedia.org/wiki/Paul_Erd%C5%91s
17. https://en.wikipedia.org/wiki/Joel_Spencer
18. Palit, A. K., & Popovic, D. (2006). *Computational intelligence in time series forecasting: Theory and engineering applications* (p. 4). London: Springer Science & Business Media. ISBN 9781846281846.

Chapter 12
Defining Threats and Critical Points for Decision-Making

Although humans have been thinking critically since the first *Homo habilis* picked up a stone tool, critical thinking as a process has only become one of the most valuable business skills in the last century. Many new decision-making strategies relying heavily on critical thinking career skills were created over the time period between 1950 and 1970, including CATWOE, PEST, and the Cause and Effect Analysis model. In this chapter, we will discuss all these processes that allow us to do critical thinking and decision-making, based on the threats against day-to-day operation and normal process in an organization or enterprises. The effectiveness of the leader is proportional to the effectiveness of the decisions the leader makes and the cascading impacts as decisions turn into action, both good and bad.

12.1 Introduction

Problem-solving is one of the leadership skills that successful business professionals and entrepreneurs are expected to have, yet many struggle with the simplest of decisions. What makes solving daily problems so natural for one person and such a struggle for the next?

The truth is even experienced decision-makers continually hone and perfect their creative problem-solving skills. In addition, there are many compelling reasons to do so. Not only do those who make better decisions have more job opportunities, how to get promoted more often, and increase their work productivity, but they are generally happier. In a recent study from the University of Chicago Booth School of Business, research found that happiness depends more on opportunities to make decisions (i.e., freedom) rather than money or connections. This means that the ability to make decisions leads to more and better opportunities for success, which improves your quality of life. In other words, the better a decision-maker you are, the happier and more successful you will be.

© Springer International Publishing AG 2017
299
B. Zohuri, M. Moghaddam, *Business Resilience System (BRS): Driven Through Boolean, Fuzzy Logics and Cloud Computation*,
DOI 10.1007/978-3-319-53417-6_12

This concept goes against what many business leaders believe—that it's what and who you know that makes you successful. In fact, how you understand and solve problems are the key to success.

Fortunately, problem-solving and decision-making are skills that can be improved upon, be studied, and be mastered. By learning specific problem-solving and decision-making techniques, you can see problems sooner and make decisions faster. This allows you to make decisions that are more confident in your job and gives you more control over the happiness and productivity in every part of your life.

For example, military decision-making process falls on upper echelon-ranking commanders. The commander is in charge of the military decision-making process and decides what procedures to use in each situation. The planning process hinges on a clear articulation of his battlefield visualization. He is personally responsible for planning, preparing, and executing operations.

From start to finish, the commander's personal role is central: his participation in the process provides focus and guidance to the staff. However, there are responsibilities and decisions that are the commander's alone (Fig. 12.1). The amount of his

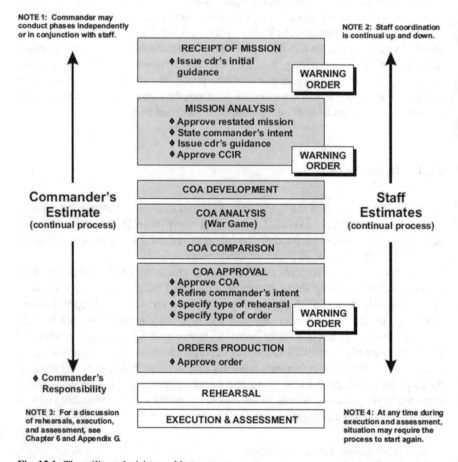

Fig. 12.1 The military decision-making process

direct involvement is driven by the time available, his personal preferences, and the experience and accessibility of the staff. The less time available, the less experienced the staff, and the less accessible the staff, generally the greater the commander involvement.

In effective decision-making processes for the Joint Force Commander (JFC), a commander has to fight the issues of future not the past events, except using the past knowledge to improve his or her critical thinking and decision-making, in order to win tactically and strategically.

In order to improve critical decision-making capabilities, joint doctrine must be changed to provide a standardized approach for the JFC, identifying key decisions and the necessary framework of supporting the critical information required by the commander to make good, timely decisions.

The effectiveness of the leader is proportional to the effectiveness of the decisions the leader makes and the cascading impacts as decisions turn into action, both good and bad. Joint doctrine defines JFC as a "general term applied to a combatant commander, sub-unified commander, or joint task force commander authorized to exercise combatant command or operational control over a joint force" [1]. The decision-making in question is for the JFC and applies to the strategic and operational levels. Since it is widely recognized that the US Armed Forces will be used with coalition armed forces, the same processes apply, whether discussing the US only or multinational forces and staffs.

As the decisions facing a commander in the vignette show, the JFC cannot rely on haphazard (i.e., lacking any obvious principle of organization) information flow to make effective decisions. The JFC's time is precious. Information presented or pushed to the JFC must be worth the JFC's time to receive and digest the information. Joint doctrine, by putting together the information from several different publications, provides Commander's Critical Information Requirements (CCIR) as the information necessary to support the JFC's decision-making.

Joint doctrine recognizes the importance of effective decision-making. Effective decision-making starts in the deliberate and crisis action planning process with well-defined decision points and corresponding CCIR. The definition of CCIR indicates that the JFC's information requirements should support the JFC's decision-making. Although it referred to an implied throughout joint doctrine, the "process" of decision-making is not defined. There is no joint doctrine that addresses the process of defining decision points or correlating CCIR to the decision points and then breaking the CCIR into subcategories that provide ownership within the JFC's staff and component organizations.

All commanders, Joint Task Force and JFC components, come from service jobs. If the service doctrines followed by key subordinates and JFC "information providers" do not follow the same decision-making processes, there will be seams in the decision-making process leading to poor-quality information flowing to the JFC in support of the CCIR. A breakdown in the process eventually leads to ineffective decisions. It is important that service doctrines synchronize decision-making processes with joint doctrine. Following is a simple example to show why it is important for service doctrine processes to match joint doctrine processes: If each service used different service-defined processes to decide targeting priorities, they would be unable to effectively identify and prioritize the targets when working together as a joint force [2].

We must determine first what joint doctrine does provide the JFC's decision-making process to analyze doctrine and provide solutions to provide Joint Force Commanders with decision-making tools not currently addressed by joint doctrine. A summary of terms and joint definitions that will be used throughout this analysis will be provided to ensure the same starting point. After definitions, the decision-making processes addressed in joint doctrine will be reviewed. Although it would be beneficial to provide the same detailed analysis of service doctrines, time and space dictate a summary of service policies with respect to decision-making processes and standardization with joint doctrine. Where gaps in definitions and joint processes are identified, solutions will be proposed. Key aspects of decision-making processes missing from the joint doctrine are standard definitions, including the key subcomponents of CCIR, linking decision points to CCIR, and how CCIR are *answered*.

In case of this military example, the following definitions are taken from the Department of Defense (DOD) Dictionary of Military and Associated Terms:

Commander's Critical Information Requirements (CCIR)—A comprehensive list of information requirements identified by the commander as being critical in facilitating timely information management, as well as the decision-making process, that affects successful mission and accomplishment. The two key subcomponents are critical friendly force information and priority intelligence requirements.

Critical information (CI)—Specific facts about friendly intentions, capabilities, and activities vitally needed by adversaries for them to plan and act effectively to guarantee failure or unacceptable consequences for friendly mission accomplishment.

Decision—In an estimate of the situation, a clear and concise statement of the line of action intended to be followed by the commander as the one most favorable to the successful accomplishment of the assigned mission.

Decision point (DP)—The point in space and time where the commander or staff anticipates making a decision concerning a specific friendly course of action. A decision point is usually associated with a specific target area of interest and is located in time and space to permit the commander sufficient lead time to engage the adversary in the target area of interest. Decision points may also be associated with friendly force and the status of ongoing operations.

Decision-support template (DST)—A graphic record of war gaming. The decision-support template depicts decision points, timelines associated with movement of forces and the flow of the operation, and other key items of information required to execute a specific friendly course of action. See Fig. 12.2.

Essential elements of friendly information (EEFI)—Key questions likely to be asked by adversary officials and intelligence systems about specific friendly intentions, capabilities, and activities so they can obtain answers critical to their operational effectiveness that is also called EEFI.

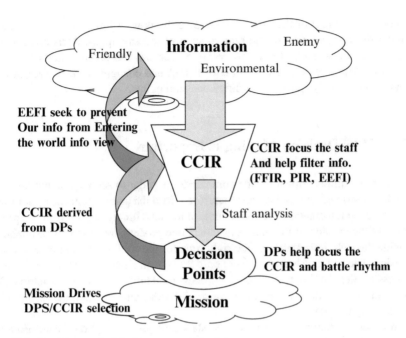

Fig. 12.2 Graphic record of decision-support template

Joint Force Commander (JFC)—A general term applied to a combatant commander, sub-unified commander, or Joint Task Force commander authorized to exercise combatant command (command authority) or operational control over a joint force. Also called a JFC.

Priority intelligence requirements (PIR)—Those intelligence requirements for which a commander has an anticipated and stated priority in the task of planning and decision-making.

One of the two key subcomponents of CCIR is friendly force information.

However, there is no definition of friendly force information. The definition of critical information describes information needed by an adversary concerning friendly forces.

As can be seen from the definitions, critical information applies to information required by the adversary instead of the JFC, so the definition of CCIR must be standard throughout joint doctrine. Since decision points may be associated with friendly forces and the status of ongoing operations, friendly force information must be defined as follows: Information about friendly forces required by the commander in support of decision-making processes [2].

The definition of CCIR mentions the decision-making process, but nowhere in the joint publications is the process of decision-making described. In fact, the decision-making process should begin while identifying decision points during crisis action planning or deliberate planning. Information required by the commander to make sound decisions must be identified in conjunction with identifying decision points. The last part of the process should be the flow of information to the commander during execution of the plan to support the anticipated decisions that the commander must make. Once the decision is made, the process starts from the beginning with a review of upcoming decisions and changing or modifying CCIR as required supporting new or modified decision points.

12.2   Why Decision-Making Is Important

Decision-making is critical for the commander in order to accomplish the mission. The decision-making process must therefore afford the commander the opportunity to receive the information necessary for him to make timely decisions in the prosecution of the mission. Without a well-understood process, the staff will be unable to provide the right information at the right time for the commander to make the right decision. The key then becomes to not only provide an efficient and well-understood process for the staff but to manage the process to identify and make available CCIR inputs in time for the commander to make decisions and act upon the information provided by the staff.

In order to manage the process, the staff must be focused on the information requirements established by the commander to make decisions. The JFC staff must be able to provide timely analysis of the information that is provided in response to CCIR. Value is added to the information as the JFC staff turns mere data into actionable knowledge for the decision-maker.

Information management is simplified by using CCIR to provide a way of taking incoming information, filtering it against a set of requirements, and applying what remains against those requirements. JFC staff efficiencies are gained by establishing an Office of Primary Responsibility (OPR) for each CCIR. Addressing an OPR in the process provides a centralized point of contact on the JFC staff for working the analysis of the information.

Consistent joint doctrine, as well as the importance of a well-articulated and understood process, is essential for commanders and their staffs to reach timely decisions. The key is to understand what must be done to change current joint doctrine so that CJTFs can form a joint staff using common, understood procedures to gain the necessary information and knowledge to make sound decisions. Thus, per reference 2 at the end of this chapter, the authors are presenting their recommendations to joint doctrine as follows:

- Establish a simple and clear process consistent throughout joint doctrine.
- Establish simple and clear definitions consistent throughout joint doctrine.

The following establishes a clear decision-making process for the JFC's staff to follow: The commander, using war-gaming techniques and the decision-support matrix (DSM) or decision-support template (DST), identifies the decision point (DP) necessary for mission accomplishment and for execution of potential branches and sequels. Once these DPs are identified, the commander then determines the applicable

CCIR for each DP. DPs are then reached and decisions made when the commander feels that the CCIR justify the need to make the decision. The process is based upon immediate commander notification when significant CCIR-related information comes in or when CCIR in combination establish the need to make a decision.

The commander utilizes CCIR to establish the priorities for information gathering and reporting. CCIR are a prioritized list of information requirements approved by the commander as critical for decision-making and are linked to the commander's DPs. They should identify opportunities and vulnerabilities that assist the commander in advising his/her higher headquarters and in supporting the JTF (see Fig. 12.1). CCIR are a tool for the commander to reduce information gaps generated by uncertainties that the commander may have concerning his own force, the threat, or the environment. Once updated, CCIR enable the commander to better understand the flow of the operation, identify risks, and make timely decisions to fulfill his intent, retain the initiative, and accomplish the mission. They assist the commander by reducing information requirements to a manageable set. More importantly, they focus the staff on the exact type and form of information the commander requires. CCIR will change as situations change and decisions are made. CCIR require continuous assessment for relevance to current and developing situations.

Instead of reacting to the threat, commanders are able to maintain tempo by controlling the flow of information necessary to attain understanding of the battle space. As events unfold, new decisions will be necessary which thereby drive changes in the CCIR. This constant state of change requires continual assessment of CCIR for relevance to current and future situations. The commander approves CCIR, but the staff recommends and manages them to assist the commander. They are updated as required by the IM plan and are tracked by the staff.

In summary, the information and process of inquiry allow developing skills in the context of decision-making and may therefore be of considerable service to a range of different readers. This chapter includes: "The Matrix of Inquiry," "The Problem of Logical Subject Matter," "Common Sense and Scientific Inquiry," "The Needed Reform of Logic," "Immediate Knowledge (i.e., Real Time): Understanding and Inference," etc.

12.3   How to Make Better Decisions and Solve Problems Faster

Any proper decision-making guide that is designed for appropriate organization or enterprise related to their business that they are conducting will give stakeholder a better understanding of what problem-solving and critical thinking entail. Not only will you learn about how to make better decisions in business, these following ideas can make you a better problem-solver at your organization or in your personal life when faced with challenges. Additionally throughout the proper guideline, it will provide you amazing online tools, videos, and resources to help you continue to learn how to make decisions better into your daily activities.

12.3.1   The Importance of Problem-Solving

As we stated at the beginning of this chapter under Sect. 12.1, problem-solving is one of the leadership skills that successful business professionals and entrepreneurs are expected to have, yet many struggle with the simplest of decisions. What makes solving daily problems so natural for one person and such a struggle for the next?

Fortunately, problem-solving and decision-making are skills that can be improved upon to be studied and to be mastered. By learning specific problem-solving and decision-making techniques, you can see problems sooner and make decisions faster. This allows you to make decisions that are more confident in your job and gives you more control over the happiness and productivity in every part of your life.

12.3.2   Critical Thinking in the Decision-Making Process

Critical thinking is the practice of methodically gathering, analyzing, and evaluating information. It is one of the most vital parts of the problem-solving and decision-making process, as it is the clearly thinking act through options that will lead to a final choice. While decision-making is the process that leads to actionable conclusions, critical thinking is the element that defines whether the choice is sound. Think about it this way: If problem-solving is the car that gets your business to its goals, critical thinking skills are the gas. See Fig. 12.3.

Although humans have been thinking critically since the first *Homo habilis* picked up a stone tool, critical thinking as a process has only become one of the most valuable business skills in the last century. John F. Dewey, the inventor of the Dewey library system and a noted educational philosopher, began touting the importance of teaching critical thinking skills in his 1938 paper, *Logic: The Theory of Inquiry*. This educational reform may have inspired the rising generation to explore the concepts more, as a resurgence of interest in the subject presented itself, between 1950 and 1970. Many new decision-making strategies (relying heavily on critical thinking career skills) were created over this time period, including CATWOE, PEST, and the Cause and Effect Analysis model.

The 5 steps of the critical thinking process are:

- **Identification** – Identify and define the problem.
- **Solution** – Design a potential solution.
- **Exploration** – Formulate a potential action plan that results in the chosen solution.
- **Action** – Take necessary steps to complete the action plan.
- **Reevaluation** – Consider the action plan and whether it solved the original problem.

Fig. 12.3 The five steps of critical thinking

Since that time, critical thinking and decision-making are synonymous business skills that are expected of corporate leaders. Still, many people don't truly understand exactly the underlying concepts that make critical thinking an effective process. There are four key structures that all are critical thinking based on:

Logic—An individual's ability to see direct relationships between causes and effects. This is one of the most important decision-making skills, as logic provides accurate predictions about what kinds of effects a potential solution will have on individuals and systems.

Truth—The unbiased data of an event. Unbiased and unemotional facts are an important part of the problem-solving process. Good critical thinking culls out these biases and focuses on the historical and documented data that will support the final conclusion.

Context—A list of extenuating pressures and factors that will or should be impacted by the final solution. Critical thinking must take into account the historical efficacy of similar solutions, the physical and abstract stressors of the decision-maker, and the assumptions or agendas of different shareholders. All of these outside elements must be considered in order to truly engage in a critical decision-making process.

Alternatives—Potential solutions not currently in use. In effective critical thinking, the individual is able to consider new ways of approaching problems that meet real-world goals and are based on accurate, unbiased data. This is the case even if alternative solutions are not used or when outside determinants are unexpected.

When you understand each of these underlying factors, you will become more aware of personal biases and be more engaged in the critical thinking process. In addition, improving your critical thinking skills leads to faster, more confident, and more productive decision-making. The fuel of critical thinking is the secret ingredient that will drive your business's success.

12.3.3   Are You Asking the Right Questions for Your Organization?

Stakeholder in any business and organization needs to set the guideline of proper thinking and decision-making in order to set the policy within the guideline.

Thought leader Clayton M. Christensen observed that business leaders often think so much about action that they fail to consider why they are acting in the first place. Unfortunately, good action isn't possible without considering the right critical thinking questions. Critical questioning allows you to clearly distinguish facts from biases, stakeholders from observers, and solutions from potential solutions. If critical thinking is the lens by which you see solutions, questioning is the telescope that gives that lens shape, structure, and purpose.

Since questioning is the means by which critical thinking and decision-making are accomplished, consider whether you truly understand what a good question looks like. A good question will result in an actionable answer, usually one that provides additional information that is helpful in reaching a final solution. But how can you formulate questions that do this?

There are a few ways to know whether the question you are asking is a good one. If you don't have good question-asking instincts, interrogate your initial question with a few of these.

1. **Is your purpose clear?**

   A good question is carefully designed to meet a particular goal. For example, instead of asking, "When can I meet with you?" a clearer questioner would ask, "Would you prefer to meet on Monday morning or Wednesday morning?" The narrower range of options encourages a quicker, more decisive answer, which can in turn be acted upon. In order to get the most actionable information possible, you need to have a distinct idea of the kinds of information you are looking for. You can then make your questions more intentional and directed as you come closer to what you are looking to know. Specific purposes of questions may include:

   - Definition: What does "work ethic" looks like in our organization?
   - Comparative: What parts of our marketing strategy are different from our competitor?
   - Causal: If we invest in this new technology, what are some potential positive and negative outcomes?
   - Evaluative: What about this product is working for our consumer? What is not?

   By knowing which types of questions to ask in each situation, you'll have a more targeted discussion that leads to actionable answers.

2. **Is the question framed correctly?**

   Even with a clearly defined purpose, the framing of the question can still help or hinder its overall effectiveness. For example, asking, "Why should we invest in a Halloween party when clown costumes are so expensive?" will not be as effective as "Why should we invest in a Halloween party when, historically, they have not improved business culture?" The first question suffers from its poor framing, as it assumes that a Halloween party must include the investment in a clown costume. Poorly framed questions can be, identified through various smaller issues, including false comparisons, false dilemmas, and ambiguity. A good question deals with only one issue at a time, and avoids bundling disparate concerns into a single blanket assessment.

3. **Is your question closed or open?**

   One of the biggest pitfalls of the questioning process is asking questions with a predefined, or "closed," set of answers. These yes or no questions don't require synthesis, analysis, or evaluation of facts. They are often asked by leaders who already have an idea of what the answer should be and have no interest in additional information. While these can be useful when only a handful of acceptable answers exist, they do not lead to creative thinking or better decision-making in management.

In contrast, an open question requires thought and evaluation to answer. These questions can open the door to outside ideas and collaboration and ultimately lead to much more productive conversations than closed questions. These questions are designed to bring additional information to light and often lead to more in-depth understanding about the problem and potential solutions.

4. **Are you following up?**

Initial questions offer a vital starting point for any critical thinking and decision-making discussion. Unfortunately, some people stop there and can be the death knell of effectiveness and efficiency. In order to get the best answers, you must engage in a series of follow-up questions to support your initial inquiry.

Consider this question: "What are some areas we can cut in order to meet our yearly budget?" On its own, it will get you some information, but may miss crucial further discussion. Questions like "Who will be affected if we cut that department?" or "What will the impact of that departmental cut be on our production processes?" will provide additional actionable information and lead to smarter, safer cuts. In fact, the highly effective Five Whys system of problem-solving is built solely upon the idea of targeted follow-up questioning.

By incorporating effective questioning into your critical thinking equation, you will get clear answers that will help you to create actionable solutions. And, as you continue to evaluate your progress, effective questioning will become one of your skills as stakeholder and decision-maker.

Even with good critical thinking and questioning skills in place, it can be difficult to maintain consistency when it comes to problem-solving. Organizations are not individuals, but instead employ an array of people with different personalities, skill sets, and strengths, which can make solving group problems virtually impossible without a clearly defined framework. For that reason, many top-level organizations choose to incorporate a standardized problem-solving methodology. Not only does this provide the consistency a business needs, but also it often leads to more focused and productive discussions. This newfound productiveness in turn leads to more actionable plans and clearly defined goals for success.

Even though these processes have mainly been designed for large organizations, organizations of any size can adapt these concepts to suit their needs. Large businesses, small businesses, and individuals can all benefit from these simple problem-solving and decision-making methods. They have proven to be effective at maintaining a structured problem-solving process regardless of the structures in which they see use.

12.3.4   Six Methods and Techniques for Problem-Solving and Decision-Making

The top six methods and techniques for problem-solving and decision-making are provided here. Although many have made variations on the 6-Step Problem-Solving Model, the only research-based version of this methodology was invented by Dr. Sidney J. Parnes and Alex Osborn in the 1950s. After working with and observing high-level advertising employees throughout the brainstorming and implementation process, Parnes and Osborn recognized that creative people go through a series of stages as they create, organize, and choose good solutions for problems. Their findings were published in 1979 under the title, *Applied Imagination: Principles and Procedures of Creative Thinking.* In their original work, the 6-Step Model was termed, "The Creative Problem-Solving (CPS) Method," and it included these key segments:

1. Objective finding
2. Fact finding
3. Problem finding
4. Idea finding
5. Solution finding
6. Acceptance finding

These six segments were further organized into three key phases of problem-solving: exploring the challenge, generating ideas, and preparing for action.

After Parnes and Osborn released these creative problem-solving techniques, many different groups and businesses adapted them to fit their needs and organizational culture, providing a consistent framework for making daily decisions. One of these popular adaptations was created by Yale University and includes an evaluative segment that provides for continual optimization of the final decision. This model also incorporates some elements from the soft stage management (SSM) model, which provides a seven-stage approach to problem-solving. The Yale adaptation has been adopted by businesses and organizations worldwide and includes these six steps of action:

1. Define the problem.
2. Determine the root cause of the problem.
3. Develop alternative solutions.
4. Select a solution.
5. Implement the solution.
6. Evaluate the outcome.

In the updated version of the CPS model, more emphasis is placed on implementation and evaluation rather than simply accepting the results of the inquiry. This provides organizational leaders with an action-based problem-solving method that has been proven through research to be consistent and adaptable for virtually any need. Still, some aspects of business work present better opportunities to use this method than others.

Large group decisions—One of the core features of the 6-Step Model is that it relies heavily on brainstorming and group problem-solving, which in turn means large groups will benefit the most from the system as presented. The more suggestions, definitions, and root cause determinations offered by participants, the wider the view of the potential problems that need to be solved becomes. In addition, when a group is the impetus for identifying and analyzing the problem at hand, members attain heightened motivation as the process reaches its final step, "preparing for action."

Comparative decision-making—Another situation in which the 6-Step Model shows its strength comes when comparing the efficacy of your organization's ideas against a competitor. The groupthink structure of the method allows for a logical discussion of potential best-case and worst-case scenarios resulting from each potential course of action. Not only is this a good thing when formulating new ideas or action plans, but it works magnificently when determining strategies to take in a competitive marketplace. The evaluative phase of the method allows for research and comparison with outside ideas and models, such as those of major competitors, which eventually will lead to a better product or idea.

Long-term restructuring—This model deals particularly well with long-term changes or processes in need of consistent evaluation and restructuring. Since the evaluation process leads back into the initial phases of defining problems and developing solutions, the method develops a circular flow that allows the user to tackle even the most daunting decision-making projects. It also adapts to the size of the project or system in which it is used, so as a small project or system gets larger and more complex, the 6-Step Model remains effective and can even be applied to individual components and subsystems as necessary.

Additional resources on the 6-Step Problem-Solving Model:

- Yale.edu—Introduction of the 6-Step Problem-Solving Model
- Utk.edu—Dealing with Problems

12.4   PEST Analysis: Political, Economic, Social, and Technological

Noted as one of the most widely used decision-making techniques, the PEST model derives from the concept that several influencing factors can affect an organization, namely, political, economic, social, and technological factors. By carefully analyzing and evaluating these factors, organizations can make more, informed decisions and have a better understanding of the long-term implications of those choices.

The PEST model of decision-making was introduced by Francis J. Aguilar, a Harvard Business professor. In 1967, he published a book including the PEST model (originally the EPST model) entitled, *Scanning the Business Environment*. Arnold Brown reorganized the acronym as STEP (strategic trend evaluation process) sometime after the book's publication, and it was adapted further by a number of authors in the 1980s into acronyms including PEST, PESTLE, and STEEPLE. It is still well known by some of these alternative nomenclatures, and each retains the core elements of the system introduced by Aguilar.

Although it was originally designed as a method for understanding the unique layout of the business arena, PEST quickly became a consistent way for leaders to understand both the internal and external pressures that affected their organizational processes and products. It can also be easily adapted for the use with acquisitions and mergers, potential investments, and marketing campaigns. After decades of its use, the PEST model has proven to be especially effective in these specific situations:

Surveying business markets—Since this was its initial function, PEST functions best as a market surveying tool. The four key elements of the model can easily be adapted to any market, regardless of size or scope. In addition, permutations of the model like PESTLE include additional pressures that help to further understand the potential marketplace, such as legal and environmental factors. This makes the PEST model perfect for political ventures, building projects, or even human resource concerns.

Evaluating strategies or markets—Another area in which the PEST model shines is the evaluation of current strategies for flaws and inconsistencies. Because the model structures itself around rigorous evaluation, it allows all members of the decision-making team to have a clear idea of the potential impacts of the chosen course of action. By adding a weighting system to each of these elements, those in the discussion can clearly see which strategies have the greatest potential for success and will meet the organization's goals. Such a system also figures in strongly when comparing markets or courses of action, as it results in data points to illustrate the projected gains and losses for each potential solution.

Large-scale change including complex elements—Finally, the model allows for a methodical consideration of various influences, so that large-scale change can be managed in advanced and intricate detail. The PEST method highlights weaknesses in potential mergers or campaigns, allows for detailed speculation about future partnerships or markets, and gives insight into the regulatory or political drawbacks for each course of action. Through applying the PEST model, it is relatively easy to create a concise checklist of items to be addressed. This makes it one of the most actionable decision-making tools for corporate-level change.

Additional resources on the PEST analysis:

- Ohiou.edu—How to Find PEST Information on a Company
- Video on PEST for Small Business
- BusinessBalls.com—PEST Resource

12.5   SWOT Analysis: Strengths, Weaknesses, Opportunities, and Threats

The SWOT model of analysis sets out to help businesses analyze their company and better understand the arenas in which they operate. In this method, the strengths, weaknesses, opportunities, and threats of a company are outlined in a grid fashion, allowing the leadership to quickly identify toxic processes and behaviors.

Albert S. Humphrey usually receives the credit for the creation of the SWOT framework, as he presented it during his work with Stanford. In reality, the concept may have originated earlier than his 1960s presentation of the concept. Several researchers, including George Albert Smith, Jr., C. Roland Christiansen, and Kenneth Andrews of the Harvard Business School, reportedly worked with a prototype of the concept during the 1950s. Their model, published in 1965 as *Business Policy, Text and Cases*, had a slightly different set of values: opportunities, risks, environment, and competition. This research likely held some sway over the Stanford research model, which Humphrey initially referred to as SOFT analysis (satisfactory, opportunity, fault, and threat). Researchers Urick and Orr changed this to SWOT by 1964, and the name stuck.

SWOT lets users evaluate potential business risks as well as rewards for business ventures on the basis of environmental pressures. Like other models, SWOT also lends itself to discourse that leads to making better decisions. Though it doesn't work very well as a standalone decision-making model, it makes an excellent supplement to another more action-based system. Some of the situations where SWOT really shines include:

Brainstorming and strategy building—SWOT lends itself to sharing and discussing potential benefits and drawbacks of a single idea or course of action. Its simple format also plays well for situations involving big picture ideas and concepts. At the planning stage, it makes large issues readily obvious, as well as illustrating key benefits for each idea. When deciding on the strategy for a particular product, plan, or business, SWOT can make an organization's position and the benefits of each situation acutely obvious. A plan that has a strong strengths–opportunities correlation will support an aggressive strategy, while a plan that has a strong weaknesses–threats connection should be approached defensively.

Business and product development—The simplicity of the SWOT matrix is perfect for easily identifying strengths and weaknesses of a business or product. This model helps encourage discussion about the competitive advantages or gaps in capabilities of a specific idea. It also helps bring to light clear threats for a course of action, such as political, technological, or environmental pressures, that must be overcome before progress can be made. And, because it is such an adaptable model, it can be used for both large-scale and small-scale problems. This flexibility makes SWOT a good choice as a standardized decision-making tool.

Gathering and organizing data—SWOT can be a good choice at the brainstorming level of creative problem-solving, but can also prove itself an excellent tool during the researching phase of a task. The simple matrix can help present and organize data in preparation for action. In addition, it can easily show where research is lacking or where more information needs to be gathered.

Additional resources on the SWOT analysis:

- St Francis.edu—What is SWOT analysis?
- Ku.edu—SWOT and Problem-Solving Resource
- QuickMBA.com—SWOT Strategy

12.5.1   Business SWOT Analysis

SWOT analysis is a useful technique for understanding your strengths and weaknesses and for identifying both the *opportunities* open to you and the *threats* you face.

What makes SWOT particularly powerful is that with a little thought, it can help you uncover opportunities that you are well placed to exploit. And by understanding the weaknesses of your business, you can manage and eliminate threats that would otherwise catch you unawares.

More than this, by looking at yourself and your competitors using the SWOT framework, you can start to construct a strategy that helps you distinguish yourself from your competitors, so that you can compete successfully in your market.

12.5.2   How to Use the Tool

Originated by Albert S Humphrey in the 1960s, the tool is as useful now as it was then. You can use it in two ways—as a simple icebreaker helping people get together to "kick off" strategy formulation or in a more sophisticated way as a serious strategy tool.

> **Tip**
> Strengths and weaknesses are often internal to your organization, while opportunities and threats generally relate to external factors. For this reason, SWOT is sometimes called internal–external analysis, and the SWOT matrix is sometimes called an IE matrix.

To help you to carry out your analysis, download and print off our free worksheet, and write down answers to the following questions in the worksheet; go to Appendix D of this book.

- **Strengths**
    - What advantages does your organization have?
    - What do you do better than anyone else?
    - What unique or lowest-cost resources can you draw upon that others can't?
    - What do people in your market see as your strengths?
    - What factors mean that you "get the sale"?
    - What is your organization's unique selling proposition (USP)?

Also, if you are having any difficulty identifying strengths, try writing down a list of your organization's characteristics. Some of these will hopefully be strengths!

When looking at your strengths, think about them in relation to your competitors. For example, if all of your competitors provide high-quality products, then a high-quality production process is not a strength in your organization's market, it's a necessity.

Weaknesses

- What could you improve?
- What should you avoid?
- What are people in your market likely to see as weaknesses?
- What factors lose you sales?

Again, consider this from an internal and external basis: Do other people seem to perceive weaknesses that you do not see? Are your competitors doing any better than you? It is best to be realistic now and face any unpleasant truths as soon as possible.

Opportunities

- What good opportunities can you spot?
- What interesting trends are you aware of?

Useful opportunities can come from such things as:

- Changes in technology and markets on both a broad and narrow scale
- Changes in government policy related to your field
- Changes in social patterns, population profiles, lifestyle changes, and so on
- Local events

Tip
A useful approach when looking at opportunities is to look at your strengths and ask yourself whether these open up any opportunities. Alternatively, look at your weaknesses and ask yourself whether you could open up opportunities by eliminating them.

- **Threats**

  - What obstacles do you face?
  - What are your competitors doing?
  - Are quality standards or specifications for your job, products, or services changing?
  - Is changing technology threatening your position?
  - Do you have bad debt or cash flow problems?
  - Could any of your weaknesses seriously threaten your business?

Tip
When looking at opportunities and threats, PEST analysis can help to ensure that you do not overlook external factors, such as new government regulations, or technological changes in your industry.

12.5.3   Further SWOT Tips

If you're using SWOT as a serious tool (rather than as a casual "warm-up" for strategy formulation), make sure you're rigorous in the way you apply it:

- Only accept precise, verifiable statements ("Cost advantage of $10/ton in sourcing raw material x", rather than "good value for money").
- Ruthlessly prune long lists of factors and prioritize them, so that you spend your time thinking about the most significant factors.
- Make sure that options generated are carried through to later stages in the strategy formation process.
- Apply it at the right level—for example, you might need to apply the tool at a product or product line level, rather than at the much vaguer whole company level.
- Use it in conjunction with other strategy tools (e.g., USP analysis and Core Competence Analysis) so that you get a comprehensive picture of the situation you're dealing with.

> **Note**
> You could also consider using the programs such as TOWS Matrix ©. This is quite similar to SWOT in that it also focuses on the same four elements of strengths, weaknesses, opportunities, and threats. However, TOWS can be a helpful alternative because it emphasizes the external environment, while SWOT focuses on the internal environment.

- **Strengths**
    - We are able to respond very quickly as we have no red tape and no need for higher management approval.
    - We are able to give really good customer care, as the current small amount of work means we have plenty of time to devote to customers.
    - Our lead consultant has strong reputation in the market.
    - We can change direction quickly if we find that our marketing is not working.
    - We have low overheads, so we can offer good value to customers.

- **Weaknesses**
    - Our company has little market presence or reputation.
    - We have a small staff, with a shallow skills base in many areas.
    - We are vulnerable to vital staff being sick and leaving.
    - Our cash flow will be unreliable in the early stages.

- **Opportunities**
    - Our business sector is expanding, with many future opportunities for success.
    - Local government wants to encourage local businesses.
    - Our competitors may be slow to adopt new technologies.

- **Threats**
  - Developments in technology may change this market beyond our ability to adapt.
  - A small change in the focus of a large competitor might wipe out any market position we achieve.

As a result of their analysis, the consultancy may decide to specialize in rapid response, good value services to local businesses and local government.

Marketing would be in selected local publications to get the greatest possible market presence for a set of advertising budget, and the consultancy should keep up to date with changes in technology where possible.

12.5.4   Key Points

SWOT analysis is a simple but useful framework for analyzing your organization's strengths and weaknesses and the opportunities and threats that you face. It helps you focus on your strengths, minimize threats, and take the greatest possible advantage of opportunities available to you.

It can be used to "kick off" strategy formulation or in a more sophisticated way as a serious strategy tool. You can also use it to get an understanding of your competitors, which can give you the insights you need to craft a coherent and successful competitive position.

When carrying out your analysis, be realistic and rigorous. Apply it at the right level and supplement it with other option–generation tools where appropriate.

12.6   FMEA Analysis

As one of the first systematic techniques for observing weaknesses in organizations, the Failure Mode and Effects Analysis (FMEA) system often sees use as a diagnostic tool for companies and other large groups. FMEA puts forth the idea that all of the elements of a structure have inevitable failure modes, which are points at which they will break down under stress or over time. The goal of FMEA, then, is to identify the probable failure mode for each component and to project the impact that these failures will have on the overall success of the plan.

The US military and surrounding industries began using this method as early as 1949 for the purpose of identifying weaknesses in potential military equipment and weapons. Adopted in the early 1960s by contractors working with the US National Aeronautics and Space Administration (NASA), FMEA helped these organizations produce parts and processes that would guarantee a high success rate for the space shuttle program. In 1967, the Society for Automotive Engineers (SAE) published a version of FMEA which, with revisions, has remained the standard failure mode model for the public aviation industry. Versions of FMEA have been used by the Automotive Industry Action Group, the US Environmental Protection Agency (EPA), and Food and Drug Administration (FDA).

Henry Ford was the first leader to widely incorporate the FMEA model to identify process weaknesses within a business. He adapted the FMEA model into two main areas: Process FMEA (PFMEA) and Design FMEA (DFMEA). PFMEA helps leaders to identify potential breakdowns of production, supply, and market failure for an organization, while engineers and other technical personnel use DFMEA to assess the ramifications of potential weaknesses and safety issues in their designs. The areas in which these two types of FMEA are most effective include:

Manufacturing and assembly processes—The initial goal of the FMEA model was to identify problems and potential failures of elements within a manufacturing process. Because of this, the FMEA model is a good choice for businesses that are heavily involved in manufacturing and production. It guides the participant through each point of the production cycle and allows him or her to foresee potential risks associated with parts, labor, and processes. Often, this results in fewer risks and elimination of unnecessary redundancies, which lead to a safer work environment and a more cost-effective business.

Business strategy—Another area in which FMEA is highly efficient is in the preparation stages of any major change. This model focuses on potential risks at every point in the new process, which motivates leaders to understand and overcome challenges long before they arise. If a clear goal or emphasis is not established before beginning the FMEA process, however, this can become overwhelming and even paralytic, encouraging stagnation within a company. By assigning a Risk Priority Number (RPN) to each failure mode element, those using this model can make it much more obvious which failure modes require immediate attention.

Customer satisfaction and safety—Both PFMEA and DFMEA can assist in bolstering the satisfaction and well-being of customers. As processes are analyzed and evaluated closely, organizations become quicker and more cost-effective, often without sacrificing the quality of the final product. Because process flaws are identified and eliminated before taking the product or process to the customer, dissatisfaction becomes much less common. The DFMEA portion of the process becomes more reliable and safer as the model is applied time, and time again, which can lead to higher employee retention and more loyal customers.

Additional resources on the FMEA method:

- ASQ.org Process Analysis with FMEA
- Video How to Become Effective in FMEA

12.7 Developing a Robust Problem Definition (CATWOE)

CATWOE helps you to look before you leap, so what we are concern with in this method are the following circumstances: What do you do when you are faced with a really big business problem? What if your employee retention was low, for example, and you wanted to know why (Fig. 12.4)?

Fig. 12.4 CATWOE. © iStockphoto erlobrown

Soft systems methodology (SSM) is a decision-making process designed to tackle real-world problems that have no formal definition or scope. In this system, users must consider six areas in order to solve these kinds of soft system problems.

SSM conceptualizes the activities or business being examined as a system, the essence of which is encapsulated in a "Root Definition."

In 1975, David Smyth, a researcher in Checkland's department, observed that SSM was most successful when the Root Definition included certain elements. These elements, remembered by the mnemonic CATWOE*, identified the people, processes, and environment that contribute to a situation, issue, or problem that you need to analyze.

CATWOE stands for:

Customers/clients	Who are they, and how does the issue affect them?
Actors	Who is involved in the situation? Who will be involved in implementing solutions? And what will impact their success?
Transformation process	What processes or systems are affected by the issue?
World view	What is the big picture? And what are the wider impacts of the issue?
Owner	Who owns the process or situation you are investigating? And what role will they play in the solution?
Environmental constraints	What are the constraints and limitations that will impact the solution and its success?

When you look at all six of these elements and consider the situation from all of these perspectives, you open your thinking beyond the issue that sits directly in front of you. By using CATWOE, the output of your brainstorming and problem-solving should be much more comprehensive, because you have considered the issue from these six, very different, perspectives.

CATWOE systematically incorporates these elements into a discussion about potential actions, looking at how these actions will influence the major players in a transition or other major problem. Originally developed by Peter Checkland and Brian Wilson, this problem-solving system has been constantly assessed and improved through continuing action research over the last 30 years. Initially, it was designed in response to the systems engineering approach to management problems. In 1966, a team of researchers at Lancaster University led by Gwilym Jenkins found that the systems engineering approach only worked when a problem could be clearly and narrowly defined. In cases wherein no clear definition was available, they found that the system was not effective for solving real and complex management problems. With Checkland and Wilson taking the lead, the SSM model was established. CATWOE was the problem-solving format that arose from their research.

CATWOE, by definition, works most effectively when it is being used to manage complex, real-world management problems. This broad approach means it can assist in solving virtually any issue that is not easily defined. Some organizational situations still lend themselves more to CATWOE than other commonly accepted models, however, in spite of this adaptability. Some common CATWOE-friendly issues include:

Identifying problems—Since the purpose of the CATWOE problem-solving method is to help define abstract problems, its ability to do so outstrips that of most other systems. Many of the day-to-day problems a manager faces are not concrete, so CATWOE can help significantly. When dealing with human resources, marketing, and workflow management, getting a clear understanding of what the problem is or how to best solve it and make decisions can feel like an impossible task. CATWOE allows leaders to consider all of the key influencers, such as people, ideologies, and environments, being impacted by the potential change or issue. This leads to a clearer understanding of the root causes that must be addressed in order to make forward progress.

Implementing solutions—The CATWOE method also presents some strong tools when preparing to take action steps. Because CATWOE focuses on considering the influencing factors, people, and environments that will be integral to a solution, this method ensures that all of those elements are in place before the implementation. CATWOE also assesses the roles each team member will play in the change, breaking individuals down into broad categories such as client, actor, or owner. Since these roles are defined in the CATWOE structure itself, each person has a better idea of how they contribute to the project's success and can in turn be easily held accountable for their responsibilities.

Organizing and aligning goals—When this problem-solving model is workshopped in a group of diverse stakeholders that includes both clients and producers, it serves to inform members about their role in the overall organization. It can also be very effective for aligning disparate worldviews and ideologies, enabling the whole team to become more focused and motivated toward a common goal. As with many of the other methodologies, CATWOE does a great job of opening discourse, but differs in that resulting action steps can't really be taken unless the group has completed the initial steps of collectively defining the problem. Unlike some other problem-solving models, CATWOE lends itself strongly to collaboration, as it uses that collaboration to feed into further action.

Additional resources on Cause and Effect Analysis:

- Washington.edu—How to Use Cause and Effect Analysis
- Video Overview—Dr. O'Loughlin on Fishbone Diagrams
- Drexel.edu—Using the CATWOE Framework
- Video—Peter Checkland @ Lancaster University—The Origins of SSM
- LifeHack.org—Find the Problem Using CATWOE

Before using CATWOE and try to solve an important problem, use the CATWOE checklist to brainstorm the various people and elements that are affected.

(A) **When to use it**

Use it when identifying the problem, to prompt thinking about what you are really trying to achieve.

Use it when seeking to implement the solution, to help consider the impact on the people involved (Table 12.1).

(B) **How to use it**

Use the areas below to stimulate thinking about the problem and/or implementing the solution.

C = Customers

- Who is on the receiving end?
- What problem do they have now?
- How will they react to what you are proposing?
- Who are the winners and losers?

A = Actors

- Who are the actors who will "do the doing," carrying out your solution?
- What is the impact on them?
- How might they react?

T = Transformation process

- What is the process for transforming inputs into outputs?
- What are the inputs? Where do they come from?

Table 12.1 When-to-use-it steps

Quick	X				Long
Logical	X				Psychological
Individual	X				Group

- What are the outputs? Where do they go to?
- What are all the steps in between?

W = World view

- What is the bigger picture into which the situation fits?
- What is the real problem you are working on?
- What is the wider impact of any solution?

O = Owner

- Who is the real owner or owners of the process or situation you are changing?
- Can they help you or stop you?
- What would cause them to get in your way?
- What would lead them to help you?

E = Environmental constraints

- What are the broader constraints that act on the situation and your ideas?
- What are the ethical limits, the laws, financial constraints, limited resources, regulations, and so on?
- How might these constrain your solution? How can you get around them?

Example
Situation: Thinking about way of putting advertisements inside cars

CATWOE:

- Customers: Advertisers, drivers, and passengers—May see things differently!
- Actors: Garage attendants, mechanics, and car washes—Extra revenue for them
- Transformation process: Putting sticker on car now in return for discounts next time
- World view: Drivers seeking every economy or fashion victims, maybe?
- Owner: Car owner—Must be persuaded of value
- Environmental constraints: Limits on marketing budget, which will be needed

(A) **How it works**

CATWOE was defined by Peter Checkland as a part of his soft systems methodology (SSM). It is a simple checklist for thinking. Like many checklists, it can be surprisingly useful when used appropriately to stimulate open thought.

12.8   Cause and Effect Analysis

In Cause and Effect Analysis, also called fishbone diagrams or Ishikawa diagrams, thinkers assess a single effect in an attempt to find its potential causes. During this four-step model, participants identify a problem, work out the involved factors, identify potential causes, and analyze the final diagram in preparation for action.

This problem-solving model was created in 1968 by University of Tokyo engineering professor Kaoru Ishikawa, although the Cause and Effect Analysis framework dates back to the 1920s. It was first included as one of the seven basic tools of quality control, which W. Edwards Deming presented to postwar Japanese engineers, including Ishikawa himself. Of these seven tools, Cause and Effect Analysis deals with critical thinking the most extensively and uses compartmentalization and categorization to define which influencers contribute to the effect in question and how.

Each industry often develops its own unique set of categories that can be used with the Ishikawa diagram. The manufacturing industry, for example, uses the six Ms (manufacturing, method, material, man power, measurement, and mother nature), while the service industry uses the five Ss (surroundings, suppliers, systems, skills, and safety). These categories are often used in conjunction with the Five Whys methodology for questioning, which can make the root causes of any effect clearer.

The Cause and Effect Analysis model has held sway for a long time, thanks to the instances in which it outperforms many newer models. The most effective implementations include:

Group decision-making—The Cause and Effect Analysis model works best with a key group of invested stakeholders, preferably from each of the main categories that the diagram will incorporate. This allows for the most in-depth analysis of the root causes of a problem from the perspective of the people who are most familiar with that aspect of the business. The Cause and Effect Analysis model also lends itself to discussion and can uncover fine details that may be closely connected with one another and in turn make analysis better. This happens most often in a group setting, where multiple members can become aware of the correlations of seemingly disparate parts of the business process.

Clearly defined problems—In complete opposition to decision-making models like CATWOE, which deal with ill-defined, nebulous issues, this model works best with concrete, tangible problems. This decision-making method starts by defining the problem, and without defining a problem clearly, the Cause and Effect Analysis model begins to break down. If the effect is vague or misunderstood by members of the team, analyzing its potential causes can be difficult. Framing is essential to the effective use of Cause and Effect Analysis, as problems like "68% employee turnover" can be much more efficiently dissected than "employees unhappy."

Complex, interrelated effects—Where this method really shines is in arenas where effects may have multiple, interrelated causes. This makes the Cause and Effect Analysis model perfect for large institutional change like mergers and acquisitions. Even on a small scale, this method does a stellar job of highlighting how seemingly unrelated processes or elements of production affect one another. Much like the PEST model, the Cause and Effect Analysis model assesses each segment of business operations that could change the outcome. This gives each stakeholder insight into the small changes that can be made within their segment and in turn helps them to understand what might make the process or product more efficient and productive.

12.9   Making Decisions Under Conditions of Risk and Uncertainty

Uncertainty is a state of having limited knowledge of current conditions or future outcomes. It is a major component of risk, which involves the likelihood and scale of negative consequences. Managers often deal with uncertainty in their work; to minimize the risk that their decisions will lead to undesired outcomes, they must develop the skills and judgment necessary for reducing this uncertainty. Managing uncertainty and risk also involves mitigating or even removing things that inhibit effective decision-making or adversely affect performance.

One cause of uncertainty is proximity: things that are about to happen are easier to estimate than those further out in the future. One approach to dealing with uncertainty is to put off decisions until data become more accessible and reliable. Of course, delaying some decisions can bring its own set of risks, especially when the potential negative consequences of waiting are great.

Dealing with this matter, we need to keep the following points under consideration:

- Uncertainty and risk are not the same thing. Whereas uncertainty deals with possible outcomes that are unknown, risk is a certain type of uncertainty that involves the real possibility of loss. Risks can be more comprehensively accounted for than uncertainty.
- Decision-making under conditions of risk should seek to identify, quantify, and absorb risk whenever possible.
- The quantity of risk is equal to the sum of the probabilities of a risky outcome (or various outcomes) multiplied by the anticipated loss as a result of the outcome.
- A firm's ability to absorb, transfer, and manage risk will often define management's risk appetite; once risks are identified and quantified, decisions may be made as to what extent risky outcomes may be tolerated.

12.9.1   Identifying Risks

Managing uncertainty in decision-making relies on identifying, quantifying, and analyzing the factors that can affect outcomes. This enables managers to identify likely risks and their potential impact. Types of risk include:

1. **Strategic risks**: These risks arise from the investments an organization makes to pursue its mission and objectives. They are often associated with competition and can include macroeconomic risks (the alignment of buyers and sellers consistent with the principles of supply and demand), transaction risks (the operational risks from merger and acquisition activity, divestitures, or partnerships), and investor relations risk (the risks associated with communicating effectively or ineffectively with the investment community).

2. **Financial risks**: These relate to potential economic losses that can result from poor allocation of resources, changes in interest rates, shifts in tax policy, increases or decreases in the price of commodities, or fluctuations in the value of currency.
3. **Operational risks**: These risks can arise due to choices about the design and use of processes to create and deliver goods and services. They can include production errors, substandard raw materials, and technology malfunctions.
4. **Legal risks**: These risks stem from the threat of litigation or ambiguity in applicable laws and regulations (including whether they are likely to change); these threats create uncertainty in the steps an organization should take to address its obligations to customers, employees, suppliers, stockholders, communities, and governments.
5. **Other risks**: Risks are very commonly associated with force majeure, or events beyond the control of the organization. These can include weather disasters, floods, earthquakes, and war or other hostilities.

12.9.2   Quantifying Risks

Once management has identified the appropriate risk category that may influence a certain decision, it may go about quantifying these risks. In other words, management will ascertain the costs incurred if a risky outcome were to happen. This can be mathematically daunting for many types of risk, especially financial risk. Generally speaking, however, risk is equal to the sum of the probabilities of a risky outcome (or various outcomes) multiplied by the anticipated loss as a result of the outcome. This is similar to performing a sensitivity analysis if the universe of outcomes is known.

12.9.3   Managing Risks

The ability of a firm to absorb, transfer, and manage risk is critical in management's decision-making process when risky outcomes are involved. This will often define management's risk appetite and help to determine once risks are identified and quantified, whether risky outcomes may be tolerated. For example, many financial risks can be absorbed or transferred through the use of a hedge, while legal risks might be mitigated through unique contract language. If managers believe that the firm is suited to absorb potential losses in the event the negative outcome occurs, they will have a larger appetite for risk given their capabilities to manage it.

12.10   What Is Threat?

To describe the threat definition as starting point, we look at the dictionary for its definition, and that is a statement of an intention to inflict pain, injury, damage, or other hostile action on someone in retribution for something done or not done or a person or thing likely to cause damage or danger. For example, hurricane damage poses a major threat to many coastal communities.

12.11   Threat Modeling

Threat modeling is a procedure for optimizing network security by identifying objectives and vulnerabilities, and then defining countermeasures to prevent or mitigate the effects of threats to the system. In this context, a threat is a potential or actual adverse event that may be malicious (such as a denial-of-service attack) or incidental (such as the failure of a storage device) and that can compromise the assets of an enterprise.

The key to threat modeling is to determine where the most effort should be applied to keep a system secure. This is a variable that changes as new factors develop and become known, applications are added, removed, or upgraded, and user requirements evolve. Threat modeling is an iterative process that consists of defining enterprise assets, identifying what each application does with respect to these assets, creating a security profile for each application, identifying potential threats, prioritizing potential threats, and documenting adverse events and the actions taken in each case.

Threat modeling is an approach for analyzing the security of an application. A structured approach enables you to identify, quantify, and address the security risks associated with an application. Threat modeling is not an approach to reviewing code, but it does complement the security code to review process. The inclusion of threat modeling in the SDLC can help to ensure that applications are being developed with security built-in from the very beginning. This, combined with the documentation produced as part of the threat modeling process, can give the reviewer a greater understanding of the system. This allows the reviewer to see where are the entry points to the application and the associated threats with each entry point. The concept of threat modeling is not new, but there has been a clear mindset change in recent years. Modern threat modeling looks at a system from a potential attacker's perspective, as opposed to a defender's viewpoint. Microsoft have been strong advocates of the process over the past number of years. They have made threat modeling a core component of their SDLC, which they claim to be one of the reasons for the increased security of their products in recent years.

When the source code analysis is performed outside of the SDLC, such as on existing applications, then the results of the threat modeling help in reducing the complexity of the source code analysis. This is done by promoting an in-depth first approach vs. breadth-first approach. Instead of reviewing all source code with equal focus, you can prioritize the security code review of components whose threat modeling has ranked with high-risk threats.

The threat modeling process can be decomposed into three high-level steps:

Step 1: Decompose the application. The first step in the threat modeling process is concerned with gaining an understanding of the application and how it interacts with external entities. This involves creating use cases to understand how the application is used, identifying entry points to see where a potential attacker could interact with the application, identifying assets, i.e., items/areas that the attacker would be interested in, and identifying trust levels which represent the access rights that the application will grant to external entities. This information is documented in the threat model document, and it is also used to produce data flow diagrams (DFDs) for the application. The DFDs show the different paths through the system, highlighting the privilege boundaries.

Step 2: Determine and rank threats. Critical to the identification of threats is using a threat-categorization methodology. A threat categorization such as STRIDE can be used or the application security frame (ASF) that defines threat categories such as auditing and logging, authentication, authorization, configuration management, data protection in storage and transit, data validation, and exception management. The goal of the threat categorization is to help identify threats both from the attacker (STRIDE) and the defensive perspective (ASF). DFDs produced in step 1 help to identify the potential threat targets from the attacker's perspective, such as data sources, processes, data flows, and interactions with users. These threats can be identified further as the roots for threat trees; there is one tree for each threat goal. From the defensive perspective, ASF categorization helps to identify the threats as weaknesses of security controls for such threats. Common threat lists with examples can help in the identification of such threats. Use and abuse cases can illustrate how existing protective measures could be bypassed, or where a lack of such protection exists. The determination of the security risk for each threat can be determined using a value-based risk model such as DREAD or a less subjective qualitative risk model based upon general risk factors (e.g., likelihood and impact).

Step 3: Determine countermeasures and mitigation. A lack of protection against a threat might indicate a vulnerability whose risk exposure could be mitigated with the implementation of a countermeasure. Such countermeasures can be identified using threat-countermeasure mapping lists. Once a risk ranking is assigned to the threats, it is possible to sort threats from the highest to the lowest risk and prioritize the mitigation effort, such as by responding to such threats by applying the identified countermeasures. The risk mitigation strategy might involve evaluating these threats from the business impact that they pose and reducing the risk. Other options might include taking the risk, assuming the business impact is acceptable because of compensating controls, informing the user of the threat, removing the risk posed by the threat completely, or the least preferable option, that is, to do nothing.

Each of the above steps is documented as they are carried out. The resulting document is the threat model for the application. This guide will use an example to help explain the concepts behind threat modeling. The same example will be used throughout each of the three steps as a learning aid. The example that will be used is a college library web site. At the end of the guide, we will have produced the threat model for the college library web site. Each of the steps in the threat modeling

12.12   Cyber Security and Threat Modeling

Application security has become a major concern in recent years. Hackers are using new techniques to gain access to sensitive data, disable applications, and administer other malicious activities aimed at the software application. The need to secure an application is imperative for use in today's world. Until recently, application security was an afterthought; developers were typically focused on functionality and features, waiting to implement security at the end of development. This approach to application security has proven to be disastrous; much vulnerability has gone undetected allowing applications to be attacked and damaged. This raises the following questions: How can application security become an integral part of the development process? How can an application design team discover and avoid vulnerabilities in their application? There are three measures that can help in discovering and avoiding security vulnerabilities:

- Have many experts contribute and share their knowledge.
- Make expert knowledge about typical vulnerabilities available to developers.
- Make system-specific knowledge available to persons searching for vulnerabilities.

To define threat modeling from Cyber Security point of view, the following statement applies. Threat modeling is a computer security optimization process that allows for a structured approach while properly identifying and addressing system threats. The process involves systematically identifying security threats and rating them according to severity and level of occurrence probability.

By identifying and rating these security threats through a solid understanding of the system or application, a security officer can logically address the threats, beginning with the most pressing.

The basis for the creation of a threat model is the development of a security specification and subsequent testing of the integrity of that specification. The process is conducted early in the design phase of a system or application and used to pinpoint the motives and methods used by an attacker to identify system threats and vulnerabilities. In other words, threat modeling involves thinking like an attacker.

Threat modeling is geared toward accomplishing the following:

- Identifying, investigating, and rating potential threats and vulnerabilities
- Identifying logical thought processes for defining the system's security

- Creating a set of standard documents that can be used to create specifications and security testing and prevent future duplication of security efforts
- Reducing threats and vulnerabilities
- Defining the overall security level of a system or application

One method being used to implement application security in the design process is threat modeling. The basis for threat modeling is the process of designing a security specification and then eventually testing that specification. The threat modeling process is conducted during application design and is used to identify the reasons and methods that an attacker would use to identify vulnerabilities or threats in the system. Threat modeling accomplishes the following: [3]

- Defines the security of an application
- Identifies and investigates potential threats and vulnerabilities
- Brings justification for security features at both the hardware and software
- Levels for identified threats
- Identifies a logical thought process in defining the security of a system
- Results in finding architecture bugs earlier and more often
- Results in fewer vulnerabilities
- Creates a set of documents that are used to create security specifications and security testing, thus preventing duplication of security efforts

By using threat modeling to identify threats, vulnerabilities, and mitigations at design time, the system development team will be able to implement application security as part of the design process.

12.13   How to Create a Threat Model?

To create a threat model, a systematic approach is required. One approach is to use data flow. By using the data flow approach, the threat modeling team is able to systematically follow the flow of data throughout the system identifying the key processes and the threats to those processes. The data flow approach uses three main steps: view the system as an adversary, characterize the system, and determine the threats. The threat modeling process outlined below is adapted from these references:

12.13.1   View the System as an Adversary

When an adversary views the application, they only see the exposed services. From the exposed services, the adversary formulates goals to attack the system.
The steps used to understand an adversary's goals are detailed below.

12.13.2   Identify the Entry/Exit Points

Entry/exit points are the places where data enters or exits the application. When identifying entry/exit points, the following data should be identified and collected:

- Numerical ID—Each entry/exit point should have a numerical ID assigned to it for cross-referencing with threats and vulnerabilities.
- Name—Assign a name to the entry/exit point and identify its purpose.
- Description—Write a description that outlines what takes place at that entry/exit point. Identify the trust levels that exist at that point.

12.13.3   Identify the Assets

Assets are the reason threats exist; an adversary's goal is to gain access to an asset. The security team needs to identify which assets need to be protected from an unauthorized user. Assets can be either physical or abstract, i.e., employee safety, company's reputation, etc. Assets can interact with other assets, and, because of this, they can act as a pass-through point for an adversary. When identifying assets, the following data should be identified and collected:

- Numerical ID—Each asset should have a numerical ID assigned to it for cross-referencing with threats and vulnerabilities.
- Name—Assign a name to the asset.
- Description—Write a description that explains why the asset needs protection.

12.13.4   Identify the Trust Levels

Trust levels are assigned to entry/exit points to define the privileges an external entity has to access and affect the system. Trust levels are categorized according to privileges assigned or credentials supplied and cross-referenced with entry/exit points and protected resources. When identifying trust levels, the following data should be identified and collected:

- Numerical ID—Each trust level should have a numerical ID to cross-reference with entry/exit points and assets.
- Name—Assign a name to the trust level.
- Description—Write a description that explains more detail about the trust level and its purpose.

12.14   Characterize the System

In order to characterize the system, background information will need to be gathered about the system. The background information will help the security team focus and identify the specific areas that need to be addressed. There are five categories of background information:

- Use scenarios
- External dependencies
- External security notes
- Internal security notes
- Implementation assumptions

12.14.1   Use Scenarios

Use scenarios describe how the system will be used or not used in terms of configuration or security goals and non-goals. A use scenario may be defined in a supported or unsupported configuration.

Not addressing use scenarios may result in a vulnerability. Use scenarios help limit the scope of analysis and validate the threat model. They can also be used by the testing team to conduct security testing and identify attack paths. The architect and end users typically identify use scenarios. When defining use scenarios, the following data should be collected:

- Numerical ID—Each use scenario should have a unique identification number.
- Description—Write a description that defines the use scenario and whether it is supported or not supported.

12.14.2   External Dependencies

External dependencies define the system's dependence on outside resources, and the security policy outside the system is being modeled. If a threat from an external dependency is disregarded, it may become a valid vulnerability. "In software systems, external dependencies often describes system wide functions such as algorithm inconsistency" [threat]. When defining external dependencies, the following data should be collected:

- Numerical ID—Each external dependency should have a unique identification number.
- Description—Write a description of the dependency.
- External security note reference—External security notes from one component can be cross-referenced with external dependencies on other components within the application.

12.14.3   External Security Notes

External security notes are provided to inform users of security and integration information for the system. External security notes can be a warning against misuse or a form of a guarantee that the system makes to the user. External security notes are used to validate external dependencies and can be used as a mitigation to a threat. However, this is not a good practice as it makes the end user responsible for security. When defining external security notes, the following data should be collected:

- Numerical ID—Each external security note should have a unique identification number.
- Description—Write a description of the note.

12.14.4   Internal Security Notes

Internal security notes further define the threat model and explain concessions made in the design or implementation of system security. When defining internal security notes, the following data should be collected:

- Numerical ID—Each internal security note should have a unique identification number.
- Description—Write a description of the security concession and justification for the concession.

12.14.5   Implementation Assumptions

Implementation assumptions are created during the design phase and contain details of features that will be developed later. When defining implementation assumptions, the following data should be collected:

- Numerical ID—Each internal implementation assumption should have a unique identification number.
- Description—Write a description of the method of implementation.

12.14.6   Modeling the System

To create a useful model, the team needs to look at the application through an adversary's eyes. Modeling diagrams are a visual representation of how the subsystems operate and work together. The type of diagramming is not important; for the purposes of this document, data flow diagrams (DFDs) will be used to model the system.

12.14.6.1 Modeling Using Data Flow Diagrams (DFDs)

DFDs are a high-level way of focusing on data and how it flows through the system. DFDs are iterative and should be organized in a hierarchy that accurately reflects the system. The six basic shapes used in a DFD are listed below with their definitions.

1. **Process**—represents a task that performs some action based on the data and should be numbered. Shown in Fig. 12.5 is a single process illustration.

2. **Multiple process**—has subprocesses and each subprocess node number should be prefixed by the parent node number, i.e., 1, 1.1, and 1.1.2. Shown in Fig. 12.6 is a multiple process illustration.

3. **External entity**—is located outside the system and interacts at a system entry/exit point. It is either the source or destination of data and may only interact with process or multiple processes. Shown in Fig. 12.7 is an external entity illustration.

4. **Data store**—place where data is saved or retrieved and it may only interact with process or multiple processes. Shown in Fig. 12.8 is a data store illustration.

5. **Data flow**—movement of data between elements. Shown in Fig. 12.9 is a data flow illustration.

6. **Trust boundary**—boundary between trust levels or privileges. Shown in Fig. 12.10 is trust boundary illustration.

To create a data flow diagram (DFD), first create an overall context diagram. This is the top level or root where the system is represented as a single multiple process or entry/exit point. Each node thereafter is a more detailed DFD describing other processes. Figure 12.11 shows an example of an overall context DFD.

Fig. 12.5 Single process illustration

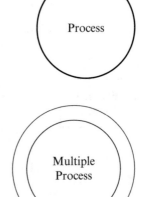

Fig. 12.6 Multiple processes illustration

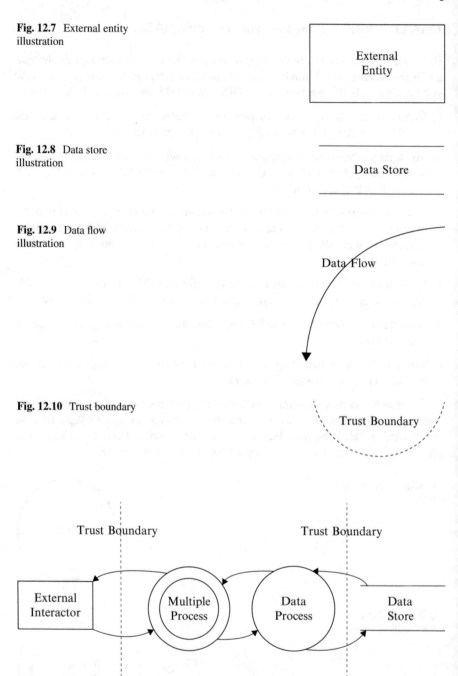

Fig. 12.7 External entity illustration

Fig. 12.8 Data store illustration

Fig. 12.9 Data flow illustration

Fig. 12.10 Trust boundary

Fig. 12.11 Data flow diagram

12.15   Determining Threats

Once the DFD is created, it is used to determine what data is supplied to a node and the goals the adversary has for the application—an adversary must supply data to launch an attack. The goals are then used within the DFD to determine the threat paths, locate the entry/exit points, and follow the data through the system. The threat path is defined as the sequence of any process nodes that perform security-critical processing [threat]. All areas where there is change or action on behalf of the data are susceptible to threats, i.e., process nodes.

12.15.1   Threat Profile

The threat profile is a security design specification for the system that describes the possible goals of the adversary and the vulnerabilities that exist as a result of those goals. Each threat in the profile must be prevented or mitigated. The threat profile consists of three main areas:

- Identify the threats.
- Investigate and analyze the threats.
- Mitigate the vulnerabilities caused by the threats.

12.15.1.1   Identifying Threats

Threat identification is a key to a secure system. Identifying threats consists of analyzing each entry/exit point and determining what critical security processing occurs at the entry/exit point and how it might be attacked. Threats are the goals of the adversary and for a threat to exist it must have a target asset. In the threat model document, the threats are cross-referenced with the assets. To identify threats or goals, ask the following questions:

How can the adversary use or manipulate the asset to:

- Modify or control the system?
- Retrieve information within the system?
- Manipulate information within the system?
- Cause the system to fail or become unusable?
- Gain additional rights?

Can the adversary access the asset:

- Without being audited?
- And skip any access control checks?
- And appear to be another user?

The next step in identifying threats is to classify the threat using the STRIDE model. The STRIDE model is documented in below:

STRIDE

- Spoofing—Allows an adversary to pose as another user, component, or other system that has an identity in the system being modeled
- Tampering—The modification of data within the system to achieve a malicious goal
- Repudiation—The ability of an adversary to deny performing some malicious activity because the system does not have sufficient proof otherwise
- Information disclosure—The exposure of protected data to a user that is not otherwise allowed access to that data
- Denial of service—Occurs when an adversary uses illegitimate means to assume a trust level with different privileges than he currently has

Classifying the threat makes it easier to understand what the threat allows an attacker to do and aids in assigning priority.

12.15.1.2 Investigating and Analyzing the Threats Using a Threat Tree

To identify vulnerable areas in the system and determine valid attack paths, the threats identified in the previous step must be analyzed to discover where the system is susceptible to the threat. One method that works well for the investigation process is to build a threat tree.

Threat trees can be expressed graphically or as text in a threat-modeling document. A threat tree consists of a root node or threat and child node(s). Each child node represents conditions needed for the adversary to find and identify the threat. Threat trees are used to determine the vulnerabilities associated with a threat. To identify a threat's vulnerabilities, begin at a node without any children and traverse it up to the root threat.

Another step in analyzing the threats is to determine the risk of the threat and the threat's conditions or child nodes by using the DREAD model. The DREAD model is documented in [secure]. When using the DREAD model, a threat-modeling team calculates security risks as an average of numeric values assigned to each of five categories.

- **Damage potential**—Ranks the extent of damage that occurs if a vulnerability is exploited.
- **Reproducibility**—Ranks how often an attempt at exploiting a vulnerability really works.
- **Exploitability**—Assigns a number to the effort required to exploit the vulnerability. In addition, exploitability considers preconditions such as whether the user must be authenticated.

- **Affected users**—A value characterizing the number of installed instances of the system that would be affected if an exploit became widely available.
- **Discoverability**—Measures the likelihood that, if unpatched, a vulnerability will be found by external security researchers, hackers, and the like.

What Is DREAD?

DREAD is part of a system for risk-assessing computer security threats previously used at Microsoft and currently used by OpenStack and many other corporations. It provides a mnemonic for risk rating security threats using five categories.

The categories are:

Damage—how bad would an attack be?

Reproducibility—how easy is it to reproduce the attack?

Exploitability—how much work is it to launch the attack?

Affected users—how many people will be impacted?

Discoverability—how easy is it to discover the threat?

The DREAD name comes from the initials of the five categories listed. It was initially proposed for threat modeling, but it was discovered that the ratings are not very consistent and are subject to debate. It was out of use at Microsoft by 2008.

When a given threat is assessed using DREAD, each category is given a rating. For example, 3 for high, 2 for medium, 1 for low, and 0 for none (rating scales running from 0 to 10 are common). The sum of all ratings for a given exploit can be used to prioritize among different exploits.

Use a scale of 1–10 to rate each category, where 1 is the least probability of occurrence and the least damage potential. Add the rating of each category and divide the total by five to get an overall risk rating for each threat. The result can further be divided into three sections to generate a high, medium, or low risk rating.

12.15.1.3 Vulnerability Resolution and Mitigation

Up to this point, the threats have been identified and analyzed. If a threat is left unresolved, it will become a vulnerability. A vulnerability is present when a threat exists, and the steps to mitigate it have not been implemented. To reduce the risk caused by threats, the team must analyze the conditions of each threat, using DREAD to assign a risk level, and identify a mitigation strategy to each condition.

Once the threat tree is completed, it can be used to identify attack paths, routes from a condition to a threat. Any attack path that is not mitigated will become a vulnerability. Figure 12.12 shows an example of a threat tree.

The threats, threat tree, vulnerabilities, and mitigations are compiled into a threat-modeling document that describes the threat profile of the system. The threat-modeling document can be used in the design process as a security design specification and in the testing process to identify the vulnerable areas of the system.

12.16   Conclusion of Threat Modeling

As the world increases its dependency on computers for critical information, the chances of applications being attacked are also increasing. Network security is no longer sufficient to secure an application. Security needs to be a part of the application design process. Implementing security during the design phase using the threat modeling process ensures that security is being designed into the application, thus decreasing the risk of an attack. Threat modeling will help you acknowledge, manage, and communicate security risks across your application, ensuring that security has been designed into the system. Threat modeling is an iterative process, and your threat model should evolve over time, changing to adapt to new threats and adjusting to changing business requirements.

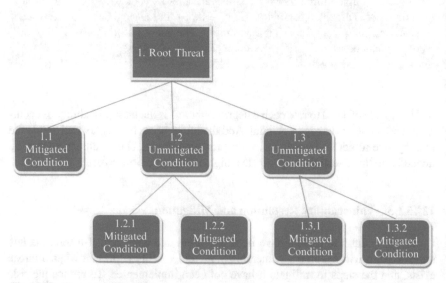

Fig. 12.12   Threat tree illustration

12.17 Enterprise Cloud Security and Governance Action Plan

Before we finish the chapter as a whole, we briefly touch upon cloud security as essential matter within enterprise and plan of action needed to be taken by information security and information technology in that organization.

The increasing complexity of cloud computing and the resulting security challenges are a force IT teams must reckon with. To succeed, system administrations need to solidify company-wide governance plans and policies.

Securing applications and data is essential for any organization, but the responsibility is not evenly distributed. IT needs to come up with specific compliance policies or principles that the rest of the organization can follow.

Public cloud removes some of the infrastructure and administrative overhead of the traditional data center, but the onus to meet cloud governance requirements still falls squarely on IT's shoulders. In the ever-shifting cloud landscape, it is important to create a governance model that resembles an ongoing process, not a product. Therefore, necessary adjustments can be made to help facilitate progress and limit any holdups [4].

Matching cloud providers to your data location, your privacy, and governance needs, as well as best practices for creating an organization-wide cloud governance strategy, are important considerations for any IT shop.

12.17.1 Cloud Security Challenges

Most businesses do not have a good grasp of what is reality and what is fiction when it comes to cloud security. According to Alert Logic's Fall 2012 State of Cloud Security Report, the variations in threat activity are not as important as where the infrastructure is located. Anything that can be accessed from outside—enterprise or cloud—has a relatively equal chance of being attacked, because attacks are opportunistic in nature, but this isn't always the case.

Web application-based attacks hit both service provider environments and on-premises environments, comprising more than 40% of the total attacks on each environment. Though these events were the most prevalent type of attack, they hit on-premises environments with much more frequency. On-premises environment users also suffered significantly more brute-force attacks compared to their counterparts in service provider environments. The 2012 report still rings true—the recent data breaches at Sony, Home Depot, and Target were unrelated to the cloud. Indeed, most attacks occur on traditional systems because those security systems are aging, and vulnerabilities have been exposed.

The importance of having effective security strategies and technologies has increased significantly. This is because cloud computing continues to grow in popularity and because the implementations become more complex and heterogeneous.

Identity and Access Management (IAM), also known as identity management, is not new, but the emergence of cloud computing has put it at center stage. Many cloud providers, such as Amazon Web Services (AWS), provide IAM as a service right out of the cloud. Others require customers to select and deploy third-party IAM systems [4].

The concept is simple: Provide a security approach and technology that allow the right individuals to access the right resources at the right times and for the right reasons. The concept follows the precept that everything and everyone get an identity, including humans, servers, devices, application programming interfaces (APIs), applications, and data. Once that verification occurs, it is just a matter of defining which identities can access other identities and creating policies that define the limits of that relationship [4].

One example would be to define and store the identity of a set of cloud-based APIs that are leveraged only by a single set of smartphones that are running an application. The APIs each have an identity, as do the smartphones, the applications, and the humans who are using the phones. An IAM service would authenticate the identity of each entity each time an entity interacts with another resource [4].

A prime example of IAM is the AWS version, which is a full-blown identity management and security system that allows users to control access to AWS cloud services. This IAM allows you to create and manage AWS users and user groups by way of permissions, which allow and disallow access to data. The benefit of Amazon's IAM is the ability to manage who can access what and in what context [4].

12.17.2   Other Players in the Game

Of course, not everyone runs AWS. Fortunately, many new IAM players are focusing cloud and usually promise to provide both identity management and single sign-on services. These players include Bitium, Centrify, Okta, OneLogin, Ping Identity, and Symplified. Each of the providers approaches cloud security and IAM differently, so you will need to review each product with regard to your specific requirements.

When selecting the right cloud security approaches, be certain to consider the following:

- The integration of cloud-based identity management solutions or other security solutions with enterprise security systems. Security should be systemic to both cloud and non-cloud systems, and you should consider ones that meet both sets of requirements.
- The design and architecture of your identity based security solution. Sometimes security services can come from your cloud provider. In many other cases, you have to select and deploy third-party security tools.

- Importance of testing, including "white hat" security tests, and they are telling, in terms of the actual effectiveness of your security systems.
- The effect on performance including, in some instances, security can slow your system to the point that it affects productivity.
- Industry and all required regulations for compliance.

12.17.3 Challenges in Governing the Cloud

Cloud governance comes in many different flavors, including service level, data level, and platform level. What is more, cloud governance and security typically work together; thus, you cannot select the right security approaches and technology without first understanding your governance strategy.

Service-level or API governance installs policies around access to services exposed by public or private clouds—those who want to access cloud services have to go through a centralized mechanism that checks to see that those who request access are appropriately authorized. This mechanism also forces compliance with predefined policies that can dictate when and how the services can be accessed. Companies that provide API/service management and governance products include Mashery and Apigee.

Data-level governance, much like service level governance, focuses on the management of both storage and data. Once again, policies are placed around data and data storage systems to define and control access.

Data-level governance, like service-level governance, focuses on the management of both storage and data. Again, policies are placed around data and data storage systems to define and control access.

Data governance is becoming more important for businesses that implement cloud computing. The Cloud Security Alliance (CSA) has a Cloud Data Governance Working Group that is defining approaches and standard technology. Perspecsys and Acaveo are among the vendors in the cloud data governance marketplace.

Platform-level governance, sometimes called a cloud management platform, is related to the management of the platforms themselves. This means placing automation services around the governance and management of a cloud platform, including provisioning and deprovisioning of cloud resources as needed by applications or data.

The objective of platform-level governance is to provide a single point of control for complex, distributed, and heterogeneous public and private cloud-based resources. This allows policies to define when and where resources are put to work and makes sure users leverage only what is necessary. The end result is that we do not overpay for subscription-based services, and the system works around issues such as outages. RightScale and ServiceMesh (now owned by CSC) are among the vendors offering platform-level governance products.

12.18   Creating Your Own Approach for Cloud Security

Your customized approach to cloud security and cloud governance requires a great deal of work to define your requirements, both business and technical. Once that is accomplished, it is easy to create a comprehensive strategy and then proceed to implement the right technology.

Most organizations continue to be concerned about the risks introduced by cloud computing. Those risks, however, are substantially less than many of the traditional systems in use today [4].

"The cloud has too many benefits to ignore, and the risks around security and governance are now solvable problems."—*David Linthicum*

12.19   Cloud Data Security Comes at a Cost

The technology needed for cloud security can be expensive, so system administration tasked with securing the cloud should prepare their CIOs for a big bill. The cost of the talent needed to create proper security architectures and approaches and then to run them effectively will set companies back.

Clouds are complex distributed systems, so what is the best way to protect them? The best cloud security model and practice is identity access management (IAM). Many cloud providers, such as Amazon Web Services (AWS), provide IAM as a service. Others require third-party IAM systems.

To ensure cloud data security, use the method and technology that enable the right individuals to access these resources at the right times and for the right reasons. This means that everything and everyone get an identity—including humans, servers, APIs, applications, data, and more. After verifying identities, define which can access other identities and create policies to define the limits of those relationships.

12.20   Explore Different Cloud Security Avenues

There are a few approaches to cloud security, including using IAM for your cloud provider, IAM software, and a third-party cloud. Cloud-based IAM system expenditures, such as those provided by AWS, are nominal. Most businesses, however, choose security options that are not tied to a single cloud provider.

The cost to run an IAM system, whether on-premises or as a service, varies. The average yearly cost is $5000 per application, so it can get expensive if you manage 1000 applications in private or public clouds and traditional systems. Everything needs to be locked up the same way; if cloud-based systems are secure, but traditional systems are not, then the system isn't completely secure.

In summary, again, indeed, *most recent attacks occur on traditional systems because those security systems are aging, and numerous vulnerabilities have been exposed.*

However, technology is not the real expense—it is the security engineers needed to build and operate effective cloud security systems that cost the most. Indeed.com reports that the average annual salary for a US worker with the words "cloud security" in his or her title is $134,000. And these talented engineers are extremely hard to find, so you'll pay even more for the best talent. Capable consultants can cost $2000 to $2500 per day.

Moving to the cloud has tremendous benefits, but security done right is costly [4].

12.21 Why It Matters?

Attackers used to favor Java exploitation, but that is no longer the case. This decrease is likely the result of several important changes in the way Web browsers evaluate and execute Java applets. Security teams can prioritize their efforts now on higher priority risks. Java users should continue to install security patches as they become available to continue guarding against potential future attacks.

Trends for the top Java exploits detected and blocked by Microsoft real-time antimalware products in the second half of 2015 are illustrated in Fig. 12.13.

Enterprise environments typically implement defense-in-depth measures, such as enterprise firewalls, that prevent a certain amount of malware from reaching users' computers. Consequently, enterprise computers tend to encounter malware at a lower rate than consumer computers. The encounter rate for consumer computers was about 2.2 times as high as the rate for enterprise computers.

Meanwhile, enterprise (domain-based) computers encountered exploits nearly as often as consumers' computers (non-domain), despite encountering less than half as much malware as non-domain computers overall. This tells CISOs that exploits are an issue for organizations and staying up to date with security updates and the latest software is their best defense. Despite these trends, you can secure your company's assets by understanding the threat landscape and devising a security strategy across all fronts, including identity and access credentials, apps and data, network devices, and infrastructure.

By adopting a proactive security stance and taking advantage of the latest in multi-factor authentication, machine learning, and analytics technologies, you can harden your company's defenses against cyber attacks and be equipped to respond in the event of a breach.

Consumer computers encounter 2x the number of threats as compared to enterprise computers. See Fig. 12.14.

Locations with the highest malware infection rates were Mongolia, Libya, the Palestinian territories, Iraq, and Pakistan, depicted in Fig. 12.15 as infection rates by country/region.

Malware is unevenly distributed around the world and each location has its own mix of threats. By studying the areas of the world that are highly impacted with malware and comparing them to the least-infected parts of the world, we can try to discover what technical, economic, social, and political factors influence regional malware infection rates. This information might help to inform future public policy that, in turn, could lead to reduced malware infection rates in highly impacted parts of the world.

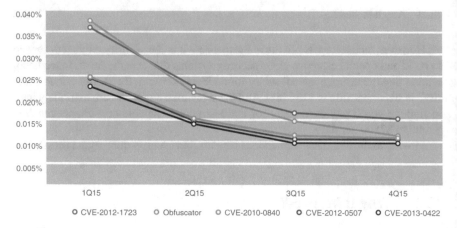

Fig. 12.13 Microsoft real-time antimalware products in the second half of 2015

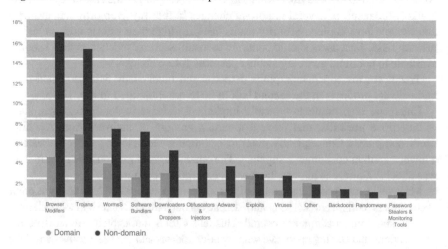

Fig. 12.14 Malware and unwanted software encounter rates for domain-based and non-domain computers

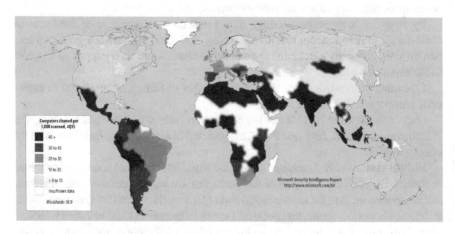

Fig. 12.15 Infection rates by country/region

If your organization only focuses on patching vulnerabilities in your most commonly used software, you are likely not managing all the vulnerabilities present in your IT environment. It is important to know if you need to take action on any of the other nearly 3000 vulnerabilities that could be in your organization's environment.

Device encryption and consistent compliance with IT rules can help reduce the odds of a breach. If you detect suspicious behavior, block and quarantine the device off the network until the threat is identified and removed.

References

1. Joint Pub 1-02, *Department of Defense Dictionary of Military and Associated Terms*. Washington, GPO, April 12, 2001 (as amended through May 7, 2002).
2. http://www.au.af.mil/au/awc/awcgate/dod/effective_decision.htm
3. Burns, S. F. GIAC Security Essentials, Version 1.4c. *Threat modeling: A process to ensure application security*. Retrieved January 5, 2005, from https://www.sans.org/reading-room/whitepapers/securecode/threat-modeling-process-ensure-application-security-1646
4. Linthicum, D. *Protecting this house: IT's role in cloud security*. TechTarget. http://searchsecurity.techtarget.com/

Chapter 13
A Simple Model of Business Resilience System

In this, chapter we can recommend and define the scope of a simple Business Resilience System (BRS) based on a simple infrastructure that one could design. As we said this is just a simple approach to give some ideas to the readers. For a more complex system and infrastructure, one needs more sophisticated design and approach to the appropriate and applicable BRS for their organization and enterprise.

13.1 Introduction

Although Business Resilience System is a must tool for today's organizations and enterprises in order to survive, maintain the integrity of day-to-day operations, and to be fully functional against an existing man-made threats such as cyber attack or natural disaster, each organization needs to look at their own need and requirements. For purpose of a demo and simple system, we can recommend a high-level requirement for BRS Demo System that is listed below. We as authors looked at a company called iJet International, Inc. and their product called Worldcue®, to build a simple BRS for Intelligent Risk Systems as a tool for a simple BRS. This helps to reduce risk of traveling abroad in today's environment of terrorist's acts against certain citizens. This will consider essential duties of the assigned and designed BRS and responsibilities. This allows the stakeholder and operator of such BRS to perform the following duties:

- Monitors global events in real time for general awareness and issues that could impact client operations
- Develops intelligence communications by category or topic
- Recognizes crisis situations and escalates information according to set protocols using independent judgment and professional training

© Springer International Publishing AG 2017   347
B. Zohuri, M. Moghaddam, *Business Resilience System (BRS): Driven Through Boolean, Fuzzy Logics and Cloud Computation*,
DOI 10.1007/978-3-319-53417-6_13

- Generates special alerts and travel notifications
- Assists in managing emergency incidents by using judgment and skills according to commonly accepted incident management best practices and client policies and procedures
- Conducts in person and/or telephone briefings with travelers who are scheduled to travel to high-risk regions
- Other duties and responsibilities as assigned by management
- Develops and issues written intelligence communications regarding global events with potential to impact staff or operations
- Recognizes crisis situations and initiates protective action using independent judgment and professional training in accordance with incident management best practices and client policies and standard operating procedures

The purpose of this chapter is to give BRS practice a high-level understanding of any services and pricing methodology. This chapter is not meant to be all encompassing but it is just a suggestion by these authors and is subject to change. In this scenario, we have chosen a simple BRS for the Worldcue® TRAVELER[1] service that is purchased for use in association with a single traveler and trip as defined by the following conditions and circumstances:

Duration—Worldcue® TRAVELER is active from the purchase date to 1 month after the end of the trip, or 13 months, whichever is shorter.

Segments—Up to eight travel legs are permitted for a single trip. On a round-trip, a return to the regional point of origin constitutes the end of the trip.

Traveler Alert Profile—The alert profile may be maintained only for the benefit of the individual traveler purchasing the Worldcue® TRAVELER service.

Travel Intelligence® Report Generation—The traveler may generate updated Travel Intelligence® Reports as desired up to a maximum of 75 times. If the maximum is reached, the last Travel Intelligence® Report generated will remain available.

Worldcue® TRAVELER Trip Changes/Cancellation

Travel Leg Changes—The traveler may make changes as needed but may not change the trip end date beyond 1 month from the original trip end date.

Trip Cancellations/Credits—As the purchase of Worldcue® TRAVELER results in the immediate activation of the service, a Worldcue® TRAVELER trip credit will be provided for cancelled trips upon request.

Worldcue® TRAVELER is the premier personal travel service offered by iJET Intelligent Risk Systems, the first intelligence and operational risk management organization for the global travel industry. Worldcue® TRAVELER uses a traveler's specific itinerary or expatriate's assignment and personal profile to deliver customized intelligence reports for over 230 countries and more than 580 cities covering entry/exit, security, health, transportation, culture, language, environment, financial, communication, and legal concerns. Worldcue TRAVELER customers receive

---

[1] Worldcue® is a service offered by iJet International, Inc.

real-time alerts before, during, and after their trip or assignment about changing conditions that might affect their itineraries or long-term stay. This constant support allows travelers and expatriates to stay safe and to circumvent potential difficulties with minimal trip/assignment disruption.

iJet Intelligent Risk Systems (www.ijet.com) specializes in providing real-time intelligence and proactive risk management services to multinational corporations and their employees. iJet's award-winning technology, modeled after government intelligence systems, continues to receive high honors for its groundbreaking innovations and efficiency. iJet's diverse and experienced intelligence analysts speak 20 languages and work 24×7 in iJet's Operations and Response Center to alert clients of changing conditions worldwide that affect their assets and employees. iJET was recognized by InfoWorld as one of the "Top 100 IT Visionaries of 2003" for its groundbreaking Worldcueý operational risk management platform and unparalleled ability to deliver objective, timely, and actionable intelligence to corporate security/risk officers, travel managers, travelers, and expatriates.

13.2 Assumptions

For the purpose of this chapter, the following assumptions were made.

- The methodology of how BRS end users will be authenticated into Worldcue®, and how portlets and services will be presented within the BRS portal framework is yet to be determined. This should be mutually agreed upon all parties involved.
- All services from a chosen vendor that are delivered to BRS clients must come from the Worldcue platform.
- All BRS end users will be authenticated by Worldcue before services are presented—regardless of final user interface location. This is to insure that our intellectual property is being protected and that the end user is only seeing the information that they should be seeing (i.e., security).
- All pushed notifications from the Worldcue platform can be "co-branded" or "private labeled" to include BRS graphics and text.
- All standard notifications will be sent out via SMTP.
- All SMS notifications will have to be mutually agreed upon and an additional message unit charge will be applied.
- ABRS will be accountable and responsible for tier-1 support for ABRS clients.
- All user licenses are priced out on an annual basis.
- Vendors like iJET should be able to offer subscription services (Daily Intelligence Briefing, Monthly Intelligence Forecast, and World Pandemic Monitor) for an annual fee, but they were not included in this proposal. We will be more than happy to provide pricing on these services if requested.
- A BRS end user will have a Worldcue license to receive any of the Intelligence, Travel Risk Management, or Asset Risk Management services from iJet.

- Worldcue user licenses, Travel Risk Management services, and Asset Risk Management services are volume discounted, paid for upfront, and are renewed annually.
- Vendors like iJET should be able to aggregate the total number of user licenses, assets, and travelers provided to all BRS clients when calculating charges. Thus, as the volume of business increases, so will BRS margins.

13.3   Access to Vendor's Intelligence or Cloud Data

Extraordinary events affecting business operations have become all too ordinary. Disruptions such as hotel bombings, new and emerging infectious diseases, and hurricanes continue to increase in number and intensity globally. Never before have corporations and government organizations been required to manage so many threats to their people and assets distributed around the world.

Provider vendor (i.e., iJet) will allow BRS' end users to have access to the following Worldcue® intelligence portlets:

- Active Alerts
- Location Intelligence
- Intelligence Briefs

    - Trip Briefs
    - Health Briefs
    - Security Briefs

- Security Assessment Ratings
- Country Risk Chronology

    - Filtering tool to search all active and archived intelligence for every country across a specified date range

Each BRS unique clients' users that have access to iJet's intelligence would have to pay an end-user license fee. The end-user license fee is volume discounted. iJet will aggregate the total number of user licenses provided to all BRS clients when calculating charges. Thus, as the number of clients increase, so will BRS margins. The price points are (these are just hypostatical figures):

• 1 user	$5000 (i.e., an assumption)
• 2–10 users	$2500 per user (i.e., an assumption)
• 11–20 users	$1500 per user (i.e., an assumption)
• 20–50 users	$1000 per user (i.e., an assumption)
• 50+ users	$750 per user (i.e., an assumption)

13.4   Access to Vendor's Analysts

BRS clients who have purchased iJet's Intelligence services will have the ability to talk to iJet's analysts about alerts and intelligence that iJET created and current global events—for a reasonable amount of time. If, during the engagement, BRS or the client would like to have iJET create custom reports, iJet will create a formal Statement of Work for the project and submit it for formal sign-off. iJet will charge an hourly blended rate for our analysts, subject matter experts, HUMINT network, advisory board members, editors, and graphic artists of:

- $200 per hour (i.e., an assumption)

13.5   Industry-Specific Analysts

If BRS would like iJet to build out an industry-specific analytical staff, we are more than capable of handling that request. For business hours coverage (9–5 Monday to Friday), this would require one (1) FTE. For 24×7 coverage, this would take five (5) FTEs. The price for a dedicated industry analyst FTE is:

- $160,000 per FTE (i.e., an assumption)

13.6   Intelligence Notification

Notifications, via SMTP methodology, of our alerts and intelligence are part of the end user's Worldcue license. There is no additional charge for this functionality. iJet will allow Worldcue licensed user access to the "User Management" portlet that would allow clients to customize their notifications settings.

13.7   Asset Notification

Protecting a dispersed set of operating assets is a challenging, yet essential, component of any organization's business resilience directive. Environmental, security, health, technology, and other international disruptions threaten facilities, employees, and the entire supply chain. When events occur, decision-makers need a comprehensive understanding of the situation, a rapid assessment of asset exposure, and the proper tools and processes to facilitate an appropriate response.

For iJet to be able to relate our intelligence to an asset, the asset will have to be loaded into the Worldcue system. As discussed previously, every iJet alert is geocoded and is given an associated threat radius. Furthermore, every asset that is loaded into Worldcue has the ability to be geocoded and given a vulnerability radius. When an alert's threat radius intersects an asset's vulnerability radius, a notification will be sent out automatically to the end user based on their notification profile (see "User Management" in the Intelligence Notification section).

There is no additional charge for Worldcue to send out a notification to a user, via SMTP methodology, for assets. However, there is a charge for each asset that is loaded into the Worldcue platform. iJet will aggregate the total number of assets provided to all BRS clients when calculating charges. Thus, as the number of assets increase, so will ABRS margins. The price points are and the asset charges are:

• 1–50	$500 per asset (i.e., an assumption)
• 51–100	$400 per asset (i.e., an assumption)
• 101–250	$300 per asset (i.e., an assumption)
• 251–500	$250 per asset (i.e., an assumption)
• 501–1000	$200 per asset (i.e., an assumption)
• 1000+	$150 per asset (i.e., an assumption)

For assets to be loaded into the Worldcue platform, based on information that is in ABRS, Accenture will have to establish a data feed of asset information from BRS to iJet. The data fields, format, timing, and transfer methodology will be mutually agreed upon by both parties.

13.8   Travel Risk Management and User Notification

It's not enough for today's decision-makers to simply ensure that traveling employees get to their destinations. Legal obligations such as Duty of Care and Duty to Warn require travel and security managers to protect the health and safety of employees traveling on behalf of their companies. When disruptions do occur, organizations quickly need an understanding of the situation, a rapid assessment of who is impacted, and the proper tools and processes to locate, communicate with, and respond to affected travelers.

If BRS would like to include iJet's traditional Travel Risk Management services in their offering, we would be able to provide those services.

iJet will allow BRS end users to have access to the following Worldcue® Travel Risk Management portlets:

• Traveler Management
• Employee Locator
• Employee Locator Reports
• World Map

There is no additional charge for Worldcue to send out a notification to individual travelers, via SMTP methodology. However, there is a charge for each trip that is loaded into the Worldcue platform.

iJet has two levels of service: TRACKER and TRAVELER. iJet will aggregate the total number of travelers provided to all ABRS clients when calculating charges. Thus, as the number of travelers increase, so will BRS margins. The price points and the prices for these services are:

TRACKER

• 1–5000	$1.73 per trip (i.e., an assumption)
• 5001–10,000	$1.44 per trip (i.e., an assumption)
• 10,001–20,000	$1.15 per trip (i.e., an assumption)
• 20,001–50,000	$0.92 per trip (i.e., an assumption)
• 50,001–100,000	$0.86 per trip (i.e., an assumption)
• 100,000+	$0.75 per trip (i.e., an assumption)

TRAVELER

• 1–5000	$3.45 per trip (i.e., an assumption)
• 5001–10,000	$2.88 per trip (i.e., an assumption)
• 10,001–20,000	$2.30 per trip (i.e., an assumption)
• 20,001–50,000	$1.96 per trip (i.e., an assumption)
• 50,001–100,000	$1.78 per trip (i.e., an assumption)
• 100,000+	$1.55 per trip (i.e., an assumption)

13.9   Hotline

Worldcue24 is a 24×7 integrated emergency hotline service capable of responding to a broad array of corporate emergencies through a single contact point from anywhere in the world. Accessing any number of a company's security, travel, medical, and other specialized response services, Worldcue24 provides emergency response notification and reporting based on each company's customized protocols. Worldcue24 is there for your people when they need it most.

Pricing for iJet's Worldcue24 services to BRS would be:

• $15,000 for each dedicated hotline number (i.e., an assumption)
• $25 per call received by iJet (i.e., an assumption)

If BRS would like to utilize SMS, 2-Way SMS, message verifications, conference bridging, etc., there would be an additional charge of:

• $13,000 for 100,000 "message units"

13.10   High-Level Requirements

The high-level requirements for BRS Demo System are listed below. Some of these listed are changes to the existing features and addition to the existing features.

13.10.1   Modification and Enhancements

1. For adding facilities, use Virtual Earth Map that should fetch the geography, location, country, and city details automatically. Add details in the database.
2. The labels in the PDP definition and master data setup screens should be configurable. This shall be applicable for location, facility, etc.
3. Response text is not necessary for "Out Perform" threshold.
4. Add the Over All Response Owner in the Process Data Point (PDP) definition screen.
5. There shall be one response owner for each of the thresholds (threshold 1, 2, 3, and 4 and BCP). These are participants in the workflow activities.
6. In the notification's email body, provide a hyperactive link to respond to the PDP event. The response text should appear in the main workflow as workflow comments.
7. Below is the new assumed screen format for configuring the master data and PDPs.

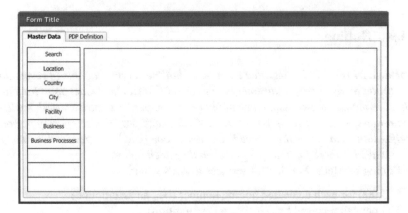

13.10.2   New Requirements

1. Response Management Unit
2. The screen format is shown below.

Response Management Unit

PDP Name	Current Threshold	Out Perform Point	Status	Response
Employee Absenteism	40	5	●	AbsenteismWorkflow

PDP Name	Current Threshold	Out Perform	Over All Owner	Status	% Complete

Responses	Owner	Status
People		
Data		
Network		

Details of Sub Workflows shall be shown on Clicking Responses in the above table

3. Provide a dependency chart for threshold and workflow definitions. We need to decide on the format of this screen and how it is used in the system.
4. Provide an interface for Notifications Setup: This screen will provide a facility to configure notifications.
5. Every time an event triggers after reaching the threshold, the ID for that incident should be of 11 chars and the format is given below:

BRSCC __ __ __ __ __ __. The first two digits/characters (CC) after "BRS" should be client code, and the rest are numeric chars.

The entire content of this chapter was based on simple design of BRS for traveling purpose around the world that companies like American Express Corporate Travel Agency could put in place for their corporate clients and this way they have a better tool and ammunition to protect their cliental base.

This BRS was designed based on a lot of assumption by these authors and iJet was chosen as an example of vendor with their Worldcue® product and their risk management tool as well. However, we as authors play a neutral ground and agnostic to any vendor. It will be the choice of enterprise to choose the right tool for their organization to implement the right Business Resilience System in place [1].

Reference

1. All the assumption was made by the authors of this book.

Chapter 14
Business Resilience System Topology of Hardware and Software

Engineers endow artifact with abilities to cope with expected anomalies. The ability may make the system robust. They are, however, a designed feature, which by definition cannot make the system "resilient." Humans at the front end (e.g., operators, maintenance people) are inherently adaptive and productive that allows them to accomplish better performances and sometimes even allows them to exhibit astonishing abilities in unexpected anomalies. However, this admirable human characteristic is a double-edged sword. Normally it works well, but sometimes it may lead to a disastrous end. Hence, a system relying on such human characteristics in an uncontrolled manner should not be called "resilient." A system should only be called "resilient" when it is tuned in such a way that it can utilize its potential abilities, whether engineered features or acquired adaptive abilities, to the utmost extent and in a controlled manner, both in expected and unexpected situations or circumstances.

14.1 Introduction

The adaptive capacity of any system is usually assessed by observing how it responds to disruptions or challenges. Adaptive capacity has limits or boundary conditions, and disruptions provide information about where those boundaries lie and how the system behaves when events push it near or over those boundaries. Resilience system in particular and being resilient in general are concerned with understanding how well the system adapts to what range or sources of variation, which means the threats that the system needs to encounter and deal with them.

This allows one to detect undesirable drops in adaptive capacity and to intervene to increase aspects of adaptive capacity. Thus, monitoring or measuring the addictiveness and resilience of a system quickly leads to a basic ambiguity. Any given incident or threat includes the system adapting to attempt to handle the disrupting

© Springer International Publishing AG 2017
B. Zohuri, M. Moghaddam, *Business Resilience System (BRS): Driven Through Boolean, Fuzzy Logics and Cloud Computation*,
DOI 10.1007/978-3-319-53417-6_14

event or variation on Service Level Agreement (SLA), governance and policy in place, per organization policy makers and stakeholders.

In this chapter, we adopt a topology using Microsoft Office SharePoint Portal Server 2007 (MOSS) that will handle the previous chapter scenario and example. This solution is a very simple approach and cost-effective with best practice approach with an effective return on investment (ROI) as well as total cost of ownership (TCO) [1].

14.2 Server Topology Design

The topology incorporates dedicated front-end Web servers.

- Windows SharePoint Services: Server(s) hosting the Windows SharePoint Services search role and database server role are protected from direct user access.
- Office SharePoint Server: Servers hosting application server roles and database server roles are protected from direct user access.
- Windows SharePoint Services: The SharePoint Central Administration site is hosted on the same server computer where the Windows SharePoint Services search role is hosted.
- Office SharePoint Server: The SharePoint Central Administration site is hosted on a dedicated application server, such as the index server.

Many permutations of topologies can be built using Microsoft Office SharePoint Portal Server 2007 (MOSS), but only a small number of these topologies are supported by Microsoft. The simplest topologies are the single-server and small-server-farm topologies, which have been covered in previous chapters.

Medium- and large-server-farm topologies can get much more complex. A *medium*-server topology requires at a minimum one front-end Web server running the search application, one index/job server running SharePoint Portal Server 2003, and one database server running Microsoft SQL Server 2007 with Service Pack 1 or later. Additional servers can be added to this topology to enable higher availability, higher capacity, or both, but you cannot have a medium farm with fewer than four servers (Fig. 14.1).

14.3 Networking Topology Design

All servers within the farm reside within a single data center, on the same vLAN.

Access is allowed through a single point of entry which is a firewall.

Windows SharePoint Services: For a more secure environment, separate the farm into three tiers (front-end Web, search, and database), separated by routers or firewalls at each vLAN boundary.Office SharePoint Server: For a more secure

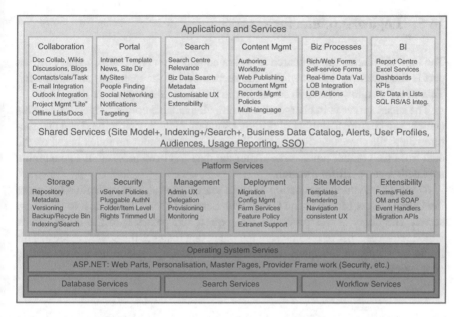

Fig. 14.1   Microsoft Office SharePoint Server 2007 Architecture (Source Microsoft)

environment, separate the farm into three tiers (front-end Web, application servers, and database), separated by routers or firewalls at each vLAN boundary.

14.4   Logical Architecture Design

- At least one zone in each Web application uses Windows authentication. This is required for the search account to crawl content within the Web application.
- Implement Web applications using host headers instead of the randomly generated port numbers that are automatically assigned. Do not use IIS host header bindings if the Web application will be hosting host header-based site collections.
- In a reverse proxy environment, consider using the default port for the public-facing network while using a non-default port on your internal network. This can help prevent simple port attacks on your internal network that assume HTTP will always be on port 80.
- Do not put an un-trusted site into the same Web application as trusted sites. This protects against intra domain scripting attacks.
- Use separate application pool accounts for central administration and each unique Web application.

14.5   Logical Architecture

The logical architecture of search is separated into two main components. The first part is the indexing and crawling. The second part is the querying of the content. This structure allows for high scalability in the physical architecture (Fig. 14.2).

The index server is in charge of crawling all available content repositories. This content can be from a number of different sources, including structured and unstructured data, file shares, and metadata. The index server builds an index based on the data from the content repositories. After the data has been indexed, the consuming application can perform a search.

When a consuming application makes a search request, the application leverages the Query Object Model or the Query Web Service to send the request to the query server. The query server processes the request and passes it to the index server, which retrieves a list of results from the index that matches the search request. This list is sent back to the query server, which returns the results to the requesting application.

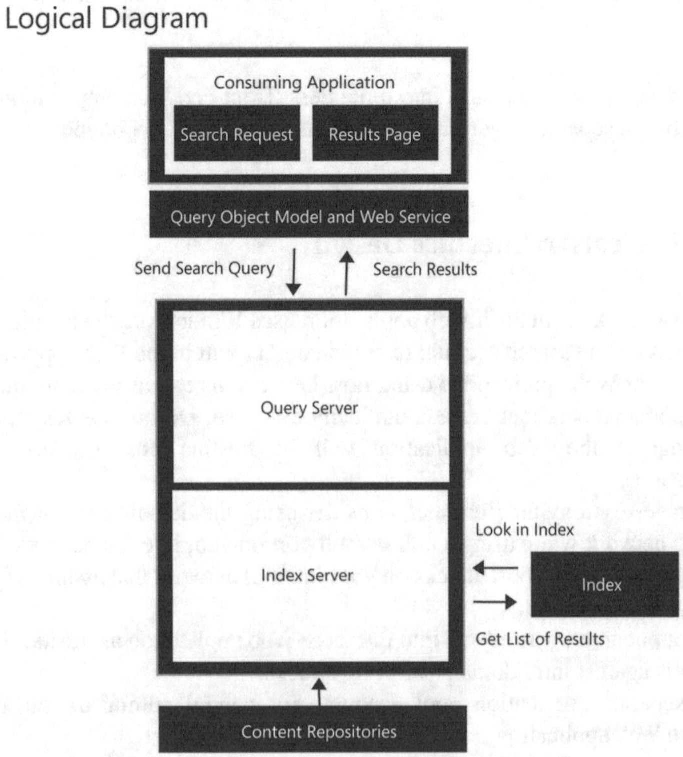

Logical Diagram

Fig. 14.2 Logical diagram

14.6   Physical Architecture

This section begins with an overview of how search can be set up on a single server, a medium-size server farm, or a large server farm. There are three main server types in the physical architecture: front-end Web server, index server, and query server. The front-end Web server is the server that handles all the user search requests. These servers access the query server returning information to the client browser. The index server performs the actual crawling and indexing of the internal and external data sources. The query server searches through the data gathered by the index server returning results to the front-end Web server for display.

14.7   Determining Hardware Requirements

Selecting an appropriate server platform is an extremely important component of a successful portal deployment. The key hardware requirements are the CPU speed, amount of RAM, and hard disk space:

- Increased CPU resources allow SharePoint Portal Server to provide an excellent experience to large numbers of users during peak usage periods.
- If there are insufficient CPU resources, users experience unacceptable server response times during peak usage periods.
- Additional RAM and hard disk space allow SharePoint Portal Server to provide improved performance.
- If there is insufficient RAM, users experience unacceptable server response times regardless of the number of active users.
- If there is insufficient hard disk space, users are not able to search for or save additional documents.

An important step in determining the server hardware requirements is establishing clear performance and scalability requirements. Unfortunately, it is frequently difficult to establish clear and detailed performance and scalability requirements for portal deployments. Most organizations find it difficult to predict the level or type of use that the site receives. To complicate matters, this level of use frequently changes and grows over time.

Most organizations are able to estimate accurately the following important deployment metrics:

1. **Number of users**. This is the total number of users that may have access to the site.
2. **Percent of active users per day**. This is the percentage of the total number of users who might use the dashboard site during any particular day. Typically, this figure ranges from 10% to 100%. This number is frequently overestimated and is commonly approximately 30%.

3. **Number of operations per active user per day**. This is the number of operations that a typical user does on the dashboard site during a typical day. An operation is an action such as viewing the home page, searching, retrieving documents, etc. This number usually ranges from 1 to 10. It is frequently possible to estimate the number of operations by analyzing the Web server log of an existing portal deployment, if one exists. Note, however, that when analyzing the Web server log, it is only necessary to consider page views, not site hits. Site hits are frequently significantly higher than the number of page views.
4. **Number of hours per day**. This is the number of hours during which most activity occurs. This number typically ranges between 10 and 24 h.
5. **Peak factor**. This is an approximate number that estimates the extent to which the peak dashboard site throughput exceeds the average throughput. This number typically ranges from 1 to 5.

These quantitative descriptions of a portal deployment can be used to estimate the required peak throughput. The following formula yields the peak throughput in operations per second (Table 14.1):

The number 360,000 is determined by:

$$100 \left(for \text{ percent conversion} \right) \times 60 \left(\text{number of minutes in an hour} \right)$$
$$\times 60 \left(\text{number of seconds in a minute} \right)$$

SharePoint Portal Server uses Hypertext Transfer Protocol (HTTP) for all communication between the client and the server. The HTTP protocol is a connectionless protocol. Therefore, it is not possible to identify the number of concurrent users. The most important measurement of server throughput is the operations per second. The following examples illustrate applying the formula to a sample deployment for three sites.

14.8   Deploying a Small Group Site

The following is an example of how you can use the preceding formula to determine the requirements for a product group portal for 400 people. The product group contributes more than 90% of the site traffic. Although there might be thousands of other users, who occasionally use the site, the number is insignificant compared to the product group usage.

Table 14.1 Portal deployment estimate for the required peak

Number of users	×	Percent of active users per day	×	Number of operations per active user per day	×	Peak factor
360,000			×	Number of hours per day		

Table 14.2 Deployment estimate

Number of users	400
Percent of active users per day	90
Number of operations per active user per day	30
Number of hours per day	12
Peak factor	5

For such a deployment, the following characteristics are reasonable (Table 14.2): These estimates yield a predicted peak throughput of 1.3 operations per second.

$$\frac{400 \times 90 \times 30 \times 5}{360,000 \times 12} = 1.25$$

Such a site could be successfully deployed with a server such as a 700 MHz Pentium III with 512 megabytes (MB) of RAM. But for our future expansion and scalability, BRS recommends the three farm approach and hardware configurations for such topology (see below).

14.9 Two-Server Farms for BRS Approach

The simplest physical server farm configuration is the two-server farm. In this topology, the Web server, query server, and index server are all located on one server, and the back-end database is located on the other server. The indexing role of this Web, query, and index server accesses content from external sources and from content databases. Because the index server and query server are on the same machine, the overall load is increased. Having separate machines for the query server and the indexer (as shown in the medium-size farm topology) would result in a faster search. This is a likely configuration for small organizations (Fig. 14.3).

14.10 Three-Server Farm (Adding the Application Layer) for BRS Approach

A recommended MOSS 2007 deployment topology is composed of three virtual layers:

- Web Front-End Layer
- Application Layer
- Database Layer

Fig. 14.3 Two-server
farms

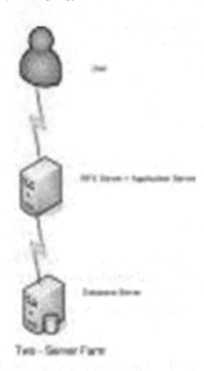

There might be cases where due to some of the factors (mentioned above), one
or more of the layers be represented by separate servers (one or more for each role
based on the redundancy requirements). The least recommended starting topology
for most of the scenarios is a two-farm server where one server acts as a database
server and the other server WFE and application server roles.

For our pilot design and proof of concept with Business Resilience System
(BRS), we should take the three-farm server approach. This topology typically
introduces another layer, and hence the three layers now are isolated and each represented by one or more servers. Each of these roles may also introduce redundancy
by adding redundant hardware and using the load-balancing techniques (Fig. 14.4).

The following Figs. 14.5 and 14.6 for Web servers and dedicated SQL Server,
including backup system for disaster and recovery system.

What is important to note is that the number of layers and not the number of servers to use in a particular farm topology differentiate the server farms depicted above.
Each of the above farm configurations is different from the other due to the introduction of roles and representation of roles at different tier levels. A three-server
farm, for instance, may also be achieved by adding a redundant database server to
the above-discussed two-server farm or by adding a redundant WFE server. So there
is much to think over for a good design and we are yet to examine the factors that
affect the number of servers in each of the server farms discussed above. I hope this
post was helpful to understand the typical server farm tiers available with MOSS. In
the near future if BRS extends its services then, we will be discussing the extension
of these baseline topologies that include redundant hardware to fulfill varied client
requirements based on availability, performance, and capacity.

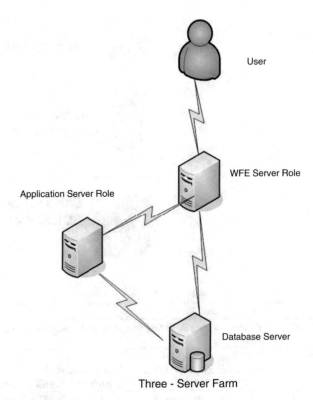

Fig. 14.4 Three-server farms

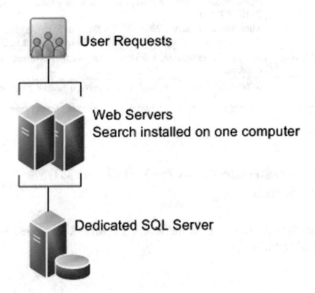

Fig. 14.5 Web server and dedicated SQL Server

Fig. 14.6 Backup system
for Web server and
dedicated SQL Server

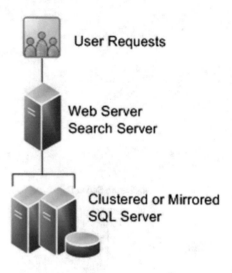

User Requests

**Web Server
Search Server**

**Clustered or Mirrored
SQL Server**

Table 14.3 MOSS and search server requirements

Component	Requirements
Operating system	Windows Server 2003 SP1 (supports 64-bit versions of Windows Server 2003)
Single-server installation	Processor: at least 2.5 GHz RAM: minimum of 1 GB, 2 GB recommended Disk: Up to 20 GB for installation, but bigger size disk is recommended
Farm deployment	Web server with a processor speed of at least 2.5 GHz, minimum 2 GB of RAM Application server with dual processors of at least 2.5 GHz, minimum 2 GB or RAM Microsoft SQL Server 2003 SP3 (or later) or SQL Server 2005 with dual processors of at least 2.5 GHz, minimum 2 GB or RAM
Internet connection	Broadband connection, 128 kbps or greater for download and activation of products
Additional components	Internet Explorer 6.0 or higher with service pack or Netscape 7.0

14.11   System Requirements for MOSS and MOSS for Search

The hardware system requirements for MOSS and MOSS for search on server are
as follows (Table 14.3):

14.12 Operating System

The operating system for this topology is listed below as:

- Server operating system is configured with the NTFS file system.
- Clocks on all servers within the farm are synchronized.

14.13 Summary

The above-suggested topology is a very high-level, uber approach without any granular and analysis approach, as a result this is not something that we can tell to implement. However, it gives you some general idea on how you can adopt your own organization need for a required Business Resilience System (BRS) that will match your enterprise requirements and day-to-day operations [1].

Reference

1. Most content of this chapter is adopted from Microsoft MOSS web site.

Chapter 15
Cloud Computing-Driven Business Resilience System

Cloud computing is an emerging commercial infrastructure and Internet-based cost-efficient computing, where information can be accessed from a Web browser by customers according to their requirement. Cloud computing in a general term is defined for anything that involves delivering hosted services over the Internet. It is based on the concept of shared computational, storage, network, and application resources provided by a third party. Knowledge is power, thus learning from experience is a fundamental way that helps individuals or organizations to improve and avoid previous mistakes.

15.1 Introduction

A number of artificial intelligence techniques such as artificial networks (AN), evolutionary computing (EC), fuzzy system (FS), case-based reasoning (CBR), and agent-based system (ABS) are few that can be mentioned. They can be applied as elements of business resilience system (BRS), in order to design and develop intelligent decision support system in general. In addition to using each intelligent technique to solve real-world problems, more effective solutions can be obtained if they are used in combination, providing that you are able to identify the right BRS for your organization and enterprise need due to the nature of your business. Indeed, hybrid paradigms combining two or more of intelligent techniques are becoming increasingly popular to deal with complex problems and new cyber threats as well as man-made thread and natural disaster that may interrupt business routine operations. Examples of hybrid paradigms include neural-fuzzy, neural-genetic, fuzzy-genetic, neural-fuzzy-genetic, fuzzy case-based reasoning (FCBR), and evolutionary case-based reasoning system (ECBRS), to name a few.

The widespread use of Internet-connected systems and distributed applications has triggered a revolution toward the adoption of pervasive and ubiquitous cloud

© Springer International Publishing AG 2017   369
B. Zohuri, M. Moghaddam, *Business Resilience System (BRS): Driven Through Boolean, Fuzzy Logics and Cloud Computation*,
DOI 10.1007/978-3-319-53417-6_15

computing environments. These environments allow users and clients to purchase computing power according to necessity, elastically adapting to different performance needs while providing higher availability. Several Web-based solutions, such as Google Docs and customer relationship management (CRM) applications, now operate in the software as a service model. Much of this flexibility is made possible by virtual computing methods, which can provide adaptive resources and infrastructure in order to support scalable on-demand sales of such applications. Virtual computing is also applied to stand-alone infrastructure as a service solution, such as Amazon Elastic Cloud Computing (EC2) and Elastic Utility Computing Architecture Linking Your Programs to Useful Systems (Eucalyptus).

Each combination brings synergy to the resulting system in such a way that the hybrid paradigm exploits the advantages of the constituent techniques and, at the same time, avoids their shortcomings. Application examples are described in Appendix A for Banking CRM and Appendix B and C PowerPoint presentation of handling data for PDP of Risk Atom. For the purpose of these presentations, we have used IBM software as solutions.

15.2   What Is Cloud and Cloud Computation

Cloud computing refers to the use, through the Internet, of diverse applications as if they were installed in the user's computer, independently of platform and location. Several formal definitions for cloud computing have been proposed by industry and academy. The following definition could be adopted in which "cloud computation" is a model for enabling convenient, on-demand network access to a shared pool of configurable computing resources (e.g., network, servers, storage, applications, and services) that can be rapidly provisioned and released with minimal management effort or service provider interaction.

There are different kinds of clouds that can be developed like private, public, and hybrid clouds. In the private cloud, the infrastructure is operated solely for an organization and managed by the organization or a third party, and it may exist on-premise or off-premise. A private cloud is supposed as a data center that supplies hosted services to a limited number of people or group of people. When a service provider uses public cloud resources to create their private cloud, the result is called a virtual private cloud. A hybrid cloud is a cloud computing environment, which is combining the benefit of different types of cloud. An organization provides and manages some resources in-house and has others provided externally in hybrid cloud.

The cloud computing is a multilayered service abstraction like Software as a Service (SaaS), Platform as a Service (PaaS), and Infrastructure as a Service (IaaS) and deployment models like private cloud, public cloud, and hybrid cloud. In the SaaS cloud model, the capability provided to the consumer is to use the provider's applications running on a cloud infrastructure, and the applications are accessible from various client devices through a Web browser [1].

Cloud computing has evolved into a business idea where Cloud Service Providers (CSPs) provide computing as a utility, which needs to be paid as per the usage. The number of CSPs has gone up in recent years so that the customer has to make a judicious choice based on various parameters such as cost, security, performance, etc. [1].

Cloud computing is being progressively adopted in different business scenarios in order to obtain flexible and reliable computing environments, with several supporting solutions available in the market. It is being based on diverse technologies, such as virtualization, utility computing, grid computing, and service-oriented architecture. Additionally, it constitutes a completely new computational paradigm; cloud computing requires high-level management routines.

Such management activities include:

(a) Service provider selection
(b) Virtualization technology selection
(c) Virtual resource allocation
(d) Monitoring and auditing in order to guarantee Service Level Agreements (SLAs)

A solution of cloud computing is composed of several elements, as clients, data center, and distributed servers, as shown in Fig. 15.1. These elements form the three parts of a solution cloud.

Each element has a purpose and has a specific role in delivering a working application based on cloud.

Cloud computing infrastructure and architecture is based on multilayers, where each layer deals with a particular aspect of making application resources available. There are two basic main layers involved: one is named a lower and the other one a higher resource layer. The lower layer comprises the physical infrastructure and is responsible for the virtualization of storage and computational resources. The higher layer provides specific services, such as:

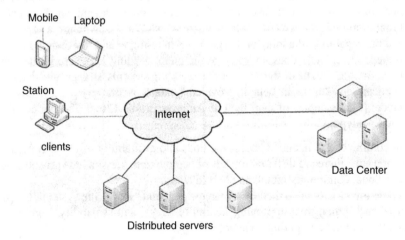

Fig. 15.1 Three elements of cloud computing solution illustration [1]

1. **Software as a Service (SaaS)**
   SaaS does provide all the functions of a traditional application; however, it provides access to specific applications through the Internet. The SaaS model reduces cost of ownership for organization and concerns with application servers, operating systems, storage, application development, etc. Hence, developers may focus on innovation, and not on infrastructure, leading to faster software system development. SaaS system reduces costs since no software licenses are required to access the applications. Instead, users access services on-demand. Since the software is mostly Web-based, SaaS allows better integration among the business units of a given organization or even among different software services. Example of SaaS is Google Docs and customer relationship management (CRM).

2. **Platform as a Service (PaaS)**
   PaaS is the middle component of the service layer in the cloud. It offers users software and services that do not require downloads or installations. PaaS provides an infrastructure with a high level of integration in order to implement and test cloud applications. The user does not manage the infrastructure, including network, server, opening system, and storage, but he or she controls deployed applications and, possibly, their configurations [2]. PaaS provides an operating system, programming languages, and application programming environments. Therefore, it enables more efficient software system implementation, as it includes tools for development and collaboration among developers. Examples of PaaS includes Microsoft Azure Service Platform (Azure), Force.com, Engine Yard, and Google App Engine.

3. **Infrastructure as a Service (IaaS)**
   IaaS is the portion of the architecture responsible for providing the infrastructure necessary for PaaS and SaaS. Its main objective is to make resources such as servers, network, and storage more readily accessible by including applications and operating systems. Thus, it offers basic infrastructure on-demand services. IaaS has a unique interface for infrastructure management; an Application Programming Interface (API) for interactions with hosts, switches, and routers; and the capability of adding new equipment in a simple and transparent manner. In general, the user does not manage the understanding hardware in the cloud infrastructure, but he or she controls the operating systems, storage, and deployed applications. The main benefit provided by IaaS is the pay-per-use business model [2]. Examples of IaaS include Amazon Elastic Cloud Computing (EC^2) and Eucalyptus.

According to the intended access methods and availability of cloud computing environments, there are different models of deployment; they include private cloud, public cloud, community cloud, and hybrid cloud.

These layers may have their own management and monitoring system, independent of each other, thus improving flexibility, reuse, and scalability. Figure 15.2 presents the cloud computing architectural layers [2].

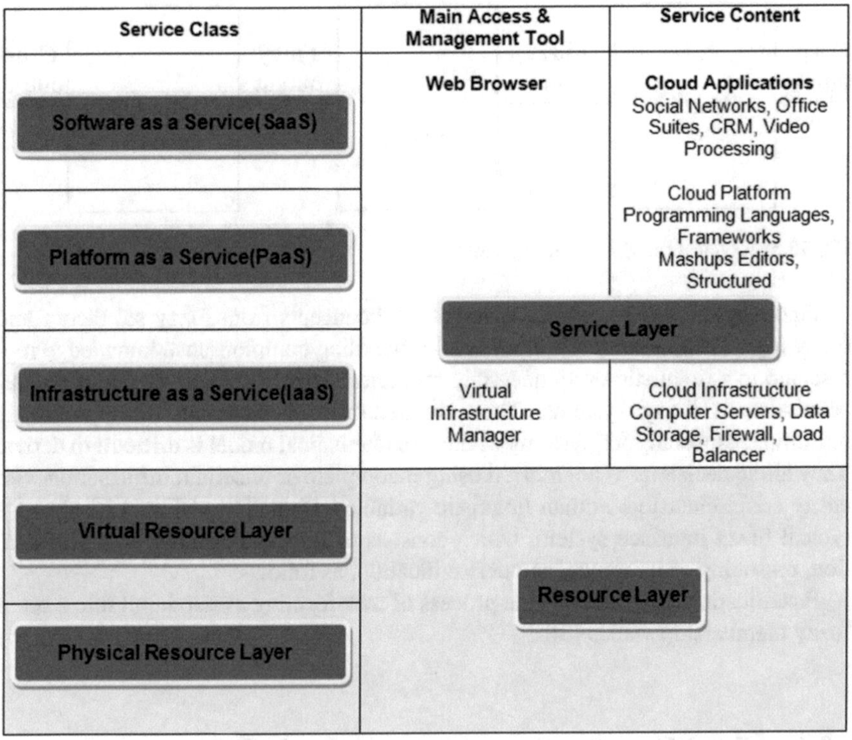

Service Class	Main Access & Management Tool	Service Content
Software as a Service(SaaS)	Web Browser	Cloud Applications Social Networks, Office Suites, CRM, Video Processing
Platform as a Service(PaaS)		Cloud Platform Programming Languages, Frameworks Mashups Editors, Structured
	Service Layer	
Infrastructure as a Service(IaaS)	Virtual Infrastructure Manager	Cloud Infrastructure Computer Servers, Data Storage, Firewall, Load Balancer
Virtual Resource Layer		
	Resource Layer	
Physical Resource Layer		

Fig. 15.2 Cloud computing architecture [2]

For more information readers should refer themselves to the article written by Vergara and Caneddo published in Computational Science and Its Applications ICCSA 2015, 15th International Conference, Banff, AB, Canada, June 22–25, 2015 Proceedings, Part I, Springer Publishing Company.

15.3   Fuzzy Logic and Fuzzy Systems

Fuzzy logic (FL), as we stated before, is a form of multivalued logic derived from fuzzy set theory to deal with human reasoning and the process of making inference and deriving decisions based on human linguistic variables in the real world. Fuzzy set theory works with uncertain and imprecise data *and/or* information. Furthermore, fuzzy sets generalize the concept of the conventional set by extending membership degree to any value between 0 and 1. Such *fuzziness* feature occurs in many real-world situations, whereby it is difficult to decide if something can be categorized exactly into a specific class or not.

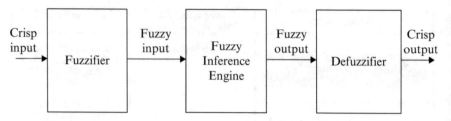

Fig. 15.3 A typical fuzzy inference system [3]

Fuzzy systems (FS), which assimilate the concepts from fuzzy set theory and fuzzy logic (FL), provide a framework for handling commonsense knowledge represented in a linguistic or an uncertain numerical form. There are two useful characteristics of fuzzy systems. They are suitable for uncertain or approximate reasoning, especially for systems where a mathematical model is difficult to derive. They allow decisions to be inferred using incomplete or uncertain information with easily comprehensible human linguistic variables. Depiction of Fig. 15.3 shows a typical fuzzy interface system, which consists of three procedures, i.e., fuzzification, reasoning or inference, and defuzzification, as follows:

Fuzzification, in general, is the process of transforming a crisp input into a set of fuzzy membership values [3].

15.4 Cloud Computation with Fuzzy Logic Concept

The fuzzy rule based assessment of the security risks and benefits of cloud computing—providing security guidance for potential and existing users and manage risk with cloud. In this section, we take advantages of three fuzzy inputs like Gracefulness, Processor Speed, and Performance used as fuzzy input to find out Trust Rating rate. Cloud-based model can be more robust, scalable, and cost-effective and would manage risk very well with the use of fuzzy logic (FL). Here, we start by introducing, cloud computing, processor speed, etc. [1].

Based on fuzzy rules, if the three fuzzy inputs like Gracefulness, Processor Speed, and Performance are used, then fuzzy output is Trust Rating rate provided by Rachna Satsangi et al. [2] as depicted in Fig. 15.4 [4].

Cloud-based model could be more robust, scalable, and cost-effective and would manage risk very well with the use of fuzzy Logic. We used three fuzzy inputs like Gracefulness, Processor Speed, and Performance as fuzzy input to find out Trust Rating rate and get better result of performance.

Old statement of expression *knowledge* is *power* has a third dimension from these authors' point of view and that is *information*. As part of Business Resilience System (BRS) infrastructure and foundation, we need to have a *Strategy of Reducing Risk* in place.

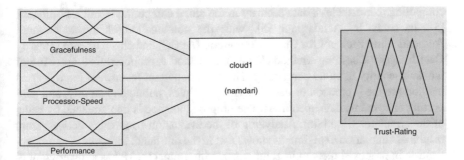

Fig. 15.4 Three fuzzy inputs like gracefulness, processor speed, and performance [4]

The information and support are provided to enable customer staff to understand how to safely and reliably use their product. Some instructions are provided to administrators and managers for setting and monitoring policies. The needful task is that users are provided with instructions on how to avoid phishing or malware attacks. Fuzzy inputs like Gracefulness, Processor Speed, and Performance are used as fuzzy input to find out Trust Rating rate [4].

As part of process, *Risk Assessment Process* should be under consideration in order to have an intelligent and autonomous BRS. The level of risk is estimated based on the likelihood of an incident scenario, mapped against the estimated negative impact. The likelihood of an incident scenario is given by a threat exploiting with a given likelihood. The likelihood of each incident scenario and the business impact were determined in consultation with the expert group contributing to this report, drawing on their collective experience. In cases where it was judged not possible to provide a well-founded estimation of the likelihood of an occurrence, the value is N/A. In many cases, the estimate of likelihood depends heavily on the cloud model or architecture under consideration [3].

15.5   Security of Data in Cloud Computing

An Efficient Way to Protect the Security of Data in Cloud Computing is a major concern of Processing Data Point (PDP) with Business Resilience System (BRS). The major factor in development limitations of cloud computing is an issue about the data security. The data security protects data versus deviation and unauthorized access. The creation acceptable level of the data security requires an attention of the principles of data security. The most important principles are confidentiality and integrity. Confidentiality protects disclosure of data of unauthorized individuals and systems; integrity ensures authenticity and information accuracy. The data security problems in all types of cloud computing services are an issue and require the use of security principles and technical mechanism safety, in order to solve user's concerns. So far, several methods have been proposed to control the security of cloud computing. Nevertheless, data security in the cloud computing is still a bottleneck [5]. In study by Azhdarpour [5], with the combination of the techniques, Authentication Protocol for Cloud Computing (APCC) and check of data integrity Data Integrity Checking Method (DICM), a method is proposed for improving of the data security in cloud computing. The most important features of the proposed method are the protection of data security principles, public auditability, and supporting of dynamic data operations. The proposed method is implemented by using of the simulator CloudSim. Analyzing of the evaluation parameters derived from results of simulation experiments shows that in the one hand, the process of authentication in proposed method in comparison with protocol SAP has a lower computational cost and communication, especially in the client side. On the other hand, the process of check integrity has the lowest cost of communication in comparison with other designs; the consumption of network bandwidth is minimizing.

The proposed method presents two schemes APCC and DICM in order to protect the confidentiality and integrity, which the most important principles of data security are considered. Performance analysis indicated that the authentication process with using APCC is more efficient and lightweight than SAP, especially the more lightweight client side. DICM is an efficient way to check the integrity of outsourced data in data centers in the cloud computing system. On the one hand, with the public auditability, it is possible that the client is not always concerned about the resource constraints and overhead of stored data integrity checking and the efficiency of storage service assess to at third party in dependent auditor. On the other hand, in DICM the public form data dynamic operations (modify, delete, and insert) is considered. This means that the index information has been removed to create tags. Accordingly, in DICM operations on each data block will not affect other block.

Therefore, it does not increase the computation cost. Moreover, DICM has the lowest cost communications in comparison with other schemes. Reducing the cost of communications represents less network bandwidth consumption and increases the speed of transfer information in outsourced data integrity checking. Both APCC and DICM schemes reduce the computational overhead on the client side. This aligns well with the idea of cloud computing which allows the user with a platform of limited performance to outsource its computational tasks to some more powerful server.

15.6   The Challenges of Cloud Integration

Heading to the Cloud with Greater Openness is part of security and risk management challenges. The cloud, of course, is all about services, so it would help if enterprises had an open platform to integrate existing service architectures with those on the cloud. That seems to be the goal of any organization's latest initiatives with a need for a BRS: a new service-oriented architecture (SOA) reference architecture and the Service-Oriented Cloud Computing Infrastructure Framework (SOCCI). The idea is to support a vendor-neutral common language to bridge the gaps that many services encounter when trying to coordinate with each other on the cloud. Through a shared terminology, services should be able to operate within a wider universe without having to recode and remap individual architectures.

Since the cloud relies on network infrastructure to drive dynamic data environments, more openness on the grid may be warranted as well. Big Switch Networks recently issued an open source version of the OpenFlow Controller as part of the software-defined networking (SDN) initiative. Dubbed Floodlight, the system monitors and maintains control data from OpenFlow-compatible switches from a server-based environment, rather than directly on the switch itself. This provides greater management centralization and more efficient distribution of network services. The system is available under an Apache 2.0 license along with Hadoop and OpenStack [6].

Open source proponents say that not only are today's platforms much more flexible and feature laden than their ancestors of the data silo age, but they also provide much greater reliability and developer support. The Internet, after all, is based largely on open source technology. If an enterprise wants to play on that field, it must adjust to someone else's turf.

In an effort to simplify cloud-based integration as part of Risk Atom PDP (see Chap. 1), we need to break it down into four basic options. The cloud-based integration is a software that runs "as a service" in the cloud. You subscribe to it or you pay as you use it. What you get in return are connectors for applications, data sources, or data files that are required foundation Risk Atom PDP.

Their four options are listed as:

1. **Application integration**: This is a good option for delivering small transactions in near real time. These integrations are typically business process oriented—that is to say, as soon as "something happens" in the originating application, for example, when a customer orders a product, it triggers the integration platform to do something—for example, submit the credit card charge to the "payments system." MuleSoft's iON is an example of this type of cloud-based integration.

2. **Data integration**: Just as with on-premise systems, data integration, typically, is done in batches, meaning you move the data between systems on a schedule, whether it is every few minutes or once a month, Tibbetts writes. You can do a number of things to the data as you move it—but it's all about moving large amounts of data. This is the approach that is usually used to make sure your Salesforce data is shared either with another application, on-premise, or in the cloud. This is the category where most cloud-based integration offerings fall, including Dell-Boomi, IBM's WebSphere Cast Iron, and Informatica.

3. **Federated data integration—aka, enterprise information integration**: This is good for fetching information for a management dashboard so you have the latest data, but you would not use it to run "big historical data reports with information from multiple places," Tibbetts says. As I understand it, this is, increasingly called data virtualization or federation. For example, Composite Software was once known as an EII solution but now markets itself as a data virtualization and federation solution and, yes, offers cloud integration.

4. **Managed file transfer**: Use this when you have files—not data, but files—that you want to manage, monitor, track, and transfer to somewhere else. It's a few evolutionary steps beyond FTP (file transfer protocol). This is where B2B integration vendors typically fit. Vendors include Hubspan, Ipswich, TIBCO, and IBM Sterling.

Those are your options; the trick, of course, is figuring out who fits where. Vendors tend to "cloud-wash" solutions these days, describing them in ways that confound and amaze. Beware if solutions promise more than the vendor does—or, really, any vendor—can deliver. Instead, focus on what's pragmatic and hard-working and, yes, maybe a little stained from hard work over the years because while cloud may be new, the integration options, it seems, are not.

References

1. Velve, A. T., & Elsenpeter, T. J. (2011). *Cloud computing—Computacão em Nuvem-Uma Abordagem Pratica*. (G. E. Mei, Trans.) (pp. 352–359). Brazil: Alta Books.
2. Jing, X., & Jian-Jun, Z. (2010). A brief survey on the security model of cloud computing. In *Engineering and Science (DCABES)* (pp. 475–478). Hong Kong: IEEE. August 2010.
3. Lim, C. P. (2010). *Handbook on decision making: Vol. 1: Techniques and applications (Intelligent Systems Reference Library)* (2010th ed.). Berlin: Springer.
4. Satsangi, R., Dashore, P., & Dubey, N. (December 2012). Risk management in cloud computing through fuzzy logic. *International Journal of Application or Innovation in Engineering & Management (IJAIEM)*, *1*(4), 144.
5. Azhdarpour, S. (2015). An efficient way to protect the security of data in cloud computing. *Journal Applied Environment and Biological Sciences, 5*(8S), 522–532.
6. Cole, A., & Lawson, L. (2011). *The challenge of cloud integration*. Enterprise Networking Planet and IT Business Edge.

Chapter 16
A General Business Resilience System Infrastructure

Knowledge is power, thus learning from experience is a fundamental way that helps individuals or organizations to improve and avoid previous mistakes. Accident Investigations (AI) and Operational Safety Reviews (OSR) are valuable for evaluating technical issues, safety management systems, and human performance and environmental conditions to prevent accidents, through a process of continuous organizational learning.

16.1 Introduction

Accidents or disasters are unexpected events or occurrences that result in unwanted or undesirable outcomes. The unwanted outcomes can include harm or loss to personnel, property, production, or nearly anything that has some inherent value. These losses increase an organization's operating cost through higher production costs, decreased efficiency, and the long-term effects of decreased employee morale and unfavorable public opinion.

Despite past accomplishments, human values and other values stimulate a continual desire to improve safety performance. Emerging concepts of system analysis, accident causation, human factors, error reduction, and measurement of safety performance strongly suggest the practicality of developing a higher order of control over hazards.

Our concern for improved preventive methods, nevertheless, does not stem from any specific, describable failure of old methods as from a desire for greater success. Many employers attain a high degree of safety, but they seek further improvement. It is increasingly less plausible that the leading employers can make further progress by simply doing more, or better, in present program. Indeed, it seems unlikely that budget stringencies would permit simple program strengthening. In addition, some scaling down in safety expenditures (in keeping with other budgets) may be necessary.

© Springer International Publishing AG 2017   379
B. Zohuri, M. Moghaddam, *Business Resilience System (BRS): Driven Through Boolean, Fuzzy Logics and Cloud Computation*,
DOI 10.1007/978-3-319-53417-6_16

Consequently, the development of new and better approaches seems the only course likely to produce more safety for the same or less money. Further, a properly executed safety system approach should make a major contribution to the organization's attainment of broader "performance goals."

With clouding technology and a new Service-Oriented Architecture (SOA) reference architecture and the Service-Oriented Cloud Computing Infrastructure Framework (SOCCI). The idea is to support design of a Business Resilience System that fits your organization needs.

With cloud computing data are floating cross networks, pretty much at the speed of light, and for human to be able to cope with such overwhelming information is just an impossible task. Thus, an intelligent BRS is a must platform for any organization in today's cyber attack world, geopolitical events around the globe, homeland security to deal with aggressive world of terrorist acts, and banking as well as financial institute to deal with their CRM and other situations.

As we saw in Chap. 1, for such BRS to be effective and cost worthy, Processing Data Point (PDP) should be able to deal with all kinds of data, coming from every direction, and filter them down to trusted level and trigger the dashboard of BRS for fast decision-making as well as action by stakeholder. For this matter, each organization and enterprise needs to look at their needs and operation to establish a right Business Resilience System.

As we also stated before, conditions of risk and uncertainty frame most decisions rendered by management and Service Level Agreement in place, by outlining the various risks that influence the decision-making process. Thus, we need to take the following points in general and ask right questions form stakeholders within organization.

- Uncertainty and risk are not the same thing. Whereas uncertainty deals with possible outcomes that are unknown, risk is a certain type of uncertainty that involves the real possibility of loss. Risks can be more, comprehensively accounted for than uncertainty.
- Decision-making under conditions of risk should seek to identify, quantify, and absorb risk whenever possible.
- The quantity of risk is equal to the sum of the probabilities of a risky outcome (or various outcomes) multiplied by the anticipated loss as a result of the outcome.
- A firm's ability to absorb, transfer, and manage risk will often define management's risk appetite; once risks are identified and quantified, decisions may be made as to what extent risky outcomes may be tolerated.

16.2   Purpose and Scope of Business Resilience System

The following is suggestion by us as authors, but you can make your own sets of questions.

The questions that you need to ask in general are that we want to understand the nature of your business, the resources you need to operate your business, the threats you face, and how you can respond to those threats. We will use this information to develop a resilience plan to monitor threats and minimize adverse impacts to your business.

The initial interview will take approximately an hour to complete (an assumption).

We will ask a series of questions to help you describe your business and risks. These questions cover three topics:

1. Business context and key performance indicators
2. Resource requirements and key capability thresholds
3. Threats, impacts, and response options

We will follow up to confirm that the information we have captured is complete and correct and to clarify any remaining questions. You may want us to talk to other members of your team to fully answer the questions we are asking.

16.2.1   Typical Use Cases for Business Resilience System

Generally speaking, you can build a use case diagram as illustrated here in Fig. 16.1, per your scope and purpose of Business Resilience System. This use case scenario allows for your programming team within IT to write an Application Programming Interfaces (API) and flowchart per your engine rules established by end users or stakeholders implementing the required Service Level Agreement (SLA) established per your organization and operation needs.

16.2.2   Typical Live Environment for Business Resilience System

Per your use, cases in above you may be able to put a hardware infrastructure in place, and we have shown a generic one here as it is depicted for a BRS that is presumably running under a hypostatical web site of www.ABRS.com domain (Fig. 16.2).

16.3   Business Resilience System Applications

In today's environment where we are living in such fast-paced and hostile environments, having a smart BRS in support of our survivability in these enjoinments is an essential requirement. Cost of dealing with any threats and recovering from it is overwhelming and devastating as well.

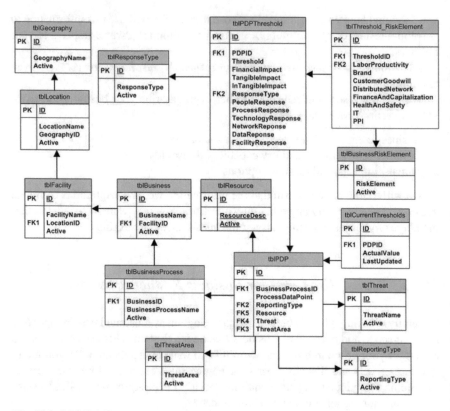

Fig. 16.1 BRS Use Cases

For example, homeland security administration can use an appropriate BRS that is built per its agenda and operation and can track all the terrorist activities around the world and in the United States, in order to predict their activities and to be able to take preventive measurements against their impacts. Industries with multi-organization across the globe could be prepared in order to deal with any unpredictable interruption in their supply chain operations so their production line does not stop. Nuclear industry can use any right BRS for their safety of their day-to-day operation so the production of electricity to the grid is not interrupted due to either terrorist activities or natural disasters.

Intelligent communities can use such system to have real-time intelligence and information around the globe due to their geopolitical changes, thus they can inform the authorities for proper and appropriate measurements to take place if need be.

At small scale any local retail stores with multiple point of sale also can use such a system to have a better and more effective supply chain, so their Key Performance Indicator (KPI) will be at their peak optimum point. E-Commerce can track any hostile takeover of their network system and stop any hacking by hackers with a right BRS in place as well.

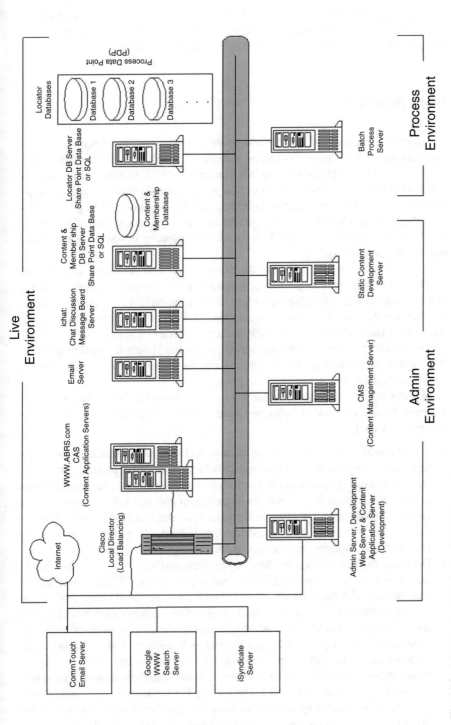

Fig. 16.2 A BRS hardware infrastructure in live mode

Application of Business Resilience System from these authors' point of view is just endless; however, the system is very cost and operational effective and allows the users of such system to be prepared against unknown events.

16.4   Summary

The adaptive capacity of any system is usually assessed by observing how it responds to disruptions or challenges. Adaptive capacity has limits or boundary conditions, and disruptions provide information about where those boundaries lie and how the system behaves when events push it near or over those boundaries. Resilience system in particular is concerned with understanding how well the system adapts and to what degree or range and sources of variation. This allows one to detect undesirable and unknown drops in adaptive capacity to intervene to increase aspects of adaptive capacity.

Thus, monitoring or measuring the adaptiveness and resilience of a system quickly leads to basic ambiguity. Any given incident includes the system adapting to attempt to handle the disrupting events or variation on textbook situations. In an episode, we observe the system stretch nearly to failure or fracture point. Hence, are there cases that fall short of breakdown success stories or anticipations of future failure stories? Moreover, if the disruption pushes the system to fracture point, do the negative consequences always indicate a brittle system, since all finite systems can be pushed eventually to a breaking point?

In summary, we have to have the following capabilities for a good and smart Business Resilience System, before we go about design and build one from end-to-end and they are [1]:

- We have to get smarter at predicting the next accident. The recombinant predictive logic that drives accident prediction today is insensitive to the pressures of normal work by normal people in normal organizations. We may not even need accident models, as the focus should be more on models of normal work. It is normal work and the slow increment drift into margins and eventually across boundaries that is a side effect of normal work in resource-constrained, adaptive organizations that seem to be an important engine behind complex system accidents.

- Detecting drift into failure that happens to seemingly safe systems, *before* breakdowns, occurs as a major role for resilience engineering. While it is critical to capture the relational dynamics and longer-term socio-organizational trends behind system failure, "drift" as such is ill-defined and not well-modeled. One important ingredient in drift is the sacrificing decision, where goals of safety and production/efficiency are weighed against one another, and managers make locally rational decisions based on criteria from inside their situated context. However, establishing a connection between individual microlevel managerial trade-offs and macro-level drift is challenging both practically and theoretically. In addition, it is probably not easy to influence sacrificing decisions so that they do not put a system on the road toward failure. Should we really micro-adjust managers' day-to-day cognitive calculus of reward?

- Detecting drift is difficult and makes many assumptions about our ability to distinguish longitudinal trends, establish the existence of safety boundaries, and warn people of their movement toward them. Instead, a critical component in estimating an organization's resilience could be a momentary charting of the distance between operations as they really go on and operations as they are imagined in the minds of managers or rule makers. This distance tells us something about the models of risk currently applied and how well (or badly) calibrated organizational decision-makers are. This, however, requires comparisons and perhaps a type of quantification (even if conceptual) that may be challenging to attain. Updating management about the real nature of work (and their deficient perception of it) also requires messengers to go beyond "reassuring" signals to management, which in turn demands a particular climate that allows the boss to hear such news.
- Looking for additional markers of resilience, what was explored in the symposium? If charting the distance between operations as imagined and as they really occur is too difficult, then an even broader indicator of resilience could be the extent to which the organization succeeds in keeping discussions of risk alive even when everything looks safe. The requirement to keep discussions of risk alive invites us to think about the role and nature of a safety organization in novel ways. Another powerful indicator is how the organization responds to failure: can the organization rebound even when exposed to enormous pressure? This closes the circle as it brings us back to theme 1, whether resilience is about effectively predicting (and then preventing) the next accident. In terms of the cognitive challenge it represents and preventing the next accident. It could conceptually be close to managing the aftermath of one: doing both effectively involves rebounding from previous understandings of the system and interpretations of the risk it is exposed to, being forced to see the gap between works as imagined and work as done, updating beliefs about safety and brittleness, and recalibrating models of risk.

To sum up this chapter, cognitively, then, predicting the next accident and mopping up the previous one demand the same types of revisionist activities and insights. Once again, throughout the symposium and echoed in the themes of the present book, this has been identified as a critical ingredient of resilience: constantly testing whether ideas about risk still match reality, updating beliefs about safety and brittleness, and recalibrating models of risk.

A main question is how to help effectively, and further work on resilience engineering system is bound to address this question.

Reference

1. Hollnagel, E., Woods, D. D., & Leveson, N. (2006). *Resilience engineering, concepts and precepts*. Aldershot: Ashgate.

Appendix A: Generic Project Planning Management

In this appendix, we will look at an application of Business Resilience System (BRS) for one simple application in a banking enterprise, where BRS is driving customer relationship management (CRM) for retaining existing customer or attracting new customers to become customer.

A.1 Introduction

Customer relationship management (CRM) is a term that refers to practices, strategies, and technologies that companies use to manage and analyze customer interactions and data throughout the customer life cycle, with the goal of improving business relationships with customers, assisting in customer retention, and driving sales growth. CRM systems are designed to compile information on customers across different channels—or points of contact between the customer and the company—which could include the company's web site, telephone, live chat, direct mail, marketing materials, and social media. CRM systems can also give customer-facing staff detailed information on customers' personal information, purchase history, buying preferences, and concerns.

Today's CRM have customer data analytics (CDA) built in them that can reap significant financial rewards for your organization's sales, marketing, and customer service departments. With so much data to contend with, companies often struggle with making sense of information from customers, public records, and external databases. Luckily, the new software in CRM evaluates the newest sales and marketing tools, making the process easier for IT managers and sales executives.

Today, many businesses such as banks, insurance companies, and other service providers realize the importance of customer relationship management (CRM) and its potential to help them acquire new customers retain existing ones and maximize their lifetime value. At this point, close relationship with customers will require a

© Springer International Publishing AG 2017
387
B. Zohuri, M. Moghaddam, *Business Resilience System (BRS): Driven Through Boolean, Fuzzy Logics and Cloud Computation*,
DOI 10.1007/978-3-319-53417-6

strong coordination between IT and marketing departments to provide a long-term retention of selected customers. This paper deals with the role of customer relationship management in the banking sector and the need for customer relationship management to increase customer value by using some analytical methods in CRM applications.

CRM is a sound business strategy to identify the bank's most profitable customers and prospects and devotes time and attention to expanding account relationships with those customers through individualized marketing, repricing, discretionary decision-making, and customized service—all delivered through the various sales channels that the bank uses. Under this case study, a campaign management in a bank is conducted using data mining tasks such as dependency analysis, cluster profile analysis, concept description, deviation detection, and data visualization. Crucial business decisions with this campaign are made by extracting valid, previously unknown, and ultimately comprehensible and actionable knowledge from large databases. The model developed here answers what the different customer segments are, who more likely to respond to a given offer is, which customers are the bank likely to lose, who most likely to default on credit cards is, and what the risk associated with this loan applicant is.

Finally, a cluster profile analysis is used for revealing the distinct characteristics of each cluster and for modeling product propensity, which should be implemented in order to increase the sales.

In this process, we present series of workflow per use cases.

A.2   Relationship Management that Pays Dividends

Turn Prospects into Clients and Harness the Power of 360° Customer and that is Equal to $$$$

- Many CRM solutions on the market today are a little more than electronic Rolodexes, offering no intelligence whatsoever. That's why we need to build our business around a **resilience relationship management platform** capable of handling our day-to-day operation with the most optimized and profitable way and in the most efficient way to maintain **customer retention** and **return on investment (ROI)** for enterprise.
- **Streamline your sales process** with **Business Resilience Services (BRS)** relationship tracking technology that intelligently centralizes multiple prospect communication streams, giving you a holistic view of how your relationship has progressed and where it's headed.
- An intelligent **BRS** not only drives an intelligent **CRM**, it also helps to have an intelligent **BCM** and **BPM** and all turn into $$$$ and increase of enterprise revenue.

A.3   Why We Need an Intelligent CRM Driven by BRS

- No industry understands or appreciates the power of numbers more than commercial banking. An intelligent CRM process and reporting helps leaders measure what is working, manage to those measurements, and identify bottlenecks before they disrupt business.
- An efficient standard report should include salesperson activity reports, opportunity-specific activity reports, weighted pipeline, and much more. Managers can also create configurable reports based on what matters to them.
- With increase in technology and functionality in present and near future smartphones, a need for some built-in mobile app functionality is table stakes at this point for any serious CRM provider, but most CRM providers are not truly optimized for any modern mobile browser yet. So we need a CRM to be able to do just that.
- That means that your marketers and sales employees can edit and access existing records, add notes, and notify colleagues of changes to records no matter where they are—with a client, at the office, on the road—instantly, if not real time but at least near real time.
- With an intelligent CRM, you and your team within Wealth Management Group will have full visibility into client and prospect records and be better prepared before, during, and after meetings.

A.4   Where to Now BI: The Future of Business Intelligence and Beyond

Accenture 2007 Survey of 1000 Middle Managers

- Managers spend up to 2 h a day searching for information.
- More than 50% of information they obtain has no value to them.
- Fifty-nine percent said they miss information that might be of value to their jobs because they cannot find it.
- Only half of all managers believe their companies do a good job in governing information distribution.

A.5   Increasing Information Volumes

- "Data is growing by a factor of 10 every 5 years, a compound annual growth rate of almost 60%" *IDC*
- This growth rate is likely to accelerate given new and evolving information generating technologies
- Examples: mobile phones, RFID tags, sensor networks, Web information
- May not be practical (or even necessary) to consolidate all of this information into a DW for *operational* decision-making

A.6   Disparity of Information

- Operational business data
  - Maintained by legacy applications, application packages, Web systems
  - Stored in transaction, event, master data stores
- Historical business information
  - Maintained by data integration software and BI applications
  - Stored in a data warehouse, data marts, data cubes
- Business content
  - Maintained by collaborative, content, operational, Web systems
  - Stored in databases, text and rich media files, Web pages, etc.

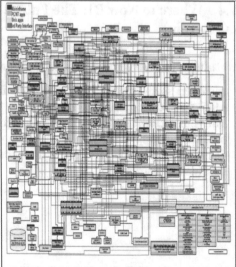

Actual Application Architecture for Consumer Electronics Company

A.7   Timeliness of Information: The Right Time

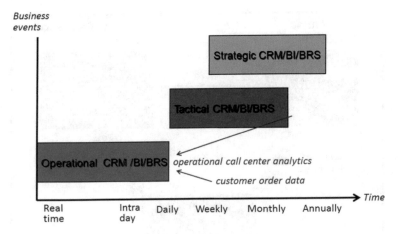

A.8   Complexity of Current Decision-Making Systems

- BI/CRM-driven decision-making is limited to users with a good knowledge of the data and combination BRS/BI technologies involved
- Less-experienced users find BI applications and tools difficult to use
- BI deployment still requires significant IT involvement

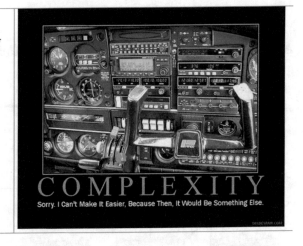

A.9 The Decision-Making Process

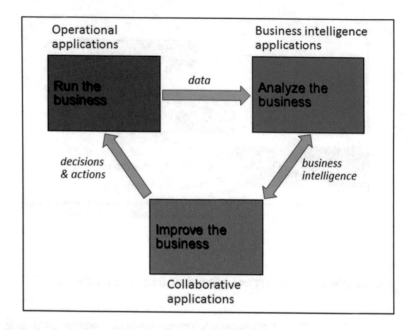

A.10 The Decision-Making Process: A More Detailed Look

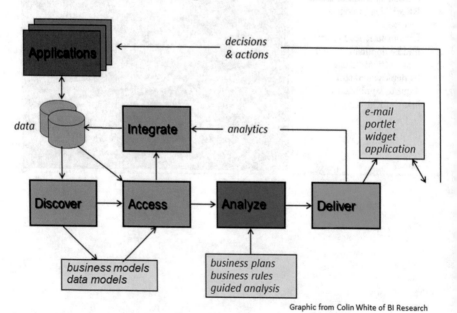

Graphic from Colin White of BI Research

A.11 Bring the Customer into Focus

- How an intelligent CRM driven by BRS using new technology is enabling banks to create a customer-centric experience that transforms customers into advocates for life.
- A 5% increase in customer retention can increase a bank's profit margin by 25–100%.
- Each year, 12 million bank customers consider switching banks.
- Ninety-six percent of unhappy customers never complain, but 91% of them leave and never come back.
- That's a dismal thought when you consider that on average a bank spends more than $400 to acquire each new customer, and a mere 5% increase in customer retention can increase a bank's profit margin by 25–100%.
- CRM for commercial banking, every customer touchpoint becomes an opportunity to connect and strengthen your existing and future relationships (B2C, C2B, B2B).
- Your entire organization can collaborate and gain visibility into the most accurate customer information.
- This connectivity empowers your organization (i.e., WMG) to engage intelligently and consistently with customers, creating lasting relationships.
- To achieve all these, a commercial banking requires a good data mining built on *fuzzy logic* infrastructure (i.e., feed from multisource vs. single one) to introduce an intelligent BRS function for their CRM variables and others (ECM, BI, KPI, MDM, etc.).
- Present CRM providers are not offering such functionality, and they are all built on *Boolean logic* database/data mining rather than fuzzy one.
- We need MDM built around fuzzy logic approach to feed right business intelligence in parallel with CRM to process right and accurate information data in hand to manage, for instance, efficient and cost-effective decision-making.
- Utilizing a fuzzy logic approach by weighing the integrity of data that is getting fed into BI and ECM from at least two directions or resources in case of mono monitoring or multidirection and resources in case of stereotype monitoring for management and stakeholder to make right call and decision near real time by flittering trusted date to them.
- In today's world where the threat is there but its definition has changed, therefore a tool like this will give the management an upper hand.

The BRS/CRM combination offering includes a portfolio of services to help organizations rapidly deploy and utilize business resilience capabilities to have a countermeasure against measure in day-to-day operation.

A.12 Business Continuity Management (BCM)

If you ask why we need such system here is why that is depicted below.

Current business continuity plans are fragmented and do not include a holistic approach to identify and avoid risk

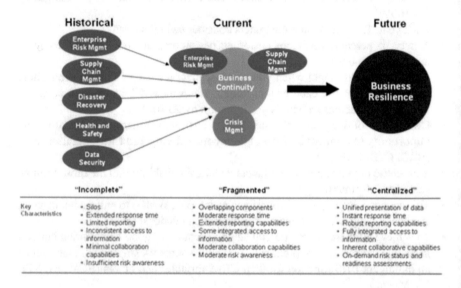

Key Characteristics	**"Incomplete"**	**"Fragmented"**	**"Centralized"**
	• Silos • Extended response time • Limited reporting • Inconsistent access to information • Minimal collaboration capabilities • Insufficient risk awareness	• Overlapping components • Moderate response time • Extended reporting capabilities • Some integrated access to information • Moderate collaboration capabilities • Moderate risk awareness	• Unified presentation of data • Instant response time • Robust reporting capabilities • Fully integrated access to information • Inherent collaborative capabilities • On-demand risk status and readiness assessments

A.13 BRS and CRM Interoperability

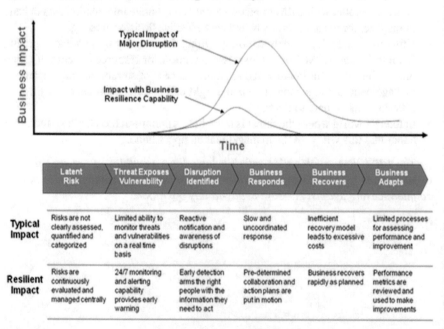

	Latent Risk	Threat Exposes Vulnerability	Disruption Identified	Business Responds	Business Recovers	Business Adapts
Typical Impact	Risks are not clearly assessed, quantified and categorized	Limited ability to monitor threats and vulnerabilities on a real time basis	Reactive notification and awareness of disruptions	Slow and uncoordinated response	Inefficient recovery model leads to excessive costs	Limited processes for assessing performance and improvement
Resilient Impact	Risks are continuously evaluated and managed centrally	24/7 monitoring and alerting capability provides early warning	Early detection arms the right people with the information they need to act	Pre-determined collaboration and action plans are put in motion	Business recovers rapidly as planned	Performance metrics are reviewed and used to make improvements

A.14 Functional Model of BRS/CRM Offering

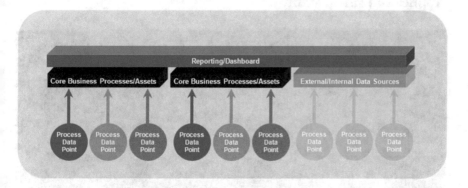

- The portal becomes the business dashboard for executives to proactively manage their organization.
- External and internal data sources (ERP, MRP, etc.) combined with the accumulation of CBPs and CBAs create a holistic view of business resilience.

A.15 Solution: A Decision Framework that Provides

- Personalized and self-service *discovery* of and *access* to business information
- An *integrated* view of an organization's business information:

  - Structured business data
  - Unstructured business content

- Easy *analysis* of business information
- *Delivery* of information via richer and more intuitive Web-based user interfaces
- *Sharing* of business information and expertise
- *Collaborative* decision-making

A.16   Business Resilience Service Driving Enterprise Application Integration

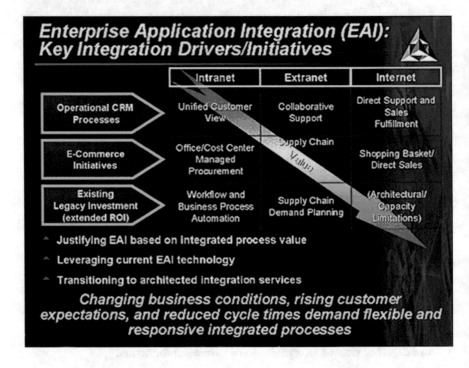

A.17   High-Impact CRM Solutions

How to use data mining with CRM in banking to segment credit card customers

- Credit cards are being used by consumers across age groups and for diverse purposes.
- 300 M credit card transaction per day is taking place globally.
- They buy different products and services according to their purchasing power, habits, standard of living, and lifestyle.
- The frequency of purchases and value of each transaction also varies; customers use credit cards for their utility bills, apparel, daily needs, and occasionally for high-value purchases.
- With e-commerce growing in popularity, having a credit card swiped instead of paying cash is increasing in popularity.
- Customers vary in terms of their payment behavior; there are some who tend to pay the full due amount, while others only pay the minimum amount and carry forward their previous balance.

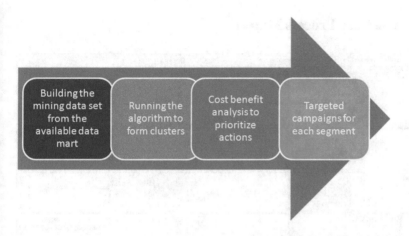

- Data mining through a banking CRM system can be a critical source for improving profitability from the credit card business. It can also effectively identify trends for cross selling with results from the segmentation exercise of other products. All data related to various segments and their related reports can be hosted on a single centralized system for analysis by numerous associated teams around the globe.

A.18 Use Cases

- **Use Case Description Documents**

  - Document interactions between system and end users
  - Identified on business process maps

- **Use Case Flow Charts**

  - Document forms, validation, decision points, and displays between system and end users from UI perspective.
  - Flow charts are function of logical engine rules with Business Rule Management System (BRMS) built-in capability/BRS.

- **Use Case Process Map**

  - Use case view of business processes

A.19   Use Case Process Map

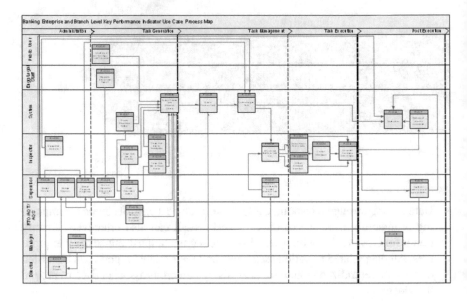

A.20   Overview of Project Plan to Implement BRS/CRM

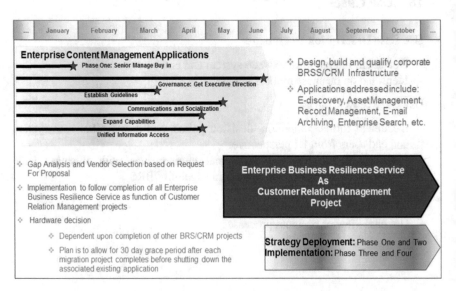

A.21   Sample of How We Can Show Timing

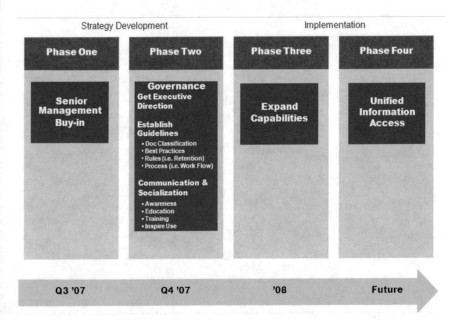

A.22   Questions and Answers

Appendix B: Information on Demand

In this appendix, we will present solution around IBM tools that deals with
information on demand, which requires as core for the Processing Data Point
(PDP) of Risk Atom (see Chap. 1).

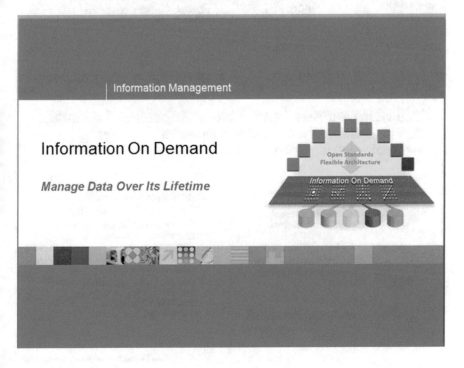

© Springer International Publishing AG 2017
B. Zohuri, M. Moghaddam, *Business Resilience System (BRS): Driven Through
Boolean, Fuzzy Logics and Cloud Computation*,
DOI 10.1007/978-3-319-53417-6

| Information Management

Are you Effectively Managing Data over its Lifetime?

"...Information is growing like crazy, we need to get control of it..."

- **Ensure accessibility, retention and compliance**
- **Reduce cost of deploying and managing data**

Information Management Software
Database Management servers
Integrated Data Management tools and solutions

Supporting Capabilities
Storage Management Solutions
Systems Management & Security
Data management consulting & implementation services

| Information Management

Data Management Must Drive Competitive Advantage
Survey: CIO's want to strengthen competitive advantage by better managing enterprise data

75% of CIO's believe they can strengthen their competitive advantage by better using and managing enterprise data.

78% of CIO's want to improve the way they use and manage their data.

...but **only 15%** believe that their data is currently comprehensively well managed.

- **Data management must drive competitive advantage**
  - Tailor application data models to support differentiable business processes
  - Service-enable data assets for business process agility
  - Empower teams to collaborate seamlessly, improve productivity from design to delivery to management

Source: Accenture CIO Data Management Survey 2007. n=167 CIOs

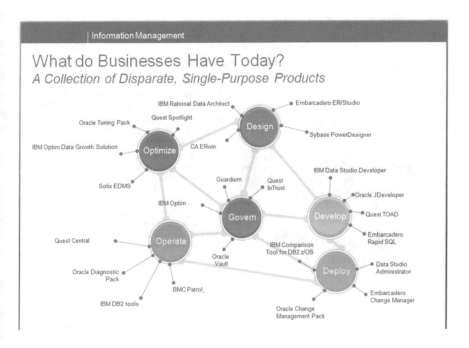

| Information Management

The Value of Integrated Data Management

- **Deliver increasing value across the lifecycle,** *from requirements to retirement*
- **Facilitate collaboration and efficiency across roles,** *via shared artifacts automation and consistent interfaces*
- **Increase ability to meet service level agreements,** *improving problem isolation, performance optimization, capacity planning, and workload and impact analysis*
- **Comply with data security, privacy, and retention policies** *leveraging shared policy, services, and reporting infrastructure*

| Information Management

Integrated Data Management Today – Powerful Capabilities

- Rational Data Architect
  - ▸ Collaborative data design to improve data quality and enterprise consistency
- Data Studio Developer and pureQuery Runtime
  - ▸ Enhance productivity up to 50% while improving code quality and providing expert-equivalent performance for Java data access
- DB2 Change Management Expert, Performance Expert, Encryption Expert, DB2 Audit Management Expert
  - ▸ Administration and security management tools
- Optim Solutions for Data Growth, Data Privacy and Test Data Management
  - ▸ Improve performance, control costs, reduce risks
  - ▸ Control data growth, streamline test data management, mask confidential data, manage data retention and destruction
  - ▸ Speed application upgrades, enable safe application retirement

Information Management

Integrated Data Management as part of an end to end system for delivering business solutions

Rational. software

Design

Govern
Standards
Models
Policies

Optimize · Develop

Tivoli. software

Operate · Deploy

WebSphere. software

Information Management

Data Management Improves Performance, Cuts Cost

TD Bank Financial Group (Canada)

Bank Financial Group

Client Profile: Banking / Finance, C$14.3 Billion Annually

Success: Data Growth

Reduced batch processing runtime for Payroll by 25%. Realized a cost savings of 50% and reduction in runtime for database utilities. Reduced database storage needs by 30% (from approximately 325 GB to 225 GB).

Corning

CORNING

Client Profile: Manufacturing, $3.7 Billion Annually

Success: Data Growth & Compliance

Improved SCM performance by 60% by archiving high volume tables, reducing overnight processing time from 6.5 hours to 2.2 hours, all while addressing new industry and regulatory standards for HR data retention.

Bausch & Lomb

Bausch&Lomb

Client Profile: Healthcare Retail, $32 Billion Annually

Success: Data Growth & Portfolio Optimization

Improved performance by 50% by archiving historical data. Reported estimated savings of $1.5 million in capacity costs, plus additional savings in hardware & software costs, by archiving to sunset several applications worldwide.

Data Management Boosts Developer Productivity

- **Challenges**
  - ▸ Executives must quickly respond to new regulatory requirements, mergers, acquisitions, and evolving customer needs
  - ▸ Developers must swiftly make changes to business-critical applications without compromising performance, availability or scalability
- **Solution**: Optim Studio
- **Key Benefits**
  - ▸ Developers can explore how workload is executed without the performance and usability challenges encountered in other tools
    - ▪ Speeds the development of high-quality applications
    - ▪ Reduces developments costs by 50 percent
  - ▸ Developers and DBAs can easily collaborate to resolve issues quickly
    - ▪ Improves developer productivity by 25 to 50 percent
- **Client Value**

  "IBM Data Studio has advanced the ease of developing, debugging and testing stored procedures by leaps and bounds. And by embedding this capability within the tool itself, rather than selling it as a separate product, we're realizing significant productivity improvements."

Success: Protecting Data Privacy

About the Client: $10 Billion Insurance Company

- **Application:**
  - ▸ Custom Insurance Applications
- **Challenges:**
  - ▸ Protecting confidential customer information required by GLB by addressing privacy vulnerabilities in the application development and testing environments
  - ▸ Creating realistic "federated" testing environments by extracting test data across complex DB2, Oracle, Informix, IMS and VSAM databases
  - ▸ Ensuring valid testing results by retaining the data integrity after sensitive information is de-identified
- **Solution:**
  - ▸ IBM® Optim™ Data Privacy Solution

- **Client Value:**
  - ▸ Mitigated risk of data breaches by implementing a consistent strategy for de-identifying sensitive data in development and testing environments
  - ▸ Improved enterprise-wide testing processes by using subsetting and transformation capabilities across applications, databases and operating systems
  - ▸ Ensured test validity by using a variety of masking techniques that preserved the data integrity, while propagating the de-identified data throughout the test environment

| Information Management

Mashups & IBM Data Management Deliver Quick Apps

- Mashup Center addresses the "quick applications" dilemma by changing the cost, information availability & TCO factors
  - ▸ Allows IT to deliver information access without losing control over the information assets
    - Slices of necessary data can be surfaced in minutes.
    - Security, governance, rate limiting protected
    - Doesn't require new silos or replication
    - Minimal technical skill required to assemble quick apps
- Mashup solutions can literally be built in hours
  - ▸ And it only gets faster from there since each mashup makes the next one easier and faster to build

Case Study

- **Business need:** Provide real-time, customizable manufacturing information for semiconductor supplier
- **Solution:** Enterprise mashups of information from semiconductor manufacturer's portal and supplier's ERP system
- **Benefits:** Reassigned one full-time employee to higher value work; significant decrease in planning and production mistakes; reduced total cost of ownership by 40%

Balance between Flexibility and Control = Immediate ROI

| Information Management

Summary – Key Values of Integrated Data Management

- Integrated Data Management Builds on Shared Artifacts
- Integrated Data Management Allows Executives To Set Policy Standards
- Integrated Data Management Allows Design Professionals To Drive Quality and Consistency
- Integrated Data Management Allows Development Professionals To Deliver Quality Software
- Integrated Data Management Allows Administrators To Deploy without Disruption
- Integrated Data Management Allows Administrators To Operate Databases to Meet SLAs
- Integrated Data Management Allows Application Allows Managers To Optimize Systems for Growth

Appendix C: Where to Now BI—The Future of Business Intelligence and Beyond

In this appendix, we will look at the future of business intelligence and beyond. This presentation was given by Dr. Claudia Imhoff of Intelligent Solutions, Inc. in 2009, where the author Zohuri was participating.

2007 Accenture Study of Middle Managers

Accenture 2007 Survey of 1,000 Middle Managers

- Managers spend up to 2 hours a day searching for information
- > 50% of information they obtain has no value to them
- 59% said they miss information that might be of value to their jobs because they can not find it
- Only half of all managers believe their companies do a good job in governing information distribution

accenture

High performance. Delivered. | Home | News | Press Kits | People | PR Contacts |

Global Home > Newsroom > Managers Say the Majority of Information Obtained for Their Work is Useless, Accenture Survey Finds

Add to Briefcase   Print Article

04 January 2007
Managers Say the Majority of Information Obtained for Their Work is Useless, Accenture Survey Finds

Information on competitors easier to find than that on other parts of their own organizations

NEW YORK and LONDON – Jan. 4, 2007 – Middle managers spend more than a quarter of their time searching for information necessary to their jobs, and when they do find it, it is often wrong, according to results of an Accenture (NYSE: ACN) survey released today.

The purpose of the online survey of more than 1,000 middle managers of large companies in the United States and United Kingdom was to uncover wide-ranging insights about the way they gather, use and analyze information.

Among the key findings: Managers spend up to two hours a day searching for information, and more than 50 percent of the information they obtain has no value to them. In addition, only half of all managers believe their companies do a good job in governing information distribution or have established adequate processes to determine what data each part of an organization needs.

Nearly three out of five respondents (59 percent) said that as a consequence of poor information distribution, they miss information that might be valuable to their jobs almost every day because it exists somewhere else in the company and they just can not find it. In addition, 42 percent of respondents said they accidentally use the wrong information at least once a week, and 53 percent said that less than half of the information they receive is valuable.

© Springer International Publishing AG 2017
B. Zohuri, M. Moghaddam, *Business Resilience System (BRS): Driven Through Boolean, Fuzzy Logics and Cloud Computation*,
DOI 10.1007/978-3-319-53417-6

2008 Accenture Study of Executives

Accenture 2008 Survey of 250 Executives

- 57% said that they don't have a beneficial, consistently updated enterprise-wide analytical capability
- 55% said their decisions rely on qualitative and subjective factors
- 61% said good data is not available for the decisions they are addressing
- 72% are working to increase business analytics usage

It's Not Getting Any Easier

- Increasing information volumes
- Number and disparity of information stores
- Information quality, accuracy, and consistency
- Timeliness of information
- Complexity of current decision-making systems
- Compliance regulations
- Acquisitions and mergers

Increasing Information Volumes

- "Data is growing by a factor of 10 every five years, a compound annual growth rate of almost 60%," *IDC*
- This growth rate is likely to accelerate given new and evolving information generating technologies
- Examples: mobile phones, RFID tags, sensor networks, web information
- ➢ May not be practical (or even necessary) to consolidate all of this information into a DW for *operational* decision making

Disparity of Information

- Operational business data
  - Maintained by legacy applications, application packages, web systems
  - Stored in transaction, event, master data stores

- Historical business information
  - Maintained by data integration software and BI applications
  - Stored in a data warehouse, data marts, data cubes

- Business content
  - Maintained by collaborative, content, operational, web systems
  - Stored in databases, text and rich media files, web pages, etc.

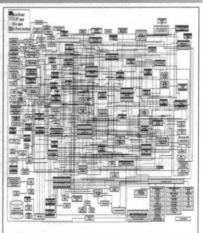

Actual Application Architecture for Consumer Electronics Company

Source: IBM

Timeliness of Information: The Right Time

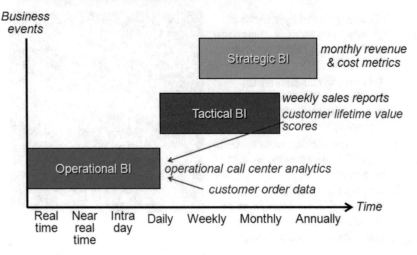

Business events

Strategic BI — *monthly revenue & cost metrics*

Tactical BI — *weekly sales reports customer lifetime value scores*

Operational BI — *operational call center analytics*

customer order data

Time →

Real time | Near real time | Intra day | Daily | Weekly | Monthly | Annually

Complexity of Current Decision-Making Systems

- BI-driven decision making is limited to users with a good knowledge of the data and BI technologies involved
- Less experienced users find BI applications and tools difficult to use
- BI deployment still requires significant IT involvement

COMPLEXITY
Sorry, I Can't Make It Easier, Because Then, It Would Be Something Else.

And Enterprises Want Even MORE!

Data Sources		Business Needs

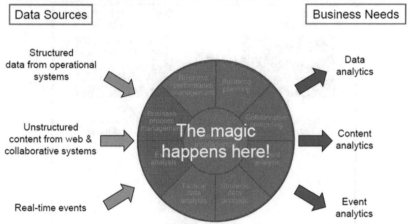

Structured data from operational systems

Unstructured content from web & collaborative systems

Real-time events

The magic happens here!

Data analytics

Content analytics

Event analytics

Solution: A Decision Framework That Provides ...

- Personalized and self-service *discovery* of and *access* to business information
- An *integrated* view of an organization's business information:
  - Structured business data
  - Unstructured business content
- Easy *analysis* of business information
- *Delivery* of information via richer and more intuitive Web-based user interfaces
- *Sharing* of business information and expertise
- *Collaborative* decision making

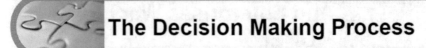

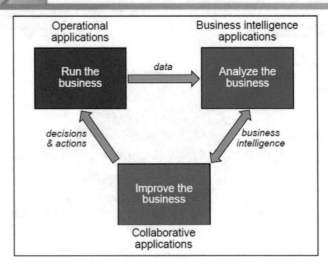

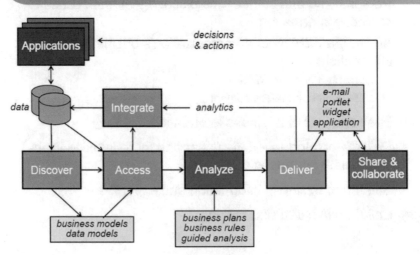

BI-Related Marketplace Directions: Technologies

- Discover
  - Search
  - Data profiling
- Access
  - Data federation
  - Data syndication
- Integrate
  - New data sources
  - Light-weight web protocols
  - Data mashups
  - Low-latency data integration

- Analyze
  - Actionable results
  - Event analytics
  - Content analytics
  - Situational analytical applications
- Deliver
  - Rich internet applications
  - Presentation mashups
  - Integration with office products
  - Advanced visualization
- Share and Collaborate
  - Integration with collaboration products
  - Social computing

BI-Related Marketplace Directions: Deployment

- Packaged Solutions
  - Operational and MDM applications
  - BI and performance management applications
- Lower Cost Solutions
  - Open source software
  - Commodity hardware
  - Appliances
  - Software-as-a-service
  - Cloud computing
- These new developments <u>extend</u> rather than replace the existing enterprise BI and DW environment

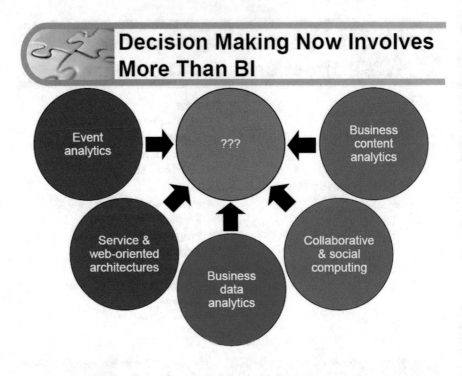

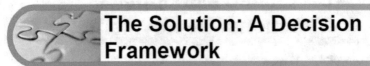

Types of Analytics

Data analytics	Event analytics	Content analytics
Intraday, daily, monthly	Real time, near real time, intraday	Intraday, daily, monthly
Static data	In-motion data	Static data
On demand	Event driven	On demand
Structured queries	Services driven	Search queries
User centric	User and application centric	User centric
Manual decision making	Manual and automated decision making	Manual decision making
Alignment to plans and budgets	Alignment to rules and expertise	Alignment to rules and expertise
Point-in-time data metrics	Continuous process and stream metrics	Point-in-time content metrics

Decision Framework: Closing the Loop

Graphic from Colin White of BI Research

Decision Framework: Information Management

Graphic from Colin White of BI Research

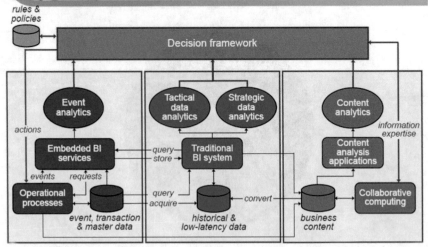

Decision Framework: Information Flow*

*See Article titled "Decision Intelligence" by Claudia Imhoff and Colin White, www.B-EYE-Network.com

Using a Decision Framework to Extend BI – 1

- Discover, Access and Integrate
  - Good enough data is okay for some analyses
  - Employ search for less experienced users to discover business data and content
  - Use low-latency DW data integration for intra-day decision making
  - Transform and integrate business content into a DW to supplement existing data
- Analyze
  - Good enough applications are okay for some analyses
  - Use content analytics to enhance data analytics
  - Use embedded BI and event analytics for agile decision making

Using a Decision Framework to Extend BI – 2

- Deliver
  - Use a services architecture to rapidly deploy on-premises and SaaS solutions
  - Use rich internet applications to improve user web interaction
  - Use web syndication and mashups to enhance delivery of results
- Share and Collaborate
  - Integrate the BI system with collaborative and office systems to provide a seamless decision making environment
  - Improve business user interaction by using social computing software to share business information, analytics and expertise

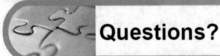

Questions?

Appendix D: SWOT Analysis Worksheet

To help you to carry out your analysis, download and print off the free worksheet page here, and write down answers to the following questions.

D.1 Worksheet

You can make up your own sheet to do your own SWOT question and answers (Fig. D.1).

© Springer International Publishing AG 2017

421

B. Zohuri, M. Moghaddam, *Business Resilience System (BRS): Driven Through Boolean, Fuzzy Logics and Cloud Computation*,
DOI 10.1007/978-3-319-53417-6

Strengths	Weaknesses
What do you do well?	What could you improve?
What unique resources can you draw on?	Where do you have fewer resources than others?
What do others see as your strengths?	What are others likely to see as weaknesses?
Opportunities	**Threats**
What opportunities are open to you?	What threats could harm you?
What trends could you take advantage of?	What is your competition doing?
How can you turn your strengths into opportunities?	What threats do your weaknesses expose you to?

Fig. D.1 Worksheet for SWOT

Index

© Springer International Publishing AG 2017
B. Zohuri, M. Moghaddam, *Business Resilience System (BRS): Driven Through
Boolean, Fuzzy Logics and Cloud Computation*,
DOI 10.1007/978-3-319-53417-6

Printed in the United States
By Bookmasters